Norovirus

Nada M. Melhem
Editor

Norovirus

Editor
Nada M. Melhem
Faculty of Health Sciences
American University of Beirut
Beirut, Lebanon

ISBN 978-3-030-27208-1 ISBN 978-3-030-27209-8 (eBook)
https://doi.org/10.1007/978-3-030-27209-8

This Springer imprint is published by the registered company Springer Nature Switzerland AG.
The registered company address is: Gewerbestrasse 11, 6330 Cham, Switzerland

Preface

Norovirus is the most common cause of acute gastroenteritis worldwide. The overall annual global burden of disease caused by noroviruses was recently estimated at more than 677 million cases and 213,515 deaths. Importantly, the majority of mortality and morbidity cases are reported among children less than 5 years old. Norovirus infection causes sudden onset of vomiting, watery and non-bloody diarrhea, abdominal cramps, fever, and malaise. An acute, self-limiting illness is usually reported among immunocompetent individuals, while severe disease outcomes are observed among immunocompromised individuals and elderly above 65 years of age.

This book is subdivided into seven chapters presenting the current available data on the burden of disease caused by noroviruses (Chap. 1); virus structure (Chap. 2); viral diversity, evolution, and selective pressure (Chap. 3); pathogenesis and clinical features of noroviruses (Chap. 4); anti-norovirus drugs (Chap. 5); immune correlates of protection (Chap. 6); and the most recent advances in the development of norovirus vaccines (Chap. 7).

Human noroviruses are estimated to result in $4.2 billion in direct costs and $60.3 billion in societal costs each year. Interestingly, high-income countries accounted for the highest estimated annual societal costs of disease ($34.7 billion) compared to low- and middle-income countries ($25.4 billion). Chapter 1 addresses the global and regional burdens of disease caused by noroviruses and presents data on the current approaches used to estimate these burdens.

Noroviruses are small, non-enveloped, positive-stranded RNA viruses belonging to the family *Caliciviridae*. Based on the capsid sequence, noroviruses were recently classified into ten genogroups (GI–GX). GI, GII, and GIV infect humans, with GI and GII noroviruses causing the majority of human infections. Based on the major viral capsid protein (VP1) sequence, GI and GII genogroups are subdivided into 9 and at least 25 genotypes, respectively. Genogroups and genotypes differ by at least 60% and 40%, respectively. A detailed overview on the virus and its structure are presented in Chap. 2.

Norovirus genogroup II genotype 4 (GII.4) is known to widely circulate and cause the majority of outbreaks. The mechanisms behind the persistence and the global circulation of GII.4 are not well understood. Chapter 3 addresses viral diversity as well as the current knowledge of evolutionary mechanisms affecting norovirus GII.4 and other genotypes. This chapter focuses on the impact of the main mechanisms suggested for norovirus evolution including antigenic variation, receptor switching, and recombination. These mechanisms are likely to differ between different strains. This chapter in addition to Chaps. 2, 6, and 7 discusses the link between susceptibility to norovirus and the presence of α1,2-linked fucose on histo-blood group antigens (HBGAs). A functional fucosyl transferase 2 (*FUT2*) gene is associated with the expression of HBGAs serving as binding receptors for human noroviruses and thus predicting susceptibility to infection. HBGAs are small carbohydrate molecules found on the erythrocytes of humans and on the surface of epithelial cells in different tissues in a wide range of vertebrate species; these molecules are also secreted in biological fluids such as saliva. The mapping of norovirus HBGA binding sites identified genotype- and strain-specific differences. The variability in these binding sites is suggested to lead to infection in different susceptible populations, thus enhancing the potential emergence of new strains.

To our knowledge, this is the first book reviewing and discussing the recent advances in the identification and development of anti-norovirus drugs (Chap. 5). Diverse molecules with therapeutic potentials and targeting different viral proteins and host cell factors were proposed. These molecules exert inhibitory activities to various steps of the norovirus life cycle including viral entry, internalization, RNA replication, translation, and polyprotein self-processing among others. Importantly, several compounds inhibiting viral binding to HBGAs were isolated with some of these compounds exhibiting a broad range of activity against GI and GII genotypes. Immunotherapeutic approaches using monoclonal antibodies and nanobodies are also discussed in this chapter. Recently, ribavirin inhibited the replication of murine norovirus in cell culture through increased mutagenesis of the replicating genomes. Two other nucleoside analogs, favipiravir (T-705) and 2CMC, also showed relative success in vivo. In addition to nucleoside analogues, Chap. 5 explores and describes other treatment strategies including protease inhibitors, anti-norovirus drug cocktails, as well as drugs targeting the gut microbiome as potential therapeutic candidates.

Recently, the Product Development for Vaccines Advisory Committee of the World Health Organization (WHO) identified norovirus as a priority disease for vaccine development. Based on an aged-structured dynamic transmission model, pediatric vaccination would avert 33% and 60% of norovirus for low vaccine efficacy and high vaccine efficacy, respectively. Currently, there are no licensed vaccines for the prevention of norovirus infection; however, the development of norovirus vaccine has been ongoing for more than a decade. The development and success of this vaccine rely on identifying and understanding the immune correlates of protection directed against noroviruses. Our current understanding of norovirus correlates of protection resulted from volunteer challenge studies and preclinical and clinical studies. Chapter 6 describes surrogate measures of immune correlates of protection including norovirus-specific antibodies and specific focus on antibodies blocking the binding of norovirus to HBGAs. Chapter 7 advances the current knowledge in relation to norovirus vaccine development whereby a number of candidates are being tested in preclinical and clinical settings. Virus-like particles

(VLPs)-based vaccine candidates expressing capsid proteins, P (the protruding domain of norovirus capsid protein) particle-based vaccines, virus replicon particles (VRPs), as well as vaccine constructs containing norovirus strains and other enteric viruses have been tested. The safety and the immunogenicity of many formulations administered intranasally, intramuscularly, or orally are described in this chapter. Until the write-up of this book, a replication-defective Adenovirus type 5 vaccine expressing the GI.1 capsid protein (Vaxart) completed two Phase I clinical trials and was reported to be safe and immunogenic. Moreover, the safety and immunogenicity of a bivalent GI and GII.4 VLP vaccine, previously evaluated in young adults and the elderly, is currently being evaluated in children in a Phase IIb study (Takeda). The development of a broad vaccine directed against predominantly circulating and newly emerging viral strains will have a significant impact on the burden of disease exerted by noroviruses. The chapters of this book present the latest on human noroviruses, their evolution, viral fitness, and preventive and control measures.

I express heartfelt appreciation to all authors for their valuable contribution and commitment extended during the different stages of processing this book. I would like to acknowledge the support of Springer Nature, its Senior Editor of Biomedicine in Vienna, Claudia Panushcka, as well as the Project Coordinator in India, Madona Samuel; both have made the preparation and production of this book a great journey.

Beirut, Lebanon Nada M. Melhem
November 2019

Contents

Global Burden of Norovirus

1

Rachel M. Burke and Aron J. Hall

1.1 Introduction

Noroviruses are small, non-enveloped, positive-stranded RNA viruses belonging to the family *Caliciviridae* [1] (Fig. 1.1). Though norovirus illness had been described as early as 1929 (termed "winter vomiting disease"), the causative agent was not identified until 1972, using Immune Electron Microscopy (IEM) on samples from a 1968 outbreak of gastroenteritis in an elementary school in Norwalk, Ohio [2]. That "prototype" virus belongs to genogroup I (GI) of the ten recognized norovirus genogroups; five genogroups (GI, GII, GIV, GVIII, and GIX) have strains that cause disease in humans [3–6]. The human disease burden caused by norovirus is large—the virus is estimated to cause nearly 20% of all acute gastroenteritis (AGE) cases worldwide [7].

Norovirus has been recognized as a predominantly winter disease; a recent review and meta-analysis was able to document this pattern worldwide [8]. Trends were clearest in the northern hemisphere, where the majority of the included studies had taken place. Similar data from Japan and Latin America confirmed winter seasonality [9, 10]. However, more data are needed especially in tropical regions for longer time frames, and in the Middle East and Northern Africa (MENA) region [11].

Noroviruses have several characteristics that enable them to be extremely successful human pathogens. They are highly contagious, with a low infectious dose (estimates range from 18 to 2800 viral particles [12, 13]). Infection leads to a generally moderate and self-limiting illness, but copious and prolonged shedding (commonly $>10^4$ copies/g of stool, up to 10^{11} copies/g, and >2 weeks in duration)

The findings and conclusions in this report are those of the authors and do not necessarily represent the official position of the US Centers for Disease Control and Prevention (CDC).

R. M. Burke (✉) · A. J. Hall
Viral Gastroenteritis Branch, Centers for Disease Control and Prevention, Atlanta, GA, USA
e-mail: RBurke@cdc.gov

N. M. Melhem (ed.), *Norovirus*,
https://doi.org/10.1007/978-3-030-27209-8_1

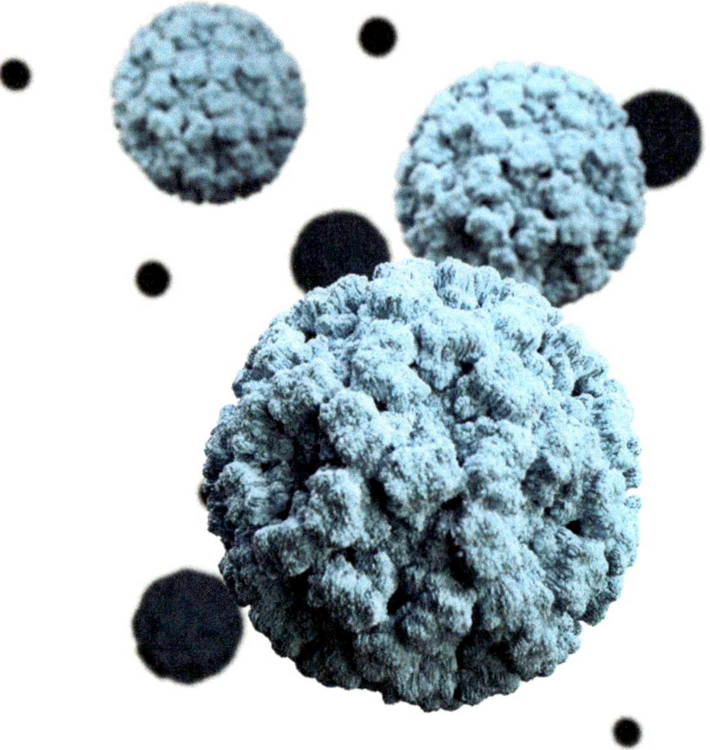

Fig. 1.1 Norovirus virion, computer-generated image. Illustrator: Alissa Eckert, MS

[14–16]. The viruses are highly environmentally stable: viruses have been recovered from environmental swabs up to 9 weeks following the onset of an outbreak, while viruses stored in groundwater have been documented to retain infectivity up to 61 days [17, 18]. Lastly, the viruses evolve rapidly, and infection provides limited immunity (estimates range from 6 months to 8 years) [19, 20]. Scientific understanding of norovirus has previously been challenged by the inability to culture the virus in vitro; however, recent advances in this area will enable new and exciting lines of research [21].

1.2 Approaches to Estimating Disease Burden Associated with Norovirus

Though noroviruses are increasingly recognized as important causes of diarrhea-associated morbidity and mortality globally, there are a number of challenges to generating estimates of the associated disease burden [22, 23]. Firstly, there are no rapid and sensitive clinical diagnostics that are in wide use; relatedly, national norovirus case-based surveillance or reporting systems are not currently in place.

However, this is rapidly changing with the increased availability and use of multi-pathogen panels in clinical practice. Nonetheless, most patients do not seek medical attention for diarrheal illness; even fewer submit stool specimens for testing. Consequently, it is often impossible to determine the etiology of AGE. Lastly, norovirus infection may not result in symptomatic AGE, or may cause an illness characterized primarily by vomiting, without prominent diarrhea.

Nonetheless, there are several approaches that can be used to estimate norovirus incidence despite these challenges. These approaches can be considered within the framework of the methodology used by the World Health Organization (WHO) Foodborne Disease Burden Epidemiology Reference Group (FERG) and the WHO Child Epidemiology Reference Group (CHERG) [24, 25]. CHERG was established in 2001 with a focus on child morbidity and mortality estimates (all causes), while FERG was established in 2013 to focus specifically on foodborne illnesses (including viral, bacterial, and parasitic causes) affecting both adults and children. Both groups use a similar methodology to generate cause-specific estimates for morbidity or mortality, which can be extended to norovirus.

This approach involves the following general steps:

1. Estimate the global incidence of all-cause diarrheal disease ("disease envelope")
2. Attribute a proportion of total diarrheal disease to noroviruses
3. Adjust the norovirus diarrhea estimate for vomiting-only cases of norovirus

For each of these steps, adjustments can be made to account for differences between age groups, disease severity (e.g., community illness, hospitalization, death), settings, and other factors of interest. The remainder of this section will discuss existing data for each of these steps and detail specific methods for generating those data, including their relative strengths and limitations.

1.2.1 Global Incidence of Diarrheal Diseases

Diarrheal diseases are an important cause of morbidity and mortality worldwide, associated with an estimated 4.5 billion illnesses and 1.7 million deaths across all ages in 2016, according to the Global Burden of Disease Study (GBD) [26, 27]. There are several approaches that may be used to calculate overall estimates of morbidity and mortality. One approach relies on systematic reviews of the literature: diarrhea surveillance results from cohort studies are translated into estimates of incidence by region and age group, incorporating country-level data as necessary (e.g., via regression) to reflect uncertainties [28–30]. This is the method that FERG has used to generate stratified morbidity and mortality estimates for the <5 and ≥ 5 age groups [24]. Another approach particularly suited to mortality estimates is to first create an overall "mortality envelope" using vital records data, and then use additional data sources and methods to attribute these deaths to various specified causes, re-scaling where necessary [31, 32]; CHERG and GBD utilized slightly different versions of this method to generate mortality estimates for children under 5 [32, 33]. Despite differences in methods, the estimates for mortality due to diarrheal illnesses in children <5 are relatively similar: CHERG estimated 0.58 million deaths in 2013, and GBD estimated 0.52 million child deaths for the same period [32, 34] (Fig. 1.2).

Though still a leading cause of infectious-disease death among children, diarrhea deaths declined by 22% in this group and by 24% across all ages (2.2 million to 1.7 million annual deaths overall) from 2006 to 2016, according to GBD estimates [26, 31, 32, 34]. These declines are likely partially attributable to general improvements in water, sanitation, hygiene, and overall economic development, as well as by increased uptake of various strategies to prevent diarrheal illnesses or mitigate their impact, including nutritional interventions, administration of zinc and oral rehydration solutions (ORS), and vaccination against rotavirus, a common cause of childhood diarrhea [36]. With increasing uptake of rotavirus vaccine worldwide, the burden of childhood diarrhea, as well as the proportion attributable to norovirus, are likely to change. For instance, increases in norovirus prevalence and decreases in rotavirus prevalence among medically attended pediatric diarrhea cases have been observed in several countries that were early introducers of rotavirus vaccine [37–39].

1.2.2 Attributing Disease Cases to Norovirus

There are several methods that can be used to attribute AGE cases to norovirus and thereby generate estimates of norovirus incidence [40]. The most common methods will be described below, and can be considered as general categories of indirect (i.e., not incorporating numbers expressly from confirmed norovirus) or direct (i.e., utilizing confirmed norovirus numbers) estimations.

1.2.2.1 Attributable Proportion Extrapolation

Similarly to the overall FERG framework, the "attributable proportion extrapolation" method of calculating norovirus incidence first calculates a "disease envelope" of overall AGE incidence, and then applies an estimated proportion attributable to norovirus. Studies using this method can take advantage of existing AGE surveillance data to estimate the disease envelope, and can incorporate other methods (e.g., based on laboratory surveillance reports, or pathogen-specific incidence reports) to estimate causes of AGE in addition to norovirus [41, 42]. One challenging aspect of this approach is choosing a proportion of disease attributable to norovirus. Although there are various studies and meta-analyses that provide overall estimates, the prevalence of norovirus among AGE cases in a given population will vary according to age group, geography, seasonality, and potentially many other factors [7, 43]. Moreover, most of the studies generating these overall estimates do not include healthy controls, who have also been shown to sometimes test positive for norovirus [44]. The relatively high prevalence of asymptomatic infection, and the long duration of shedding, contribute to the overall challenge of attributing norovirus as the etiological agent even among AGE cases that test positive for norovirus [22].

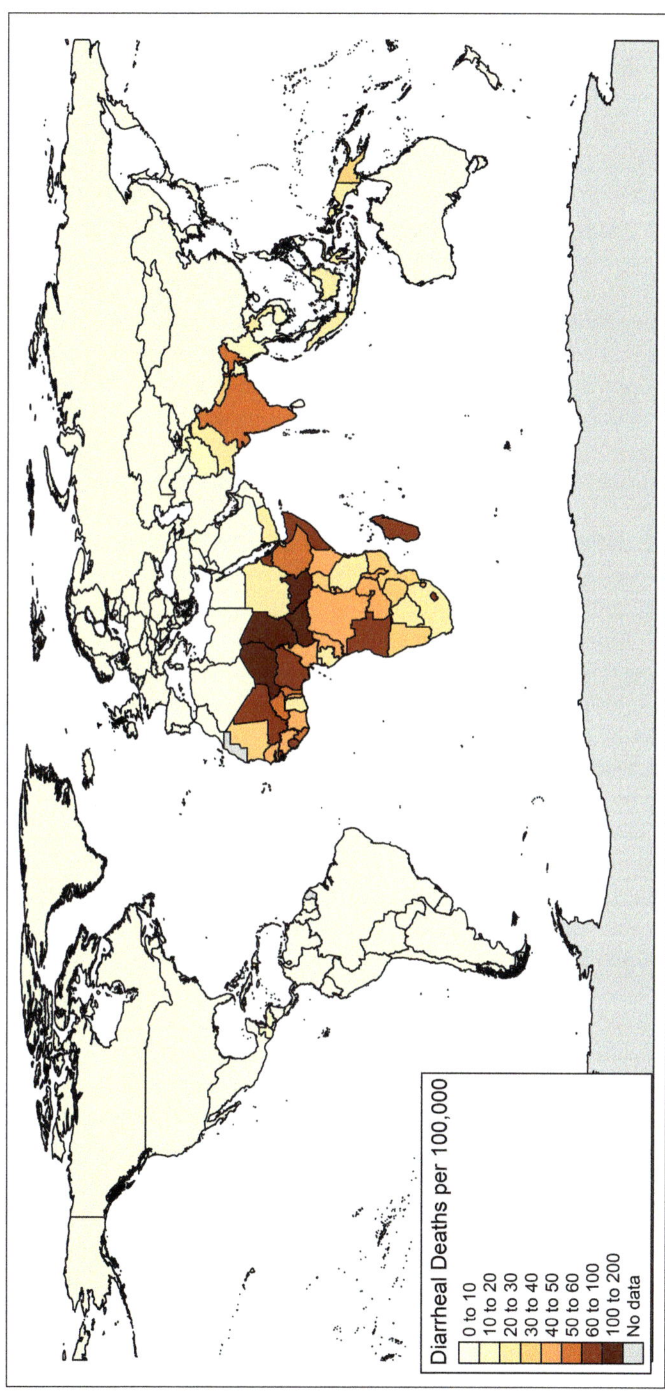

Fig. 1.2 All-ages diarrheal mortality rates by country, 2017, as estimated by the Global Burden of Disease 2017 study [35]

1.2.2.2 Indirect Attribution from Regression Modeling

Another "indirect" approach that has been used to estimate norovirus disease burden utilizes regression modeling. Specifically, time-series regression models are constructed using (nationally) representative administrative datasets with information on causes of and outcomes (e.g., hospitalizations) associated with AGE. Cause-unknown outcomes are modeled as a function of the monthly outcomes with known causes; parameters from these models are then used to allocate cause-unknown AGE cases to different causes, with the remainder (model residual) assumed to represent norovirus after accounting for seasonality [45–47]. This method can be subject to biases that could cause under- or over-estimation of the burden. For instance, the method assumes that all residual seasonality is due to norovirus, which may overestimate the burden if other diseases with similar seasonality also make up the residual. However, this approach also tends to assume that there is 1 month of every year with no norovirus-associated outcomes, which can cause an underestimate given that norovirus is known to circulate year-round. However, this approach has the advantage of making use of existing databases with broad representativeness.

1.2.2.3 Lab-Confirmed Population-Based Surveillance

The ideal approach to calculating norovirus incidence would be to use population-based active surveillance with laboratory confirmation. In this approach, cases are systematically sampled from a known catchment area, allowing the direct calculation of norovirus incidence. Examples include the Sensor Study in the Netherlands and the Infectious Intestinal Disease (IID) studies in the United Kingdom (Table 1.1). The Sensor study enrolled persons registered at sentinel general practices and followed them weekly for 6 months [48]. Similarly, the IID studies also enrolled persons served by a sample of general practices, but followed them for up to a year [49]. These studies are able to capture community incidence of norovirus, in addition to care-seeking behavior and rates of norovirus-associated clinical visits and hospitalizations. However, for these particular studies, recruitment based on registration or utilization of general practitioners could still result in bias if these persons are not representative of the overall population in the area. Of course, as with any other cohort study, active surveillance studies such as these are also subject to potential bias from loss to follow-up. Active surveillance can also be conducted in the healthcare setting in order to capture norovirus-associated medical visits and hospitalizations; Payne et al. utilized data from the New Vaccine Surveillance Network, which includes several hospital sites across the U.S., to make estimates of norovirus burden in children <5 years [38]. A disadvantage of active surveillance systems is their high resource cost, which precludes their usage in many countries.

Passive surveillance, in which routinely submitted stools are tested for pathogens, can also be used to estimate norovirus burden, if the catchment population is known. Results can also be adjusted where necessary for healthcare utilization rates (which may come from a separate sub-study). Examples of the use of passive surveillance systems (Kaiser Permanente, Veterans Affairs) are reported in Hall et al. [50] and Grytdal et al. [51, 52]. Passive surveillance studies can be limited by the fact that they utilize specimens collected through routine care, and clinicians may elect not to

Table 1.1 Incidence estimates of norovirus-associated outcomes

Norovirus-associated outcome	Methodology	Location and scope	Annual rate per 10,000 population (uncertainty bounds)	References
Deaths	Passive surveillance	Germany (national)	0.005 (0.001–0.011)	Bernard et al. [53]
	Attributable proportion extrapolation	Australia (national)	0.008 (NR)	Gibney et al. [54]
	Attributable proportion extrapolation	Netherlands (national)	0.040 (0.020–0.070)	Verhoef et al. [55]
	Indirect attribution from regression	US (national)	0.027 (0.023–0.031)	Hall et al. [46]
	Attributable proportion extrapolation	US (national)	0.019 (0.011–0.029)	Scallan et al. [42]
Hospitalizations	Active surveillance	Guatemala (sub-national): Highlands	0.6 (NR)	Bierhoff et al. [56]
		Guatemala (sub-national): Lowlands	1.7 (NR)	
	Indirect attribution from regression	Taiwan (national)	6.7 (6.4–7.1)	Burke et al. [57]
	Indirect attribution from regression	Japan (national)	13 (9–20)	Chang et al. [58]
	Passive surveillance	Germany (national)	3.4 (2.6–4.1)	Kowalzik et al. [59]
	Indirect attribution from regression	UK (England; national)	7.1 (6.2–8.0)	Verstraeten et al. [60]
	Indirect attribution from regression	Canada (national)	2.1 (1.6–2.6)	Morton et al. [61]
	Indirect attribution from regression	US (national)	0.3–0.9 (range over 5 years)	Karve et al. [62]
	Attributable proportion extrapolation	Netherlands (national)	1.2 (0.5–2.0)	Verhoef et al. [55]

(continued)

Table 1.1 (continued)

Norovirus-associated outcome	Methodology	Location and scope	Annual rate per 10,000 population (uncertainty bounds)	References
	Indirect attribution from regression	US (national)	2.4 (NR)	Lopman et al. [47]
	Attributable proportion extrapolation	US (national)	1.9 (1.1–2.9)	Scallan et al. [42]
Emergency Department visits	Indirect attribution from regression	US (national)	2.9–5.5 (range over 5 years)	Karve et al. [62]
	Indirect attribution from regression	US (national)	13.5 (8.0–18.9)	Gastanaduy et al. [45]
Outpatient visits	Active surveillance	Guatemala (sub-national): Highlands	12 (NR)	Bierhoff et al. [56]
		Guatemala (sub-national): Lowlands	36 (NR)	
	Indirect attribution from regression	Japan (national)	389 (269–558)	Chang et al. [58]
	Active surveillance	Kenya (sub-national): Urban	194 (183–206)	Shioda et al. [63]
		Kenya (sub-national): Rural	262 (238–291)	
	Indirect attribution from regression	UK (England; national)	49 (45–55)	Verstraeten et al. [60]
	Active surveillance	China (sub-national)	150 (140–160)	Yu et al. [64]
	Passive surveillance	US (Kaiser Permanente; (sub-national)	56 (53–59)	Grytdal et al. [51]
	Active surveillance	US (Kaiser Permanente; (sub-national)	50 (43–57)	Hall et al. [65]
	Indirect attribution from regression	US (national)	3.4–9.5 (range over 5 years)	Karve et al. [62]
	Indirect attribution from regression	US (national)	57 (40–74)	Gastanaduy et al. [45]

	Attributable proportion extrapolation	Netherlands (national)	9.2 (5.0–15.0)	Verhoef et al. [55]
	Active surveillance	UK (national)	21 (14–30)	Tam et al. [49]
	Passive surveillance	US (sub-national)	64 (36–120)	Hall et al. [50]
	Active surveillance	UK (national)	54 (48–60)	Phillips et al. [66]
	Active surveillance	Germany (sub-national)	63 (29–107)	Karsten et al. [67]
Total illnesses	Active surveillance	Guatemala (sub-national): Highlands	62 (NR)	Bierhoff et al. [56]
		Guatemala (sub-national): Lowlands	164 (NR)	
	Active surveillance	Peru (sub-national)	650 (540–780)	Romero et al. [68]
	Active surveillance	Kenya (sub-national): Urban	412 (387–439)	Shioda et al. [63]
		Kenya (sub-national): Rural	964 (873–1072)	
	Active surveillance	China (sub-national)	890 (820–970)	Yu et al. [64]
	Passive surveillance	US (sub-national)	689 (577–771)	Grytdal et al. [51]
	Attributable proportion extrapolation	Australia (national)	976 (NR)	Gibney et al. [54]
	Attributable proportion extrapolation	Canada (national)	1040 (924–1163)	Thomas et al. [41]
	Attributable proportion extrapolation	Netherlands (national)	380 (264–544)	Verhoef et al. [55]
	Active surveillance	UK (national)	470 (391–565)	Tam et al. [49]
	Passive surveillance	US (sub-national)	650 (370–1200)	Hall et al. [50]
	Attributable proportion extrapolation	US (national)	698 (430–1028)	Scallan et al. [42]
	Active surveillance	UK (national)	450 (380–520)	Phillips et al. [66]

test samples where viral disease is suspected. Further, these studies may be limited in their generalizability outside the catchment population.

1.2.3 Adjustment for Vomiting-Only Presentations of Norovirus

As mentioned above, norovirus is clinically characterized by a high prevalence of vomiting, particularly among children, and often to a much greater extent than other viruses causing AGE [69–75]. Indeed, norovirus infection can sometimes lead to a vomiting-only illness, which would not be detected in standard surveillance platforms for AGE, since these typically require diarrhea as part of the case definition. For this reason, estimates of norovirus incidence should be adjusted upward by what may be termed a "vomiting inflation factor" (VIF). The VIF can be calculated as follows:

$$VIF = \frac{1}{(1 - proportion\ of\ vomiting\ only\ norovirus\ illnesses)}$$

Multiplying the incidence estimate by the VIF will generate an adjusted overall incidence estimate. For example, a WHO-commissioned meta-analysis generated a VIF of 1.19, suggesting that accounting for vomiting-only presentations of norovirus would increase incidence estimates by nearly 20%. This figure was then used in the FERG conclusions to show a potential range of illness numbers [24].

1.3 Estimates of Norovirus Disease Burden

1.3.1 Overall

In a 2015 analysis using a combined methodology (national data for low-mortality countries and CHERG for all other countries), the overall disease burden of norovirus was estimated to encompass 685 million (95% UI: 491–1123 million) cases and 212,489 (95% UI: 160,595–278,420) deaths annually [33]. Estimates of the incidence of norovirus-related medical outcomes vary, but population-based studies suggest an annual incidence of 100–1000 norovirus illnesses, 5–400 norovirus-associated outpatient visits, 1–13 norovirus hospitalizations, and 0.005–0.04 norovirus deaths per 10,000 persons [40–42, 45–47, 49–51, 53–64, 66–68, 65, 76] (Table 1.1). Given that norovirus prevalence among AGE has been shown to vary by setting and geography [7], these numbers are not necessarily representative of global norovirus burden. It is reassuring that estimates applying different methods to the same geography are generally similar, at least within an order of magnitude.

The setting of norovirus illness can serve as a proxy for disease severity. Community-acquired norovirus rarely results in death, but norovirus acquired in a healthcare facility or other institution (particularly among the elderly) is more likely to result in severe outcomes [77]. Outbreaks in institutional settings can also have high economic costs [78, 79].

1.3.2 By Geography

The burden of norovirus has been shown to vary substantially by geography, with a recent analysis by WHO region estimating the lowest annual rate of illness (534.3/ 10,000; 95% UI: 155.0–2392.4) in the South-East Asia region (SEAR), and the highest annual rate of illness (1820.3/ 10,000; 95% UI: 102.9–3121.2) in the Eastern Mediterranean region (EMR), which encompasses countries in the Arabian Peninsula, northeastern Africa, and southwestern Asia [33]. The estimated death rates, however, were highest in SEAR and the African region (AFR), both estimated at 0.7 deaths/10,000 persons annually. Nonetheless, because nationally representative estimates are only available in developed-country settings, and because healthcare-seeking behaviors vary across geographies (for instance, given the presence or absence of national healthcare plans), it is difficult to make generalizations about differences in incidence rates by geography.

1.3.3 By Age

Comparing norovirus incidence estimates by age group, it is clear that young children suffer the highest burden of community-acquired illness, and also tend to show greater incidence of seeking care or being hospitalized, sometimes an order of magnitude above adults [51, 57–59, 61, 62, 64, 66, 80] (Table 1.2). Estimates of hospitalization rates from population-based studies range from 5 to 150 per 10,000 child-years, with the highest rates in settings with national healthcare systems [38, 47, 56–62, 81–83]. Outpatient visit rate estimates range from 60 to 1600 per 10,000 child-years, with the highest estimates in Japan (~1600), Kenya (~850), and China (~750) [38, 45, 51, 56, 58, 60, 62–64, 66, 80, 81, 84]. Estimates of total illnesses are also high among children, ranging from ~200 up to 10,000 per 10,000 child-years in some age groups [37, 51, 56, 63, 64, 66, 68, 76, 80, 85–88].

Another group at high risk for medical care and severe outcomes is the elderly [45–47, 53, 57, 59, 61, 62, 82, 89] (Table 1.3), though not all studies show a substantially higher incidence in this group when compared to younger adults [51, 58, 66, 80]. Differences tend to be most notable for hospitalizations, where estimated rates can be 4–19 times as high among the elderly as among younger adults and older children [47, 57–62, 89].

1.3.4 By Genogroup

Human norovirus disease is caused by noroviruses in five genogroups: GI, GII, GIV, GVIII, and GIX, comprising multiple genotypes [4, 6]. Non-GII outbreaks are mainly caused by GI viruses, while GIV is rarely associated with outbreaks [90, 91]. Global pandemic norovirus has invariably been caused by a GII.4 strain, as different variants evolve and replace the previously circulating strains [92]. However, the recent emergence of GII.17 in Asia [93] and of recombinant GII.P16–GII.2

Table 1.2 Incidence estimates of norovirus-associated outcomes in pediatric populations

Norovirus-associated outcome	Methodology	Location and scope	Age group	Annual rate per 10,000 population (uncertainty bounds)	References
Deaths	Indirect attribution from regression	US (national)	0–59 mo.	0.013 (0.009–0.017)	Hall et al. [46]
Hospitalizations	Active surveillance	Guatemala (sub-national): Highlands	0–59 mo.	5.1 (NR)	Bierhoff et al. [56]
		Guatemala (sub-national): Lowlands	0–59 mo.	10.5 (NR)	
	Indirect attribution from regression	Taiwan (national)	0–59 mo.	63.7 (61.5–66.1)	Burke et al. [57]
	Indirect attribution from regression	Japan (national)	0–59 mo.	48 (39–56)	Chang et al. [58]
	Passive surveillance	Germany (national)	0–11 mo.	34.8–53.1 (range over 6 years)	Kowalzik et al. [59]
			12–23 mo.	41.8–52.0 (range over 6 years)	
	Active surveillance	Colombia (sub-national)	0–11 mo.	106 (NR)	Lopez-Medina et al. [81]
			12–23 mo.	102 (NR)	
			24–59 mo.	20 (NR)	
	Indirect attribution from regression	UK (England)	0–59 mo.	33 (24–44)	Verstraeten et al. [60]
	Passive surveillance	Hong Kong (sub-national)	0–59 mo.	147.5 (NR)	Chan et al. [82]
	Active surveillance	Israel (sub-national)	0–59 mo.	10.4 (9.5–11.3)	Leshem et al. [83]
	Indirect attribution from regression	Canada (national)	0–59 mo.	8.4 (5.6–11.2)	Morton et al. [61]
	Indirect attribution from regression	US (national)	0–59 mo.	1.0–5.7 (range over 5 years)	Karve et al. [62]

	Active surveillance	US (sub-national)	0–59 mo.	7.2 (5.8–8.7)	Payne et al. [38]
	Indirect attribution from regression	US (national)	0–59 mo.	9.4 (NR)	Lopman et al. [47]
Emergency Department visits	Active surveillance	Israel (sub-national)	0–59 mo.	9.9 (9.0–10.7)	Leshem et al. [83]
	Indirect attribution from regression	US (national)	0–59 mo.	9.4–28.4 (range over 5 years)	Karve et al. [62]
	Indirect attribution from regression	US (national)	0–59 mo.	38.3 (24.2–52.4)	Gastañaduy et al. [45]
	Active surveillance	US (sub-national)	0–59 mo.	140.7 (NR)	Payne et al. [38]
Outpatient visits	Active surveillance	Guatemala (sub-national): Highlands	0–59 mo.	63 (NR)	Bierhoff et al. [56]
		Guatemala (sub-national): Lowlands	0–59 mo.	221 (NR)	
	Indirect attribution from regression	Japan (national)	0–59 mo.	1569 (1325–1792)	Chang et al. [58]
	Active surveillance	Colombia (sub-national)	0–11 mo.	208 (NR)	Lopez-Medina et al. [81]
			12–23 mo.	1086 (NR)	
			24–59 mo.	218 (NR)	
	Active surveillance	Kenya (sub-national): Urban	0–59 mo.	860 (792–942)	Shioda et al. [63]
		Kenya (sub-national): Rural	0–59 mo.	743 (609–947)	
	Indirect attribution from regression	UK (England)	0–59 mo.	340 (275–426)	Verstraeten et al. [60]
	Active surveillance	China (sub-national)	0–24 mo.	740 (540–930)	Yu et al. [64]
	Active surveillance	Italy (sub-national)	0–59 mo.	151 (124–182)	Dona et al. [84]

(continued)

Table 1.2 (continued)

Norovirus-associated outcome	Methodology	Location and scope	Age group	Annual rate per 10,000 population (uncertainty bounds)	References
	Passive surveillance	US (sub-national)	0–59 mo.	256 (NR)	Grytdal et al. [51]
	Active surveillance	UK (national)	0–59 mo.	144 (85–245)	O'Brien et al. [80]
	Indirect attribution from regression	US (national)	0–59 mo.	23.2–87.4 (range over 5 years)	Karve et al. [62]
	Indirect attribution from regression	US (national)	0–59 mo.	233 (156–310)	Gastañaduy et al. [45]
	Active surveillance	US (sub-national)	0–59 mo.	319 (NR)	Payne et al. [38]
	Active surveillance	UK (national)	0–59 mo.	320 (260–380)	Phillips et al. [66]
Total illness	Active surveillance	Guatemala (sub-national): Highlands	0–59 mo.	207 (NR)	Bierhoff et al. [56]
		Guatemala (sub-national): Lowlands	0–59 mo.	495 (NR)	
	Active surveillance	Guatemala (sub-national)	<18 years	140 (NR)	Olson et al. [76]
	Active surveillance	Peru (sub-national)	<2 years	6240 (4630–8260)	Romero et al. [68]
			2–4 years	1140 (680–1790)	
	Active surveillance	Kenya (sub-national): Urban	0–59 mo.	1751 (1604–1925)	Shioda et al. [63]
		Kenya (sub-national): Rural	0–59 mo.	2223 (1818–2843)	
	Active surveillance	China (sub-national)	0–11 mo.	2030 (1380–2790)	Yu et al. [64]
			12–23 mo.	1470 (920–2080)	
			24–59 mo.	250 (170–350)	
	Passive surveillance	US (national)	0–59 mo.	1521 (NR)	Grytdal et al. [51]
	Active surveillance	UK (national)	0–59 mo.	1426 (998–2039)	O'Brien et al. [80]

	Indirect attribution from regression	Netherlands (national)	0–48 mo.	390 (NR)	Enserink et al. [85]
	Active surveillance (birth cohort)	Ecuador (sub-national)	<36 mo.	1700 (1400–2100)	Lopman et al. [86]
	Active surveillance	Peru (sub-national)	12–24 mo.	11,000 (NR)	Zambruni et al. [87]
	Active surveillance	Nicaragua (sub-national)	0–59 mo.	2290 (1890–2780)	Becker-Dreps et al. [37]
	Active surveillance (birth cohort)	Peru (sub-national)	0–5 mo.	2880 (2040–4080)	Saito et al. [88]
			6–11 mo.	7920 (2520–9600)	
			12–24 mo.	8400 (7320–10,080)	
	Active surveillance	UK (national)	0–59 mo.	2140 (1590–2770)	Phillips et al. [66]

Table 1.3 Incidence estimates of norovirus-associated outcomes in geriatric populations

Norovirus-associated outcome	Methodology	Location and scope	Age group	Annual rate per 10,000 population (uncertainty bounds)	References
Deaths	Indirect attribution from regression	US (national)	65+	0.20 (0.17–0.23)	Hall et al. [46]
	Passive surveillance	Germany (national)	75+	0.05 (0.01–0.11)	Bernard et al. [53]
Hospitalizations	Indirect attribution from regression	Taiwan (national)	65+	8.4 (7.9–9.0)	Burke et al. [57]
	Indirect attribution from regression	Japan (national)	65+	13 (9–20)	Chang et al. [58]
	Passive surveillance	Germany (national)	85+	11.8–20.9 (range over 6 years)	Kowalzik et al. [59]
	Indirect attribution from regression	UK (England)	65+	17 (14–20)	Verstraeten et al. [60]
	Passive surveillance	Hong Kong (sub-national)	85+	58.1 (NR)	Chan et al. [82]
	Passive surveillance	US (Veterans Affairs; sub-national)	>65	8 (including HAI) (NR)	Grytdal et al. [52]
	Indirect attribution from regression	Canada (national)	65–84	6.0 (4.9–7.1)	Morton et al. [61]
			85+	24.4 (20.2–28.7)	
	Indirect attribution from regression	US (national)	65–74	1.5–3.0 (range over 5 years)	Karve et al. [62]
			75–84	2.4–9.8 (range over 5 years)	
			85+	5.5–10.1 (range over 5 years)	
	Indirect attribution from regression	US (national)	65–74	4.7 (NR)	Lopman et al. [47]
			75–84	9.2 (NR)	
			85+	18.5 (NR)	

	Indirect attribution from regression	UK (national)	65+	1.0–4.3 (range over 6 years)	Haustein et al. [89]
Emergency Department visits	Indirect attribution from regression	US (national)	65–74	2.4–7.0 (range over 5 years)	Karve et al. [62]
			75–84	4.6–10.5 (range over 5 years)	
			85+	8.0–17.4 (range over 5 years)	
	Indirect attribution from regression	US (national)	65+	14.5 (8.8–20.2)	Gastañaduy et al. [45]
	Indirect attribution from regression	England (national)	65+	11.7 (10.4–13.0)	Haustein et al. [89]
Outpatient visits	Indirect attribution from regression	Japan (national)	65+	245 (107–445)	Chang et al. [58]
	Indirect attribution from regression	UK (England; national)	65+	39 (33–45)	Verstraeten et al. [60]
	Active surveillance	China (sub-national)	65+	260 (200–320)	Yu et al. [64]
	Passive surveillance	US (sub-national)	>65	79 (NR)	Grytdal et al. [51]
	Active surveillance	UK (national)	65+	21 (11–40)	O'Brien et al. [80]
	Active surveillance	UK (national)	65+	37 (27–47)	Phillips et al. [66]
	Passive surveillance	US (Veterans Affairs; sub-national)	>65	20 (NR)	Grytdal et al. [52]
	Indirect attribution from regression	US (national)	65–74	3.4–11.0 (range over 5 years)	Karve et al. [62]
			75–84	8.9–22.7 (range over 5 years)	
			85+	8.1–15.0 (range over 5 years)	
	Indirect attribution from regression	US (national)		54 (36–72)	Gastanaduy et al. [45]
Total illnesses	Active surveillance	UK (national)	65+	277 (196–391)	O'Brien et al. [80]
	Active surveillance	China (sub-national)	65+	780 (610–950)	Yu et al. [64]

strains in Asia [94–98] and Europe [99, 100] highlights geographical differences in norovirus molecular epidemiology, and raises the possibility that GII.4 will not dominate future outbreaks worldwide [93].

Noroviruses associated with endemic (sporadic) disease have a similar genotype distribution. A 2013 review of sporadic pediatric norovirus cases found that, while GII was still by far the most common genogroup worldwide, the distribution of genogroups did vary by region [101]. In some countries, the prevalence of GI was approximately one third. This study also reflected a potential decrease in the role of GII.4 in sporadic pediatric disease in Asia. Studies including adults have also shown a predominance of GII noroviruses, though prevalence varies by region, and a potential emergence of GII.17 in some parts of Asia [102–108].

1.3.4.1 GII.4 Noroviruses

GII viruses are associated with 90% of outbreak samples in the United States [90] and over 50% of samples in outbreaks worldwide [91]. Globally, GII.4 has been associated with the majority of outbreaks, particularly winter outbreaks in most geographies [109]. GII.4 viruses evolve rapidly, and new antigenically distinct strains emerge every few years, escaping population immunity and causing pandemics [110, 111].

In terms of other epidemiologic differences, GII.4 is less likely than other genotypes to be associated with foodborne outbreaks, and more likely to be associated with healthcare outbreaks and other outbreaks characterized by person-to-person spread [112, 113]; however, recent analysis suggests that among non-healthcare outbreaks and after controlling for other variables, GII.4 may be more likely to be foodborne (as opposed to person-to-person or other transmission modes) [114]. This genotype also tends to cause more severe outcomes as compared to other genotypes: incidence rate ratios (IRR) for hospitalization and mortality rates comparing GII.4 to non-GII.4 outbreaks have ranged 1.4–9.4 and 2.8–3.1, respectively [114, 115]. In several studies, GII.4 has been shown to cause a more severe illness presentation, with more vomiting, when compared to other genotypes [116, 117]; however, in a more recent analysis, symptoms were not significantly associated with genotype once adjusting for transmission, setting, and seasonality [114]. Additional multivariable analyses from other surveillance systems will be useful to further understand the relationships among genotype, setting, and transmission.

1.3.4.2 Emerging Non-GII.4 Noroviruses

Beginning in winter 2014/2015, a novel GII.17 strain (GII.17 Kawasaki) emerged and appeared to escape population immunity in Asia, causing widespread outbreaks in China and Japan, and replacing GII.4 in some locations [93, 103, 118–120]. Though not predominant, this strain has also been detected in Europe and North America [93, 118, 121, 122]. In contrast to GII.4, which typically has the greatest effect on children and the elderly, GII.17 Kawasaki has been documented to cause severe illness across the age spectrum [123]. Given its rapid emergence and outbreak potential, this is a strain that is being actively monitored.

Novel recombinant GII viruses continue to emerge, affecting adults as well as children, and across different geographies [94–97, 99, 100, 122, 124, 125]. These genotypes are also being monitored, given their potential to escape population immunity and cause large outbreaks. Further analysis is needed to elucidate the relative role of polymerase versus capsid changes in the genome in evoking population immunity, as well as to clarify any difference in clinical or epidemiologic characteristics of these recombinants.

1.4 Economic Burden of Norovirus

Bartsch et al. recently developed a computational simulation model to estimate the economic burden of norovirus globally [126]. Based on this model, norovirus was associated with $4.2 billion (95% UI: $3.2–5.7 billion) in direct costs and $60.3 billion (95% UI: $44.4–83.4 billion) in societal costs each year [126]. Productivity losses, approximately half of which were attributable to deaths, made up the bulk of overall costs. All costs varied across age groups and geography (Fig. 1.3).

Estimated annual costs associated with norovirus, generated in the study described above, were highest in high-income countries as compared to low- and middle-income countries (U$34.7 billion, 95% UI: $24.3–49.6 billion; vs. $25.4 billion, 95% UI: $19.4–35.2 billion) [126]. The lowest per-episode costs were seen in AFR ($50), while the highest per-episode costs were seen in the Americas (>$150). These differences are likely reflective of overall differences in health systems and healthcare-seeking behavior, and should be interpreted cautiously given that the model relied on assumptions to extrapolate from countries with existing data to countries without data.

Just as incidence of norovirus medical outcomes varies by age, so does the associated cost of illness. Despite the fact that the modeled average cost per norovirus episode was highest in the elderly, children under 5 accounted for the majority of overall annual societal costs due to norovirus—$39.8 billion (95% UI: $27.2–$58.1 billion), as compared to $20.4 billion (95% UI: $16.9–$25.4 billion) for all other age groups combined [126].

While it is clear that norovirus illnesses cause substantial economic burden across geographies, data gaps remain. As norovirus vaccines get further in development, additional data on cost of illness will be important to be able to assess cost-effectiveness of potential vaccination programs.

1.5 Critical Data Gaps

Although norovirus research has expanded substantially in the last decade, critical knowledge gaps remain [7, 22]. The bulk of population-based norovirus burden studies take place in highly developed countries, which have different circulating norovirus strains [91], patterns of care-seeking behavior and associated costs [126], and prevalence of norovirus among AGE cases [7]. While several developing-

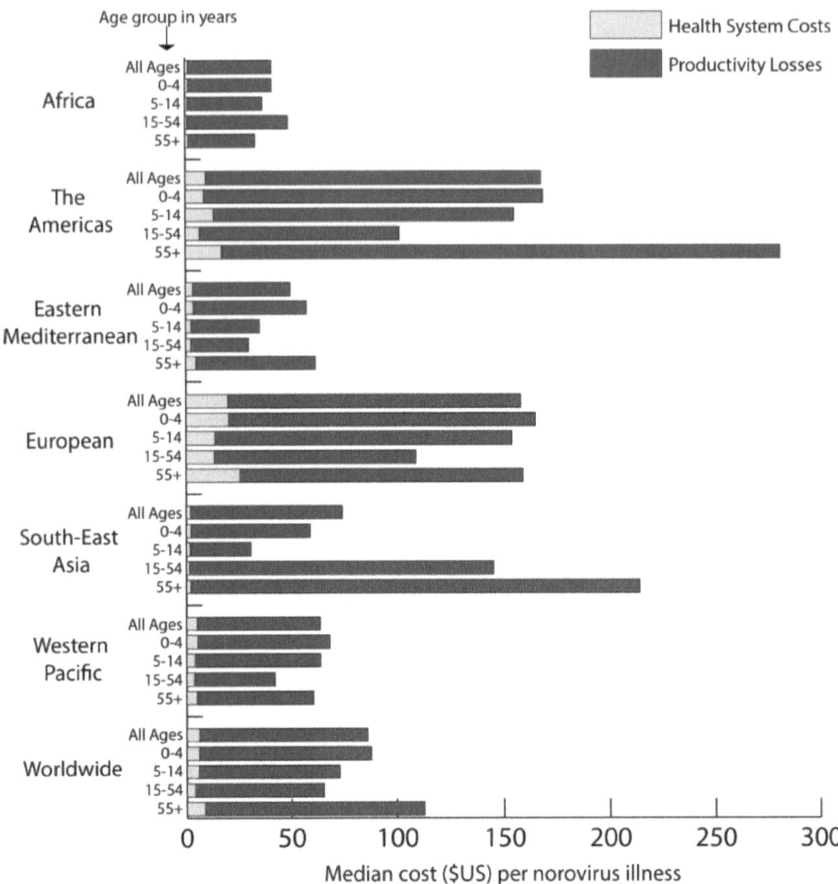

Fig. 1.3 Adapted from Bartsch et al. Median health system and productivity lost cost per norovirus illness by age group for each region and globally (total bar represents societal cost). Results based on simulation model

country setting birth cohorts have recently been undertaken and provide important data [86, 88], norovirus transmission dynamics and outcomes are known to be different across age groups. Thus, in addition to further birth cohort studies, there is a need for population-based incidence studies that cover the entire age spectrum and represent different geographies. For instance, most of the population-based incidence studies take place in developed settings in North America, Europe, or Asia; a few studies have been conducted in South America, but these are generally restricted to a single site. Population-based studies are especially sparse in Africa and the Middle East, though norovirus also circulates and causes AGE in those regions [7, 8, 11].

Another challenge to estimating norovirus disease burden is the high prevalence of asymptomatic infections and the long period of shedding, such that norovirus is often detected in the stool of healthy controls [22]. This makes it difficult to pinpoint exactly when norovirus is the disease-causing agent, complicating estimates of AGE burden attributable to norovirus. One possibility that has been explored is to use Ct value (cycle threshold from real time polymerase chain reaction, RT-PCR) as a proxy of viral load, with lower Ct values corresponding to higher viral loads, in turn serving as proxy for disease-causing status of norovirus; however, this method has shown varying levels of success, and Ct cut-offs will vary by genotype, population, and even laboratory [127–129]. Better technology or methodology will be necessary to improve distinction between asymptomatic versus disease-causing norovirus infection.

1.6 Additional Areas of Research

The field of human norovirus research was long limited by the lack of a viable method of in vitro culture [21]. However, in 2016, scientists from Baylor College of Medicine reported the successful culture of human norovirus in human intestinal enteroids (HIEs), also termed "miniguts" [21]. The ability to culture norovirus in vitro opens up a number of research possibilities. Most relevant to global norovirus burden include the ability to better understand the mechanisms of norovirus evolution, the impact of various host factors on norovirus infection and shedding, and the infectiousness of shed virus following therapeutic interventions.

An additional area of ongoing norovirus research is vaccine development given the widespread burden of norovirus [22]. Clinical research on norovirus vaccines will likely go hand in hand with research on norovirus burden in different populations: a better characterization of the disease burden will inform and promote vaccine development, and vaccine studies can also provide information on norovirus burden. Indeed, the existence of norovirus vaccines in development may well spur investment in enhanced surveillance for norovirus in preparation for studies of vaccine impact.

1.7 Conclusions

Norovirus is a globally important cause of AGE, associated with an estimated annual 685 million (95% UI: 491–1123 million) cases and 212,489 (95% UI: 160,595–278,420) deaths worldwide [33]. Although norovirus occurs across the age spectrum, young children and elderly persons are most affected, with the elderly most likely to suffer the most severe outcomes [45–47, 51, 66, 80, 82]. Norovirus is ubiquitous worldwide, though outcomes vary [32]. Although GII.4 strains have historically predominated, with new variants emerging every few years to cause global pandemics, recent surveillance suggests that other strains, such as GII.17 or GII.2/GII.P16–GII.2 recombinants, may replace GII.4 in some geographies [91–93, 109].

Despite the clear role that norovirus plays in the global burden of AGE, several challenges have historically existed to generating a valid estimate of norovirus incidence. These challenges have included lower rates of healthcare-seeking behavior and specimen submission, limited use of norovirus diagnostics in clinical practice, a lack of national surveillance systems, and a significant proportion of asymptomatic and prolonged shedding among currently or previously infected individuals. Nonetheless, several methods have been used to successfully generate incidence estimates. However, key gaps remain in data from developing countries, particularly from adult populations. More data are needed to better understand asymptomatic and prolonged shedding, and the impact this may have on norovirus attribution to AGE.

New technologies such as norovirus culture in HIEs will enable further research into norovirus evolution, infection and transmission dynamics, and effectiveness of therapeutic interventions. Further areas of exciting research include vaccine development, which will likely complement increased surveillance activities.

References

1. Robilotti E, Deresinski S, Pinsky BA. Norovirus. Clin Microbiol Rev. 2015;28(1):134–64.
2. Kapikian AZ. The discovery of the 27-nm Norwalk virus: an historic perspective. J Infect Dis. 2000;181(Suppl 2):S295–302.
3. Kroneman A, Vega E, Vennema H, Vinje J, White PA, Hansman G, et al. Proposal for a unified norovirus nomenclature and genotyping. Arch Virol. 2013;158(10):2059–68.
4. Zheng DP, Ando T, Fankhauser RL, Beard RS, Glass RI, Monroe SS. Norovirus classification and proposed strain nomenclature. Virology. 2006;346(2):312–23.
5. Vinje J. Advances in laboratory methods for detection and typing of norovirus. J Clin Microbiol. 2015;53(2):373–81.
6. Chhabra P, de Graaf M, Parra GI, Chan MC, Green K, Martella V, Wang Q, White PA, Katayama K, Vennema H, Koopmans MPG, Vinjé J. Updated classification of norovirus genogroups and genotypes. J Gen Virol. 2019;100(10):1393–406.
7. Ahmed SM, Hall AJ, Robinson AE, Verhoef L, Premkumar P, Parashar UD, et al. Global prevalence of norovirus in cases of gastroenteritis: a systematic review and meta-analysis. Lancet Infect Dis. 2014;14(8):725–30.
8. Ahmed SM, Lopman BA, Levy K. A systematic review and meta-analysis of the global seasonality of norovirus. PLoS One. 2013;8(10):e75922.
9. Thongprachum A, Khamrin P, Maneekarn N, Hayakawa S, Ushijima H. Epidemiology of gastroenteritis viruses in Japan: prevalence, seasonality, and outbreak. J Med Virol. 2016;88 (4):551–70.
10. da Silva PT, Peiro JR, Mendes LC, Ludwig LF, de Oliveira-Filho EF, Bucardo F, et al. Human norovirus infection in Latin America. J Clin Virol. 2016;78:111–9.
11. Kreidieh K, Charide R, Dbaibo G, Melhem NM. The epidemiology of norovirus in the Middle East and North Africa (MENA) region: a systematic review. Virol J. 2017;14(1):220.
12. Teunis PF, Moe CL, Liu P, Miller SE, Lindesmith L, Baric RS, et al. Norwalk virus: how infectious is it? J Med Virol. 2008;80(8):1468–76.
13. Atmar RL, Opekun AR, Gilger MA, Estes MK, Crawford SE, Neill FH, et al. Determination of the 50% human infectious dose for Norwalk virus. J Infect Dis. 2014;209(7):1016–22.
14. Teunis PF, Sukhrie FH, Vennema H, Bogerman J, Beersma MF, Koopmans MP. Shedding of norovirus in symptomatic and asymptomatic infections. Epidemiol Infect. 2015;143(8):1710–7.

15. Aoki Y, Suto A, Mizuta K, Ahiko T, Osaka K, Matsuzaki Y. Duration of norovirus excretion and the longitudinal course of viral load in norovirus-infected elderly patients. J Hosp Infect. 2010;75(1):42–6.
16. Murata T, Katsushima N, Mizuta K, Muraki Y, Hongo S, Matsuzaki Y. Prolonged norovirus shedding in infants <or=6 months of age with gastroenteritis. Pediatr Infect Dis J. 2007;26 (1):46–9.
17. Kotwal G, Cannon JL. Environmental persistence and transfer of enteric viruses. Curr Opin Virol. 2014;4:37–43.
18. Seitz SR, Leon JS, Schwab KJ, Lyon GM, Dowd M, McDaniels M, et al. Norovirus infectivity in humans and persistence in water. Appl Environ Microbiol. 2011;77(19):6884–8.
19. Hall AJ. Noroviruses: the perfect human pathogens? J Infect Dis. 2012;205(11):1622–4.
20. Simmons K, Gambhir M, Leon J, Lopman B. Duration of immunity to norovirus gastroenteritis. Emerg Infect Dis. 2013;19(8):1260–7.
21. Ettayebi K, Crawford SE, Murakami K, Broughman JR, Karandikar U, Tenge VR, et al. Replication of human noroviruses in stem cell-derived human enteroids. Science. 2016;353 (6306):1387–93.
22. Lopman BA, Steele D, Kirkwood CD, Parashar UD. The vast and varied global burden of norovirus: prospects for prevention and control. PLoS Med. 2016;13(4):e1001999.
23. Yen C, Hall AJ. Editorial commentary: challenges to estimating norovirus disease burden. J Pediatric Infect Dis Soc. 2013;2(1):61–2.
24. WHO estimates of the global burden of foodborne diseases: foodborne disease burden epidemiology reference group 2007-2015. Geneva, Switzerland: World Health Organization; 2015.
25. Lanata CF, Fischer-Walker CL, Olascoaga AC, Torres CX, Aryee MJ, Black RE, et al. Global causes of diarrheal disease mortality in children <5 years of age: a systematic review. PLoS One. 2013;8(9):e72788.
26. GBD 2016 Causes of Death Collaborators. Global, regional, and national age-sex specific mortality for 264 causes of death, 1980-2016: a systematic analysis for the Global Burden of Disease Study 2016. Lancet. 2017;390(10100):1151–210.
27. GBD 2016 Disease and Injury Incidence and Prevalence Collaborators. Global, regional, and national incidence, prevalence, and years lived with disability for 328 diseases and injuries for 195 countries, 1990-2016: a systematic analysis for the Global Burden of Disease Study 2016. Lancet. 2017;390(10100):1211–59.
28. Fischer Walker CL, Perin J, Aryee MJ, Boschi-Pinto C, Black RE. Diarrhea incidence in low- and middle-income countries in 1990 and 2010: a systematic review. BMC Public Health. 2012;12:220.
29. Walker CL, Rudan I, Liu L, Nair H, Theodoratou E, Bhutta ZA, et al. Global burden of childhood pneumonia and diarrhoea. Lancet. 2013;381(9875):1405–16.
30. Walker CL, Black RE. Diarrhoea morbidity and mortality in older children, adolescents, and adults. Epidemiol Infect. 2010;138(9):1215–26.
31. Liu L, Johnson HL, Cousens S, Perin J, Scott S, Lawn JE, et al. Global, regional, and national causes of child mortality: an updated systematic analysis for 2010 with time trends since 2000. Lancet. 2012;379(9832):2151–61.
32. Global Burden of Disease Working Group. Global, regional, and national life expectancy, all-cause mortality, and cause-specific mortality for 249 causes of death, 1980-2015: a systematic analysis for the Global Burden of Disease Study 2015. Lancet. 2016;388(10053):1459–544.
33. Pires SM, Fischer-Walker CL, Lanata CF, Devleesschauwer B, Hall AJ, Kirk MD, et al. Aetiology-specific estimates of the global and regional incidence and mortality of diarrhoeal diseases commonly transmitted through food. PLoS One. 2015;10(12):e0142927.
34. Liu L, Oza S, Hogan D, Perin J, Rudan I, Lawn JE, et al. Global, regional, and national causes of child mortality in 2000-13, with projections to inform post-2015 priorities: an updated systematic analysis. Lancet. 2015;385(9966):430–40.

35. GBD Compare Data Visualization [Internet]. IHME, University of Washington. 2016. http://vizhub.healthdata.org/gbd-compare/. Accessed 1 May 2017.
36. Bhutta ZA, Das JK, Walker N, Rizvi A, Campbell H, Rudan I, et al. Interventions to address deaths from childhood pneumonia and diarrhoea equitably: what works and at what cost? Lancet. 2013;381(9875):1417–29.
37. Becker-Dreps S, Bucardo F, Vilchez S, Zambrana LE, Liu L, Weber DJ, et al. Etiology of childhood diarrhea after rotavirus vaccine introduction: a prospective, population-based study in Nicaragua. Pediatr Infect Dis J. 2014;33(11):1156–63.
38. Payne DC, Vinje J, Szilagyi PG, Edwards KM, Staat MA, Weinberg GA, et al. Norovirus and medically attended gastroenteritis in U.S. children. N Engl J Med. 2013;368(12):1121–30.
39. McAtee CL, Webman R, Gilman RH, Mejia C, Bern C, Apaza S, et al. Burden of norovirus and rotavirus in children after rotavirus vaccine introduction, Cochabamba, Bolivia. Am J Trop Med Hyg. 2016;94(1):212–7.
40. Hall AJ, Lopman BA, Payne DC, Patel MM, Gastanaduy PA, Vinje J, et al. Norovirus disease in the United States. Emerg Infect Dis. 2013;19(8):1198–205.
41. Thomas MK, Murray R, Flockhart L, Pintar K, Pollari F, Fazil A, et al. Estimates of the burden of foodborne illness in Canada for 30 specified pathogens and unspecified agents, circa 2006. Foodborne Pathog Dis. 2013;10(7):639–48.
42. Scallan E, Hoekstra RM, Angulo FJ, Tauxe RV, Widdowson MA, Roy SL, et al. Foodborne illness acquired in the United States—major pathogens. Emerg Infect Dis. 2011;17(1):7–15.
43. Patel MM, Widdowson MA, Glass RI, Akazawa K, Vinje J, Parashar UD. Systematic literature review of role of noroviruses in sporadic gastroenteritis. Emerg Infect Dis. 2008;14(8):1224–31.
44. Rouhani S, Penataro Yori P, Paredes Olortegui M, Siguas Salas M, Rengifo Trigoso D, Mondal D, et al. Norovirus infection and acquired immunity in 8 countries: results from the MAL-ED study. Clin Infect Dis. 2016;62(10):1210–7.
45. Gastanaduy PA, Hall AJ, Curns AT, Parashar UD, Lopman BA. Burden of norovirus gastroenteritis in the ambulatory setting—United States, 2001-2009. J Infect Dis. 2013;207(7):1058–65.
46. Hall AJ, Curns AT, McDonald LC, Parashar UD, Lopman BA. The roles of *Clostridium difficile* and norovirus among gastroenteritis-associated deaths in the United States, 1999-2007. Clin Infect Dis. 2012;55(2):216–23.
47. Lopman BA, Hall AJ, Curns AT, Parashar UD. Increasing rates of gastroenteritis hospital discharges in US adults and the contribution of norovirus, 1996-2007. Clin Infect Dis. 2011;52(4):466–74.
48. de Wit MA, Koopmans MP, Kortbeek LM, Wannet WJ, Vinje J, van Leusden F, et al. Sensor, a population-based cohort study on gastroenteritis in the Netherlands: incidence and etiology. Am J Epidemiol. 2001;154(7):666–74.
49. Tam CC, Rodrigues LC, Viviani L, Dodds JP, Evans MR, Hunter PR, et al. Longitudinal study of infectious intestinal disease in the UK (IID2 study): incidence in the community and presenting to general practice. Gut. 2012;61(1):69–77.
50. Hall AJ, Rosenthal M, Gregoricus N, Greene SA, Ferguson J, Henao OL, et al. Incidence of acute gastroenteritis and role of norovirus, Georgia, USA, 2004-2005. Emerg Infect Dis. 2011;17(8):1381–8.
51. Grytdal SP, DeBess E, Lee LE, Blythe D, Ryan P, Biggs C, et al. Incidence of norovirus and other viral pathogens that cause acute Gastroenteritis (AGE) among Kaiser permanente member populations in the United States, 2012-2013. PLoS One. 2016;11(4):e0148395.
52. Grytdal SP, Rimland D, Shirley SH, Rodriguez-Barradas MC, Goetz MB, Brown ST, et al. Incidence of medically-attended norovirus-associated acute gastroenteritis in four Veteran's Affairs Medical Center populations in the United States, 2011-2012. PLoS One. 2015;10(5):e0126733.
53. Bernard H, Hohne M, Niendorf S, Altmann D, Stark K. Epidemiology of norovirus gastroenteritis in Germany 2001-2009: eight seasons of routine surveillance. Epidemiol Infect. 2014;142(1):63–74.

54. Gibney KB, O'Toole J, Sinclair M, Leder K. Disease burden of selected gastrointestinal pathogens in Australia, 2010. Int J Infect Dis. 2014;28:176–85.
55. Verhoef L, Koopmans M. W VANP, Duizer E, Haagsma J, Werber D, et al. The estimated disease burden of norovirus in The Netherlands. Epidemiol Infect. 2013;141(3):496–506.
56. Bierhoff M, Arvelo W, Estevez A, Bryan J, McCracken JP, Lopez MR, et al. Incidence and clinical profile of norovirus disease in Guatemala, 2008-2013. Clin Infect Dis. 2018;67 (3):430–6.
57. Burke RM, Shih SM, Yen C, Huang YC, Parashar UD, Lopman BA, et al. Burden of severe norovirus disease in Taiwan, 2003-2013. Clin Infect Dis. 2018;67(9):1373–8.
58. Chang CH, Sakaguchi M, Weil J, Verstraeten T. The incidence of medically-attended norovirus gastro-enteritis in Japan: modelling using a medical care insurance claims database. PLoS One. 2018;13(3):e0195164.
59. Kowalzik F, Binder H, Zoller D, Riera-Montes M, Clemens R, Verstraeten T, et al. Norovirus gastroenteritis among hospitalized patients, Germany, 2007-2012. Emerg Infect Dis. 2018;24 (11):2021–8.
60. Verstraeten T, Cattaert T, Harris J, Lopman B, Tam CC, Ferreira G. Estimating the burden of medically attended norovirus gastroenteritis: modeling linked primary care and hospitalization datasets. J Infect Dis. 2017;216(8):957–65.
61. Morton VK, Thomas MK, McEwen SA. Estimated hospitalizations attributed to norovirus and rotavirus infection in Canada, 2006-2010. Epidemiol Infect. 2015;143(16):3528–37.
62. Karve S, Krishnarajah G, Korsnes JS, Cassidy A, Candrilli SD. Burden of acute gastroenteritis, norovirus and rotavirus in a managed care population. Hum Vaccin Immunother. 2014;10 (6):1544–56.
63. Shioda K, Cosmas L, Audi A, Gregoricus N, Vinje J, Parashar UD, et al. Population-based incidence rates of diarrheal disease associated with norovirus, sapovirus, and astrovirus in Kenya. PLoS One. 2016;11(4):e0145943.
64. Yu J, Ye C, Lai S, Zhu W, Zhang Z, Geng Q, et al. Incidence of norovirus-associated diarrhea, Shanghai, China, 2012-2013. Emerg Infect Dis. 2017;23(2):312–5.
65. Hall A, Naleway A, Esterberg E, Moua K, Donald J, Debess E, et al. Incidence of medically-attended acute gastroenteritis and norovirus infection across the age spectrum. IDWeek; October 10, 2015; San Diego, 2015.
66. Phillips G, Tam CC, Conti S, Rodrigues LC, Brown D, Iturriza-Gomara M, et al. Community incidence of norovirus-associated infectious intestinal disease in England: improved estimates using viral load for norovirus diagnosis. Am J Epidemiol. 2010;171(9):1014–22.
67. Karsten C, Baumgarte S, Friedrich AW, von Eiff C, Becker K, Wosniok W, et al. Incidence and risk factors for community-acquired acute gastroenteritis in north-west Germany in 2004. Eur J Clin Microbiol Infect Dis. 2009;28(8):935–43.
68. Romero C, Tinoco YO, Loli S, Razuri H, Soto G, Silva M, et al. Incidence of norovirus-associated diarrhea and vomiting disease among children and adults in a community cohort in the Peruvian Amazon Basin. Clin Infect Dis. 2017;65(5):833–9.
69. Rockx B, De Wit M, Vennema H, Vinje J, De Bruin E, Van Duynhoven Y, et al. Natural history of human calicivirus infection: a prospective cohort study. Clin Infect Dis. 2002;35 (3):246–53.
70. Kaplan JE, Gary GW, Baron RC, Singh N, Schonberger LB, Feldman R, et al. Epidemiology of Norwalk gastroenteritis and the role of Norwalk virus in outbreaks of acute nonbacterial gastroenteritis. Ann Intern Med. 1982;96(6 Pt 1):756–61.
71. Kawada J, Arai N, Nishimura N, Suzuki M, Ohta R, Ozaki T, et al. Clinical characteristics of norovirus gastroenteritis among hospitalized children in Japan. Microbiol Immunol. 2012;56 (11):756–9.
72. de Andrade JS, Rocha MS, Carvalho-Costa FA, Fioretti JM, Xavier Mda P, Nunes ZM, et al. Noroviruses associated with outbreaks of acute gastroenteritis in the State of Rio Grande do Sul, Brazil, 2004-2011. J Clin Virol. 2014;61(3):345–52.

73. Barreira DM, Ferreira MS, Fumian TM, Checon R, de Sadovsky AD, Leite JP, et al. Viral load and genotypes of noroviruses in symptomatic and asymptomatic children in Southeastern Brazil. J Clin Virol. 2010;47(1):60–4.
74. Wu TC, Liu HH, Chen YJ, Tang RB, Hwang BT, Yuan HC. Comparison of clinical features of childhood norovirus and rotavirus gastroenteritis in Taiwan. J Chin Med Assoc. 2008;71 (11):566–70.
75. Lopman BA, Reacher MH, Vipond IB, Sarangi J, Brown DW. Clinical manifestation of norovirus gastroenteritis in health care settings. Clin Infect Dis. 2004;39(3):318–24.
76. Olson D, Lamb MM, Lopez MR, Paniagua-Avila MA, Zacarias A, Samayoa-Reyes G, et al. A rapid epidemiological tool to measure the burden of norovirus infection and disease in resource-limited settings. Open Forum Infect Dis. 2017;4(2):ofx049.
77. Kambhampati A, Koopmans M, Lopman BA. Burden of norovirus in healthcare facilities and strategies for outbreak control. J Hosp Infect. 2015;89(4):296–301.
78. Zingg W, Colombo C, Jucker T, Bossart W, Ruef C. Impact of an outbreak of norovirus infection on hospital resources. Infect Control Hosp Epidemiol. 2005;26(3):263–7.
79. Lopman BA, Reacher MH, Vipond IB, Hill D, Perry C, Halladay T, et al. Epidemiology and cost of nosocomial gastroenteritis, Avon, England, 2002-2003. Emerg Infect Dis. 2004;10 (10):1827–34.
80. O'Brien SJ, Donaldson AL, Iturriza-Gomara M, Tam CC. Age-specific incidence rates for norovirus in the community and presenting to primary healthcare facilities in the United Kingdom. J Infect Dis. 2016;213(Suppl 1):S15–8.
81. Lopez-Medina E, Parra B, Davalos DM, Lopez P, Villamarin E, Pelaez M. Acute gastroenteritis in a pediatric population from Cali, Colombia in the post rotavirus vaccine era. Int J Infect Dis. 2018;73:52–9.
82. Chan MC, Leung TF, Chung TW, Kwok AK, Nelson EA, Lee N, et al. Virus genotype distribution and virus burden in children and adults hospitalized for norovirus gastroenteritis, 2012-2014. Hong Kong Sci Rep. 2015;5:11507.
83. Leshem E, Givon-Lavi N, Vinje J, Gregoricus N, Parashar U, Dagan R. Differences in norovirus-associated hospital visits between Jewish and Bedouin children in Southern Israel. Pediatr Infect Dis J. 2015;34(9):1036–8.
84. Dona D, Mozzo E, Scamarcia A, Picelli G, Villa M, Cantarutti L, et al. Community-acquired rotavirus gastroenteritis compared with adenovirus and norovirus gastroenteritis in Italian children: a pedianet study. Int J Pediatr. 2016;2016:5236243.
85. Enserink R, van den Wijngaard C, Bruijning-Verhagen P, van Asten L, Mughini-Gras L, Duizer E, et al. Gastroenteritis attributable to 16 enteropathogens in children attending day care: significant effects of rotavirus, norovirus, astrovirus, Cryptosporidium and Giardia. Pediatr Infect Dis J. 2015;34(1):5–10.
86. Lopman BA, Trivedi T, Vicuna Y, Costantini V, Collins N, Gregoricus N, et al. Norovirus infection and disease in an Ecuadorian birth cohort: association of certain norovirus genotypes with Host FUT2 secretor status. J Infect Dis. 2015;211(11):1813–21.
87. Zambruni M, Luna G, Silva M, Bausch DG, Rivera FP, Velapatino G, et al. High prevalence and increased severity of norovirus mixed infections among children 12-24 months of age living in the suburban areas of Lima. Peru J Pediatric Infect Dis Soc. 2016;5(3):337–41.
88. Saito M, Goel-Apaza S, Espetia S, Velasquez D, Cabrera L, Loli S, et al. Multiple norovirus infections in a birth cohort in a Peruvian Periurban community. Clin Infect Dis. 2014;58 (4):483–91.
89. Haustein T, Harris JP, Pebody R, Lopman BA. Hospital admissions due to norovirus in adult and elderly patients in England. Clin Infect Dis. 2009;49(12):1890–2.
90. Vega E, Barclay L, Gregoricus N, Shirley SH, Lee D, Vinje J. Genotypic and epidemiologic trends of norovirus outbreaks in the United States, 2009 to 2013. J Clin Microbiol. 2014;52 (1):147–55.
91. Chen SY, Chiu CH. Worldwide molecular epidemiology of norovirus infection. Paediatr Int Child Health. 2012;32(3):128–31.

92. Siebenga JJ, Vennema H, Renckens B, de Bruin E, van der Veer B, Siezen RJ, et al. Epochal evolution of GGII.4 norovirus capsid proteins from 1995 to 2006. J Virol. 2007;81(18):9932–41.
93. de Graaf M, van Beek J, Vennema H, Podkolzin AT, Hewitt J, Bucardo F, et al. Emergence of a novel GII.17 norovirus – end of the GII.4 era? Euro Surveill. 2015;20(26) https://doi.org/10.2807/1560-7917.es2015.20.26.21178.
94. Ao Y, Wang J, Ling H, He Y, Dong X, Wang X, et al. Norovirus GII.P16/GII.2-associated gastroenteritis, China, 2016. Emerg Infect Dis. 2017;23(7):1172–5.
95. Lu J, Fang L, Sun L, Zeng H, Li Y, Zheng H, et al. Association of GII.P16-GII.2 recombinant norovirus strain with increased norovirus outbreaks, Guangdong, China, 2016. Emerg Infect Dis. 2017;23(7):1188–90.
96. Thongprachum A, Okitsu S, Khamrin P, Maneekarn N, Hayakawa S, Ushijima H. Emergence of norovirus GII.2 and its novel recombination during the gastroenteritis outbreak in Japanese children in mid-2016. Infect Genet Evol. 2017;51:86–8.
97. Liu LT, Kuo TY, Wu CY, Liao WT, Hall AJ, Wu FT. Recombinant GII.P16-GII.2 norovirus, Taiwan, 2016. Emerg Infect Dis. 2017;23(7):1180–3.
98. Iritani N, Kaida A, Abe N, Sekiguchi J, Kubo H, Takakura K, et al. Increase of GII.2 norovirus infections during the 2009-2010 season in Osaka City, Japan. J Med Virol. 2012;84(3):517–25.
99. Bidalot M, Thery L, Kaplon J, De Rougemont A, Ambert-Balay K. Emergence of new recombinant noroviruses GII.p16-GII.4 and GII.p16-GII.2, France, winter 2016 to 2017. Euro Surveill. 2017;22(15) https://doi.org/10.2807/1560-7917.ES.2017.22.15.30508.
100. Niendorf S, Jacobsen S, Faber M, Eis-Hubinger AM, Hofmann J, Zimmermann O, et al. Steep rise in norovirus cases and emergence of a new recombinant strain GII.P16-GII.2, Germany, winter 2016. Euro Surveill. 2017;22(4) https://doi.org/10.2807/1560-7917.ES.2017.22.4.30447.
101. Hoa Tran TN, Trainor E, Nakagomi T, Cunliffe NA, Nakagomi O. Molecular epidemiology of noroviruses associated with acute sporadic gastroenteritis in children: global distribution of genogroups, genotypes and GII.4 variants. J Clin Virol. 2013;56(3):185–93.
102. Yi J, Wahl K, Sederdahl BK, Jerris RR, Kraft CS, McCracken C, et al. Molecular epidemiology of norovirus in children and the elderly in Atlanta, Georgia, United States. J Med Virol. 2016;88(6):961–70.
103. Zhang P, Chen L, Fu Y, Ji L, Wu X, Xu D, et al. Clinical and molecular analyses of norovirus-associated sporadic acute gastroenteritis: the emergence of GII.17 over GII.4, Huzhou, China, 2015. BMC Infect Dis. 2016;16(1):717.
104. Park K, Yeo S, Jeong H, Baek K, Kim D, Shin M, et al. Updates on the genetic variations of norovirus in sporadic gastroenteritis in Chungnam Korea, 2009-2010. Virol J. 2012;9:29.
105. Wang X, Du X, Yong W, Qiao M, He M, Shi L, et al. Genetic characterization of emergent GII.17 norovirus variants from 2013 to 2015 in Nanjing, China. J Med Microbiol. 2016;65(11):1274–80.
106. Lim KL, Eden JS, Oon LL, White PA. Molecular epidemiology of norovirus in Singapore, 2004-2011. J Med Virol. 2013;85(10):1842–51.
107. Lim KL, Hewitt J, Sitabkhan A, Eden JS, Lun J, Levy A, et al. A multi-site study of norovirus molecular epidemiology in Australia and New Zealand, 2013-2014. PLoS One. 2016;11(4):e0145254.
108. Ifantidou AM, Kachrimanidou M, Markopoulou S, Kansouzidou A, Malisiovas N, Papa A. Molecular epidemiology of noroviruses in Northern Greece, 2005-2006. J Med Virol. 2015;87(1):170–4.
109. Siebenga JJ, Vennema H, Zheng DP, Vinje J, Lee BE, Pang XL, et al. Norovirus illness is a global problem: emergence and spread of norovirus GII.4 variants, 2001-2007. J Infect Dis. 2009;200(5):802–12.

110. Zheng DP, Widdowson MA, Glass RI, Vinje J. Molecular epidemiology of genogroup II-genotype 4 noroviruses in the United States between 1994 and 2006. J Clin Microbiol. 2010;48(1):168–77.
111. Zhang J, Shen Z, Zhu Z, Zhang W, Chen H, Qian F, et al. Genotype distribution of norovirus around the emergence of Sydney_2012 and the antigenic drift of contemporary GII.4 epidemic strains. J Clin Virol. 2015;72:95–101.
112. Verhoef L, Hewitt J, Barclay L, Ahmed SM, Lake R, Hall AJ, et al. Norovirus genotype profiles associated with foodborne transmission, 1999-2012. Emerg Infect Dis. 2015;21(4):592–9.
113. Bitler EJ, Matthews JE, Dickey BW, Eisenberg JN, Leon JS. Norovirus outbreaks: a systematic review of commonly implicated transmission routes and vehicles. Epidemiol Infect. 2013;141(8):1563–71.
114. Burke RM, Shah MP, Wikswo ME, Barclay L, Kambhampati A, Marsh Z, et al. The norovirus epidemiologic triad: predictors of severe outcomes in US norovirus outbreaks, 2009-2016. J Infect Dis. 219(9):2018, 1364–1372.
115. Desai R, Hembree CD, Handel A, Matthews JE, Dickey BW, McDonald S, et al. Severe outcomes are associated with genogroup 2 genotype 4 norovirus outbreaks: a systematic literature review. Clin Infect Dis. 2012;55(2):189–93.
116. Friesema IH, Vennema H, Heijne JC, de Jager CM, Teunis PF, van der Linde R, et al. Differences in clinical presentation between norovirus genotypes in nursing homes. J Clin Virol. 2009;46(4):341–4.
117. Huhti L, Szakal ED, Puustinen L, Salminen M, Huhtala H, Valve O, et al. Norovirus GII-4 causes a more severe gastroenteritis than other noroviruses in young children. J Infect Dis. 2011;203(10):1442–4.
118. Parra GI, Green KY. Genome of Emerging Norovirus GII.17, United States, 2014. Emerg Infect Dis. 2015;21(8):1477–9.
119. Parra GI, Squires RB, Karangwa CK, Johnson JA, Lepore CJ, Sosnovtsev SV, et al. Static and evolving norovirus genotypes: implications for epidemiology and immunity. PLoS Pathog. 2017;13(1):e1006136.
120. Matsushima Y, Ishikawa M, Shimizu T, Komane A, Kasuo S, Shinohara M, et al. Genetic analyses of GII.17 norovirus strains in diarrheal disease outbreaks from December 2014 to March 2015 in Japan reveal a novel polymerase sequence and amino acid substitutions in the capsid region. Euro Surveill. 2015;20(26) https://doi.org/10.2807/1560-7917.es2015.20.26.21173.
121. Chan MCW, Hu Y, Chen H, Podkolzin AT, Zaytseva EV, Komano J, et al. Global spread of norovirus GII.17 Kawasaki 308, 2014-2016. Emerg Infect Dis. 2017;23(8):1359–4.
122. van Beek J, de Graaf M, Al-Hello H, Allen DJ, Ambert-Balay K, Botteldoorn N, et al. Molecular surveillance of norovirus, 2005-16: an epidemiological analysis of data collected from the NoroNet network. Lancet Infect Dis. 2018;18(5):545–53.
123. Chan MC, Lee N, Hung TN, Kwok K, Cheung K, Tin EK, et al. Rapid emergence and predominance of a broadly recognizing and fast-evolving norovirus GII.17 variant in late 2014. Nat Commun. 2015;6:10061.
124. Cannon JL, Barclay L, Collins NR, Wikswo ME, Castro CJ, Magana LC, et al. Genetic and epidemiologic trends of norovirus outbreaks in the United States from 2013 to 2016 demonstrated emergence of novel GII.4 recombinant viruses. J Clin Microbiol. 2017;55(7):2208–21.
125. Lu QB, Huang DD, Zhao J, Wang HY, Zhang XA, Xu HM, et al. An increasing prevalence of recombinant GII norovirus in pediatric patients with diarrhea during 2010-2013 in China. Infect Genet Evol. 2015;31:48–52.
126. Bartsch SM, Lopman BA, Ozawa S, Hall AJ, Lee BY. Global economic burden of norovirus gastroenteritis. PLoS One. 2016;11(4):e0151219.
127. Phillips G, Lopman B, Tam CC, Iturriza-Gomara M, Brown D, Gray J. Diagnosing norovirus-associated infectious intestinal disease using viral load. BMC Infect Dis. 2009;9:63.

128. Elfving K, Andersson M, Msellem MI, Welinder-Olsson C, Petzold M, Bjorkman A, et al. Real-time PCR threshold cycle cutoffs help to identify agents causing acute childhood diarrhea in Zanzibar. J Clin Microbiol. 2014;52(3):916–23.
129. Trang NV, Choisy M, Nakagomi T, Chinh NT, Doan YH, Yamashiro T, et al. Determination of cut-off cycle threshold values in routine RT-PCR assays to assist differential diagnosis of norovirus in children hospitalized for acute gastroenteritis. Epidemiol Infect. 2015;143 (15):3292–9.

The Virus

2

Christopher Ruis, Lucy Thorne, and Judith Breuer

2.1 Norovirus Classification and Genome Organization

Noroviruses belong to the *Caliciviridae*, a family of icosahedral small non-enveloped positive-sense single-stranded RNA viruses [1]. There are currently five accepted genera within the *Caliciviridae*: *Norovirus*, *Sapovirus*, *Lagovirus*, *Nebovirus* and *Vesivirus* (Fig. 2.1). Three additional genera within the family have been suggested, but are yet to be approved: *Recovirus*, *Valovirus* and chicken calicivirus [2, 3]. The *Caliciviridae* family has a broad host range and includes important human gastrointestinal pathogens in the Norovirus and Sapovirus genera [1]. Important animal pathogens in the family include feline calicivirus (FCV, a vesivirus) that typically causes respiratory disease in cats and rabbit hemorrhagic disease virus (RHDV, a lagovirus) that causes an often fatal hemorrhagic disease in rabbits.

The norovirus genome is a linear single-stranded positive-sense RNA molecule of 7.5–8.3 kilobases (kb) in length and is bound at the 5′ end by the virally encoded protein VPg (Virus Protein, genome linked) and is polyadenylated at the 3′ end [1, 4, 5] (Fig. 2.2). The norovirus genome typically encodes three open reading frames (ORFs) and contains very short 5′ and 3′ untranslated regions (UTRs) (Fig. 2.2). ORF1 is more than 5 kb in length and encodes a ~200 kilodalton (kDa) polyprotein

C. Ruis (✉)
Division of Infection and Immunity, University College London, London, UK

Department of Medicine, University of Cambridge, Cambridge, UK
e-mail: cr628@cam.ac.uk

L. Thorne
Division of Infection and Immunity, University College London, London, UK

J. Breuer
Division of Infection and Immunity, University College London, London, UK

Department of Microbiology, Virology and Infection Control, Great Ormond Street Hospital for Children, London, UK

N. M. Melhem (ed.), *Norovirus*,
https://doi.org/10.1007/978-3-030-27209-8_2

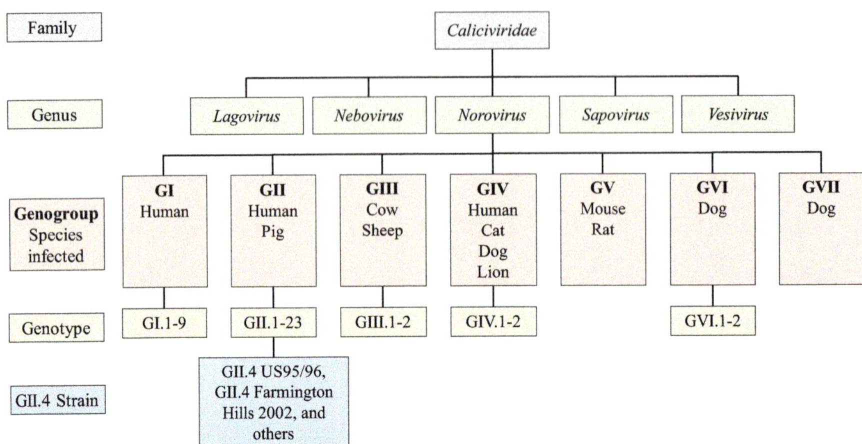

Fig. 2.1 Classification of the Caliciviridae Family. The hierarchical classification of the Caliciviridae family is shown, with Norovirus being one of the five accepted genera within the family. There are seven genogroups (GI-GVII) within the Norovirus genus, each of which is associated with infection of different species. Five of these genogroups are further divided into multiple genotypes. The GII.4 genotype has caused the majority of human cases and outbreaks since the mid-1990s and is further divided into strains that are named by their year of emergence and the geographical location in which they were first identified. The genotype classification shown here is based on the VP1 capsid; due to the presence of intergenic recombination, the RdRp is classified separately

that is cleaved into six non-structural proteins (NS) by the virally encoded protease (Pro/NS6), these are: p48 (murine norovirus homologue NS1/2), nucleoside triphosphatase NTPase (NS3), p22 (NS4), VPg (NS5), protease (Pro, NS6) and RNA-dependent RNA polymerase (RdRp, NS7) (Fig. 2.2). ORF2 is roughly 1.8 kb in length and encodes the major capsid protein viral protein 1 (VP1), 180 copies of which are incorporated into the viral particle [6]. VP1 is divided into the shell (S) and protruding (P) domains, with the P domain being further subdivided into two subdomains P1 and P2 [6, 7] (Fig. 2.2). The S domain forms the interior of the viral particle, while the P2 domain is surface exposed [6, 7]. ORF3 is approximately 0.6 kb in length and encodes VP2, a small basic protein that is incorporated into the interior surface of the viral particle through interactions with the VP1 shell domain [8, 9]. ORFs 2 and 3 are also produced as a ~2.4 kb subgenomic RNA that is identical to the corresponding region in the full length genome (Fig. 2.2). Murine noroviruses contain a fourth ORF translated from an alternative reading frame within ORF2 and encoding virulence factor 1 (VF1), an antagonist of the innate immune response [10] (Fig. 2.2).

Until recently [11, 12], human noroviruses (HuNoVs) could not be cultured in vitro and classification has therefore been based on sequence and phylogenetic analysis [13–15]. Classification is based on the RdRp and VP1 capsid regions (Fig. 2.2) and was originally based on percentage amino acid similarity [13]. However, this has more recently been demonstrated to be unreliable with the ever

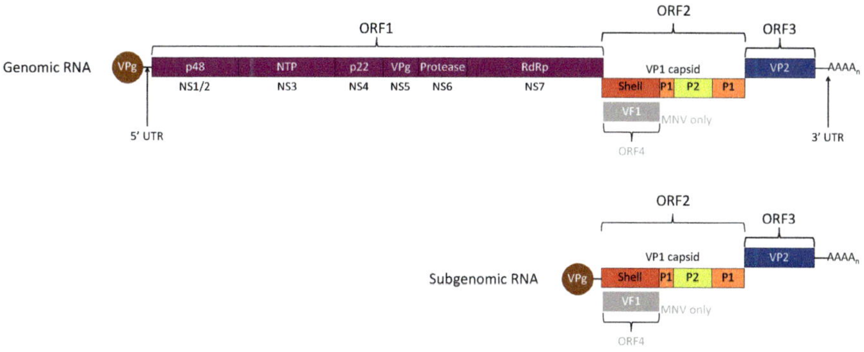

Fig. 2.2 Norovirus genome organization. The norovirus genome ranges between 7.3 and 7.7 kb in length, depending on the genogroup, and typically encodes three open reading frames. ORF1 encodes a non-structural polyprotein that is cleaved by the virally encoded protease into six proteins. These proteins are named differently in human and murine noroviruses; the protein name in a human norovirus is shown along the genome and the protein name in murine norovirus is shown below the genome. ORF2 encodes the VP1 capsid protein that is divided into the S domain and two protruding subdomains P1 and P2. Two distinct regions in ORF2 that flank the sequence encoding the P2 domain encode the P1 domain. ORF3 encodes a small basic protein VP2. VP1 and VP2 are also encoded as a subgenomic RNA, the sequence of which is identical to the corresponding region of the full length genomic RNA. The 5′ end of the genomic and subgenomic RNAs are covalently bound by VPg (NS5) and the 3′ end is polyadenylated. The genomic RNA contains very short 5′ and 3′ UTRs that contain conserved RNA secondary structures important in viral replication. The murine norovirus genome contains ORF4 that is translated from an alternate reading frame within ORF2 and encodes VF1

increasing availability of sequence data [14]. Therefore classification is now based on phylogenetic clustering [14]. Recombination frequently occurs close to the ORF1-ORF2 boundary, necessitating a dual-typing system where the RdRp and VP1 are genotyped separately [14–18], for example a GII.P21-GII.3 virus has a genogroup II genotype 21 RdRp and a genogroup II genotype 3 VP1.

Until the write-up of this book, noroviruses were divided into seven genogroups (GI–GVII), several of which can be further divided into genotypes [15] (Fig. 2.1). Recently, GVIII-GX were advanced for an updated classification [19]. With classification based on VP1, there are 9 recognized genotypes within GI, 23 in GII, 2 in GIII, 2 in GIV, 1 in GV, 2 in GVI and 1 in GVII [15] (Fig. 2.1). The VP1 genotypes GI.3, GI.5, GI.6, GI.7, GII.3 and GII.6 are further divided into subclusters [15]. The different genogroups are associated with infection of different species (Fig. 2.1). Human norovirus infections are predominantly associated with genogroups GI and GII, with a small number of infections caused by GIV.1. However, the majority of human cases and outbreaks are caused by a single genotype within genogroup GII, called GII.4 [20, 21], which has also caused six pandemics since the mid-1990s [20, 22]. The GII genotypes GII.11, GII.18 and GII.19 infect pigs [15, 23–25]. GIII is associated with infection of cows and sheep, while GIV.2 is associated with infection of cats and dogs, with one reported infection of a lion [15, 25–31]. Murine norovirus (MNV) in genogroup GV was first detected as a lethal agent in STAT1 knockout mice and has since been recognized as a common pathogen infecting mice in research facilities [32, 33], in addition to strains detected in wild mice and rats [34, 35]. The GVI genogroup is associated with infection of cats and dogs, while GVII so far only contains canine noroviruses [31, 35–41]. Infections of additional species have also been identified, including infection of harbour porpoise [42], bats [43–45] and sea lions [46], further demonstrating the wide species tropism of noroviruses.

The rapid evolution of HuNoVs within the GII.4 genotype has necessitated the further subdivision of this genotype into strains (Fig. 2.1), each of which is named by the geographical location and year in which the strain was first detected [20, 21]. Each of the six GII.4 pandemics was caused by a distinct strain: US95/96, Farmington Hills 2002, Hunter 2004, Den Haag 2006, New Orleans 2009 and Sydney 2012 [14, 20, 22]. There is also a number of epidemic GII.4 strains that exhibit lower prevalence or a more localised geographical distribution, including Lanzhou 2002, Kaiso 2003, Asia 2003, Yerseke 2006, Apeldoorn 2007 and Osaka 2007 [14, 47, 48]. Additional GII.4 strains that circulated prior to the first GII.4 pandemic have also been identified, including CHDC 1970s, Tokyo 1980s, Bristol 1992 and Camberwell 1994 [14, 20, 49, 50].

2.2 The Viral Particle

The norovirus particle is ~38 nm in diameter and is formed from 90 dimers of the VP1 capsid protein and a small number of copies of VP2 [6, 9]. The particle exhibits $T = 3$ icosahedral symmetry. This icosahedral structure requires the VP1 capsid protein to adapt to three quasi-equivalent positions, with subunits in each position referred to as A, B and C. The curvature needed to form a closed shell requires two types of dimer interaction: C/C dimers have a 'flat' conformation and are important for forming the shell while A/B dimers have a 'bent' conformation and extend further from the surface [6, 51]. The S domain forms the icosahedral shell of the capsid and when expressed alone can form smooth icosahedral particles, while the P domain forms dimeric interactions that stabilize the viral particle [6, 7, 52]. The S domain is formed from the N-terminal 221 residues (residue numbers relative to the GII.4 VP1) and is connected to the P domain by a flexible hinge approximately ten residues in length [6, 7] (Fig. 2.3). This hinge contributes to the formation of the different curvatures in the A/B and C/C dimers by facilitating the rotation of the S domain relative to the P domain [6]. Residues 50–221 in the S domain form an eight-stranded antiparallel β sandwich, which is a fold common to many viral capsid proteins. Two alpha helices are also found within the S domain and are located between the second and third β-strands and between the fourth and fifth β-strands of the β-sandwich, respectively [6]. The P domain is formed from residues 222–540 (residue numbers relative to the GII.4 VP1) and structures of multiple genotypes have demonstrated that the P2 subdomain (residues 275–417) is an insertion in the P1 subdomain (residues 222–274 and 418–540) (Fig. 2.3) [6, 7, 53, 54]. The P1 subdomain contains two twisted β-sheets, while the P2 domain contains an

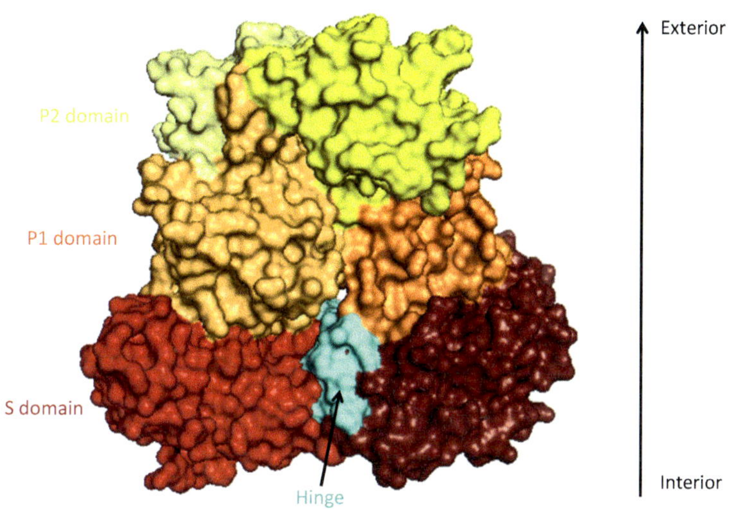

Fig. 2.3 The structure of the VP1 capsid protein. The viral particle is formed from 90 homodimers of the VP1 capsid protein and a small number of VP2 molecules that bind to the interior face of the particle through interactions with the VP1 S domain. Here, a single VP1 dimer is shown with the monomers represented in different shades. The S domain (red) is found on the interior of the viral particle and forms smooth icosahedral particles when expressed alone. The P1 (orange) and P2 (yellow) domains form inter-monomer contacts that stabilize the particle. The P2 domain is found on the exterior of the viral particle where it is thought to be targeted by host humoral immunity. The S domain is connected to the P1 domain by a flexible hinge (cyan). The structure of the Norwalk virus (GI.1) VP1 is shown (PDB: 1IHM) [6]

antiparallel β-barrel [6, 7]. There is only a single α-helix in the P domain and the overall content of secondary structure in the P domain is low [7].

Upon expression in vitro, VP1 self-assembles to form virus like particles (VLPs) that are highly antigenically similar to native virions but do not contain the viral genome [55, 56]. While VP2 is not required for VLP formation, it increases VLP stability and size homogeneity [57]. Additionally, VP2 has been demonstrated to protect VP1 from disassembly and from degradation by the host proteases trypsin and pancreatin [57]. The interaction of VP1 with VP2 is mediated by an isoleucine (I52 in GI.1 Norwalk virus) within an IDPWI motif in the VP1 S domain that is highly conserved across genogroups [9]. Two recombinant expression systems have been developed that enable the production of large quantities of VLPs: the baculovirus expression system [58] and the Venezuelan equine encephalitis (VEE) replicon system [56]. One advantage of the VEE expression system is that it employs mammalian cells rather than insect cells used in the baculovirus expression system and this may result in more physiological post-translational modifications [56]. VLPs produced by these systems have been useful for characterizing aspects of norovirus biology in the absence of HuNoV cell culture and reverse genetic systems until recently [11, 12, 59]. Example applications include characterizing viral antigenicity [60, 61], investigating attachment factor-binding [60] and serving as a platform for vaccine development [62].

2.3 Capsid Immunogenetic Epitopes

The exposure of the VP1 P2 domain on the surface of the viral particle (Fig. 2.3) suggests that this region is the dominant target of host humoral immune responses. A recent study demonstrated that the VP1 proteins of non-GII.4 HuNoV genotypes remain 'static' at the amino acid level for long periods [21]. In contrast, extensive work has shown that the GII.4 VP1 evolves rapidly at the amino acid level, with most variation mapping to the P2 domain [21, 60, 61, 63–66]. This sequence change has been demonstrated to result in antigenic change through time with the use of mouse monoclonal antibodies [60, 66], human monoclonal antibodies [61, 67], mouse polyclonal sera [60] and human polyclonal sera [60, 68, 69] raised against the VP1 protein from different GII.4 strains. Antibody binding to a VLP is not indicative of neutralization of infection and, in the absence of a HuNoV culture system until recently [11, 12], a surrogate neutralization assay was developed that measures the ability of monoclonal antibodies or sera to block the binding of a VLP to a histo-blood group antigen (HBGA) binding partner [60]. We discuss interactions between norovirus and HBGAs in more details below, but HBGAs are thought to act as attachment factors for noroviruses. Therefore, prevention of norovirus-HBGA interactions may prevent infection of target cells. Antibodies that block the interaction between a VLP and HBGA are termed blockade antibodies and the presence of such antibodies in sera correlates with protection from HuNoV infection in humans [70] and chimpanzees [71]. This surrogate neutralization assay can distinguish VLPs too similar to be separated by antibody binding alone [60].

Immunological epitopes within the GII.4 capsid were initially proposed based on rapidly evolving sites on the surface of the viral particle [61, 64]. Five such epitopes have been identified and are termed epitopes A–E (Fig. 2.4) [61]. Epitopes A, D and E are classified as blockade epitopes as mutations within these epitopes can alter binding of blockade antibodies. These epitopes have been experimentally verified by switching these regions between viral strains [61, 68, 69, 72, 73]. Epitope A is a conformational epitope consisting of residues 294, 296–298, 368, 372 and 373 (residue numbers relative to post-Farmington Hills 2002 GII.4 strains) and has been verified as a blockade epitope in multiple pandemic GII.4 strains [61, 68, 69, 72]. It is thought to be immunodominant and likely consists of several overlapping epitopes [68, 69]. Epitope D consists of residues 393–395 and variation within this epitope can alter HBGA binding [60, 61]. Epitope E consists of residues 407, 412 and 413 and was initially identified as a GII.4 Farmington Hills 2002 blockade epitope [61, 73]. Additionally, residues 355–357 likely contribute to epitope E [73]. While epitopes B and C have been proposed due to their high variability and surface location, they have not yet been experimentally verified [61]. Epitope B consists of residues 333 and 382, while epitope C consists of residues 340 and 376 [61]. Each of these five epitopes likely includes additional nearby residues [61]. The presence of

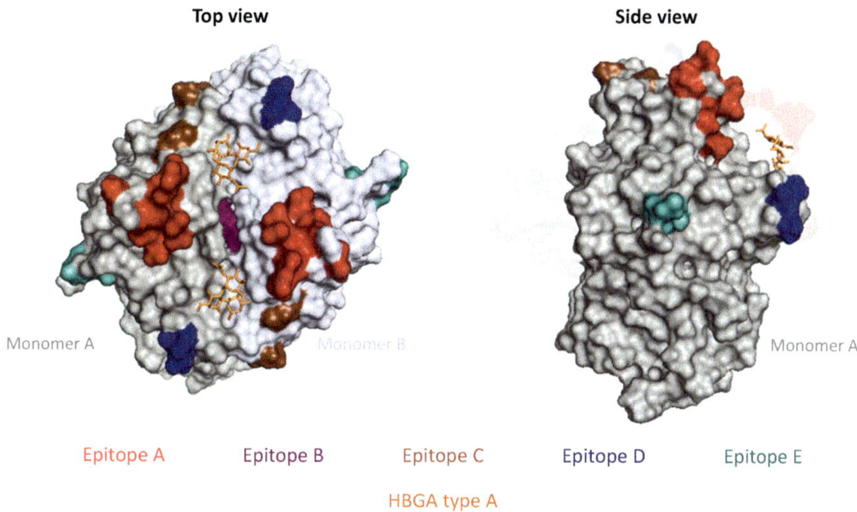

Fig. 2.4 Epitopes within the GII.4 VP1 capsid. The GII.4 VP1 capsid is shown from a top view and a side view with blockade epitope A (red), putative epitope B (purple), putative epitope C (brown), blockade epitope D (blue) and blockade epitope E (teal). The residues not currently known to contribute to an immune epitope within the GII.4 VP1 capsid are coloured in different shades in the different monomers. In the side view, one monomer has been made transparent for clarity. The structure of a GII.4 Sydney 2012 P domain is shown (PDB: 4WZT) [53]; the HBGA type A bound by the P2 domain within this structure is labelled in orange

at least one more non-blockade epitope within the GII.4 capsid has been mooted [68].

A human monoclonal antibody (NVB 71.4) isolated by Lindesmith et al. [61] was demonstrated to block the interaction of a diverse set of GII.4 VLPs with their HBGA binding partner [67]. These VLPs included viruses collected between 1987 and 2012 and this recognition therefore suggests the presence of a conserved epitope [67]. While the residues bound by NVB 71.4 are yet to be mapped, access to this epitope is conformation and temperature dependent and is regulated by residues E316, R484 and K493 (and possibly other residues) [67]. Site 310 is also likely to regulate access to this so called 'universal epitope' [69].

2.4 Host Genetic Susceptibility

Early volunteer studies conducted with the GI.1 Norwalk virus suggested that not all individuals were inherently susceptible to infection [74]. Extensive studies have suggested that susceptibility and resistance to certain norovirus genotypes is in part determined by the HBGAs expressed on the surface of cells within the gastrointestinal tract [12, 75–82]. HBGAs are small carbohydrate molecules found on the erythrocytes of humans and some other species of great ape [83]. Additionally, these molecules are expressed on the surface of epithelial cells in different tissues in

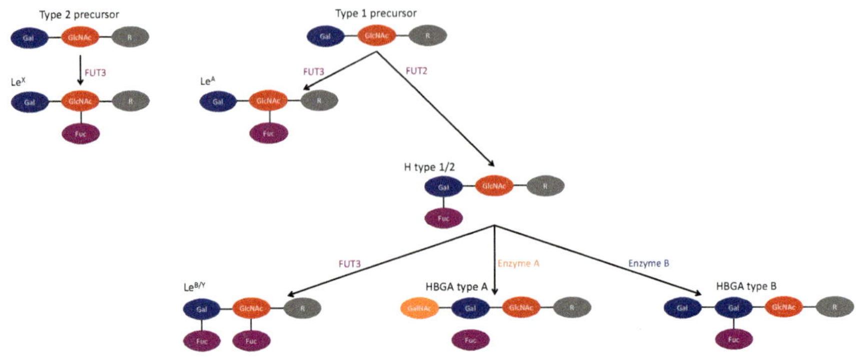

Fig. 2.5 The HBGA biosynthesis pathway. HBGAs are complex carbohydrates formed by the addition of sugar moieties onto precursor molecules. The synthesis of HBGAs begins with the type 1 and type 2 precursor disaccharides that differ only by the linkage between the Gal and GlcNAc sugars; the type 1 precursor has a β1,3 linkage while the type 2 precursor has a β1,4 linkage. The enzymes FUT2, FUT3, enzyme A and enzyme B catalyse the addition of sugar moieties to the carbohydrate chain and are shown next to the step in the synthesis pathway they catalyze, in the same color as the sugar group they add. FUT2 adds a fucose moiety onto Gal of the type 1 precursor, a step that is essential for the downstream production of the HBGA type A, HBGA type B, Le^B and Le^Y HBGAs. Individuals with a functional FUT2 enzyme are termed secretors, while mutations in this enzyme characterize the non-secretor and weak secretor phenotypes. The HBGAs can be attached to other structures as a part of glycoprotein or glycolipid chains and the other structures are represented by the R group. The Le^A and Le^X and the Le^B and Le^Y HBGAs differ from one another only by the specific linkages between sugar groups. A single structure is therefore shown for Le^B and Le^Y for simplicity. *Gal* galactose, *GlcNAc* *N*-acetylglucosamine, *GlaNAc* N-acetylgalactosamine, *Fuc* fucose

a wide range of vertebrate species and are also secreted into biological fluids such as saliva [51, 84]. Noroviruses use HBGAs from the A/B/H and Lewis antigens as attachment factors and bind to these molecules in a genotype- and strain-specific manner [60, 81]. The expression of these HBGAs on the surface of gastrointestinal cells is under the control of fucosyl- and glycosyltransferases encoded by the *FUT2*, *FUT3* and *ABO* genes (Fig. 2.5) [7, 53]. The synthesis of HBGAs begins with a disaccharide precursor molecule formed of galactose and *N*-acetylglucosamine moieties [85]. This precursor can be modified by the FUT3 enzyme to form the Lewis A (Le^A) or Le^X antigen, depending on the exact linkage within the precursor (Fig. 2.5). Alternatively, the FUT2 enzyme adds a fucose moiety onto the type 1 precursor molecule to form HBGA type H, which can be further modified by FUT3 or enzymes encoded by the *ABO* gene to form Le^B, Le^Y, HBGA type A and HBGA type B (Fig. 2.5). Polymorphisms in the genes encoding enzymes in this synthesis pathway result in differential HBGA expression in saliva and on the surface of gastrointestinal cells. These polymorphisms vary greatly in frequency depending on ethnicity [75, 86].

Of particular importance for norovirus infection are several inactivating mutations in the polymorphic *FUT2* gene, including the G428A nonsense mutation that is homozygous in roughly 20% of the European and North American populations [79, 82, 87]. The FUT2 enzyme is required for the expression of several HBGAs on the surface of gastrointestinal cells (Fig. 2.5) [75, 85]. Therefore, individuals who do not encode a functional FUT2 enzyme are termed non-secretors and can only express Le^A and Le^X, providing they possess a functional FUT3 enzyme. Conversely, individuals with a functional FUT2 enzyme are termed secretors and can express additional HBGAs depending on their genotype at the *FUT3* and *ABO* loci [51, 81] (Fig. 2.5). Non-secretor individuals account for 5% to more than 50% of different populations worldwide [86] and initially drew attention in the norovirus field due to their resistance to the GI.1 Norwalk virus [75]. This resistance has been observed in volunteer studies and studies of natural infections and correlates with the inability of Norwalk virus to bind to non-secretor HBGAs in vitro [75, 85, 88]. However, Norwalk virus does exhibit in vitro binding to the secretor HBGA types A and H and Le^B [89]. While the non-secretor phenotype is very rare in a number of Asian populations, a different *FUT2* mutation resulting in the weak secretor phenotype is present in 15–20% of individuals [79, 86]. The weak secretor phenotype results in low level expression of secretor HBGAs.

Further characterization has revealed additional HuNoV genotypes and strains that exhibit specific host susceptibility patterns. In addition to resistance to Norwalk virus infection, multiple studies have demonstrated that non-secretors are highly resistant to infection with specific strains within the globally dominant GII.4 genotype [78, 80, 87, 90–92]. In support of this, specific strains within GII.4 are unable to replicate in human intestinal enteroid cultures (HIEs, described in more detail below) derived from non-secretor individuals [12]. However, in vitro binding of certain GII.4 strains to non-secretor HBGAs [53, 93] and non-secretor Lewis positive saliva (from non-secretor individuals with a functional FUT3 enzyme) [94] has been demonstrated and cases of symptomatic and asymptomatic GII.4 infection of non-secretors have been reported [80, 87, 95]. Therefore, specific GII.4 strains are thought to be capable of infecting non-secretors. Additionally, non-secretor individuals are known to be susceptible to other genotypes including GI.3 [96, 97] and GII.2 [98]. GI.3 infection of both secretors and non-secretors was identified in water-borne [96] and food-borne [97] outbreaks and correlates with the ability of GI.3 VLPs to bind Le^A in vitro [99]. In the case of GII.2, a challenge study found no association between secretor status and susceptibility to infection with this genotype [98]. Interestingly, the GII.3 genotype was demonstrated to be capable of infecting non-secretor Lewis positive children [100] and could replicate in some, but not all, HIEs derived from non-secretor individuals [12] but did not exhibit binding to non-secretor saliva [100]. It has been suggested that infection of non-secretor individuals by a usually secretor-specific strain could be facilitated by the presence of enteric bacteria expressing HBGAs to promote attachment to target cells [11]. However, another explanation is that certain strains may have evolved to use alternative attachment factors. The presence of alternate attachment factors is supported by the observation that the GII.4 Hunter 2004 pandemic strain binds only weakly to all tested HBGAs in vitro [61, 77]. While the specific molecules that may act as additional attachment factors are not fully understood, heparan sulfate and gangliosides are proposed binding partners of human noroviruses [101, 102]. Specifically, three different genogroup GII strains bound to cells in a heparin sulfate-dependent manner, while two genogroup GI VLPs bound only weakly [101]. Gangliosides are glycosphingolipids that contain one or more sialic acid moieties. The ganglioside GM3 is expressed on a variety of cell types and has been demonstrated to bind to VLPs from the GI.3 and GII.4 genotypes [102]. A GII.4 VLP also bound to galactosylceramide, a molecule that could facilitate cellular entry through its clustering into lipid microdomains [103]. Genotypes that only infect non-secretors have not yet been identified; genotypes that are capable of infecting non-secretors also infect secretors. However, it is possible that the non-secretor phenotype may increase susceptibility to certain genotypes [86], although the mechanisms underlying this association have not been explored. Weak secretor individuals often express both secretor and non-secretor HBGAs concurrently and so are likely to be susceptible to a large range of norovirus genotypes.

Specific populations in, for example, the Philippines, Saudi Arabia and Tanzania have more than 50% non-secretors and norovirus prevalence would therefore be predicted to be lower in these regions, although this is yet to be tested [86]. Conversely, populations with a high percentage of secretors could be hypothesized to have higher norovirus infection rates and disease burdens. Given the differential prevalence of HBGA receptors in different populations, it has recently been proposed that populations with a large diversity of HBGAs would contain a greater diversity of norovirus genotypes [86]. This is supported by a study in Burkina Faso, a population in which a broad range of HBGA receptors are expressed [80]. Fourteen different norovirus genotypes were detected which is more than are typically found in similar studies carried out in Europe or North America [80, 86]. However, further investigation is required to examine potential confounding factors including socio-economic status.

2.5 Cell Tropism of Noroviruses

Until recently, there was no reproducible and robust cell culture system for HuNoVs and MNVs were the only members of the genus that could be cultivated in vitro [12, 33]. This has greatly hampered understanding of the viral life cycle, host-pathogen interactions and the mechanisms used by noroviruses to infect the intestinal tract. MNVs replicate efficiently in primary and cultured murine macrophages and B cells, primary dendritic cells and cultured T cells, suggesting a tropism for haematopoietic cells [11, 33, 104, 105]. This tropism is supported by the detection of MNV antigen in cells that morphologically resemble macrophages and dendritic cells in vivo [104, 106] and by the identification of MNV genomes and non-structural protein in Peyer's patch B cells of infected mice [11]. A recent study identified positive-strand and negative-strand MNV RNA in macrophages, dendritic cells, B cells and T cells in the gut-associated lymphoid tissue of immunocompetent mice, suggesting productive infection of these cell types in vivo [105]. In vivo replication in B cells is also supported by a reduction in MNV load in B cell-deficient mice [11]. It was proposed that noroviruses could gain access to immune system cells through transcytosis across intestinal epithelial cells [107, 108]. This transcytosis was demonstrated with MNV in vitro and is mediated by microfold (M) cells which are intestinal cells that sample particulate antigen [107, 108]. The depletion of M cells from mice reduced, but did not abolish, MNV titers in the intestine [109], suggesting the existence of additional mechanisms by which MNV can cross the epithelial cell barrier and/or infection of additional cell types accessible without transcytosis. Recently, Lee et al. demonstrated infection of intestinal epithelial cells (IECs) by the persistent MNV strain CR6 using a highly sensitive flow cytometry assay double staining for NS1/2 and NS6/7 [110]. They demonstrated that CR6 establishes persistent infection within a very small number of IECs in wild-type mice, with both non-structural proteins and MNV genomes detected in such cells. However, the acute MNV strain CW3 did not establish infection in IECs. It was subsequently revealed that the IECs targeted by CR6 are tuft cells, a rare type of specialized IEC that sense and initiate immune responses against intestinal worms and parasites [111]. Tuft cells proliferate in response to the type 2 interferons interleukin (IL)-4 and IL-25 and thus IL-4 and IL-25 promote MNV infection. Persistent MNV infection is only established within ~1.4% of tuft cells, which equates to ~100 cells per mouse [111]. Interestingly, the ability of CR6 to establish persistent infection is dependent on NS1/2; a chimeric virus containing the CR6 backbone with the CW3 NS1/2 did not infect IECs and did not establish persistent infection. The defective growth of this chimera in wild-type mice was rescued in IFN-λ receptor 1-knockout mice, suggesting that the CR6 NS1/2 evades IFN-λ-mediated antiviral immunity to enable establishment of a persistent infection [110]. The mechanism through which this evasion occurs is still under investigation.

The cell tropism of HuNoVs has been unclear which has presented a major hurdle in the development of a HuNoV in vitro propagation system. After the identification of MNV and its immune cell tropism, research efforts focused largely on whether HuNoV may infect the same cell types. A similar tropism was supported by the detection of HuNoV antigens in cells resembling dendritic cells and macrophages in chimpanzee and immunodeficient mouse models [71, 112], although these models do not recapitulate symptomatic human norovirus infections. Additionally, HuNoV antigen was identified in intestinal lamina propria cells from a human biopsy [113]. However, efforts to culture HuNoVs in vitro in macrophages and dendritic cells derived from peripheral blood mononuclear cells have been unsuccessful [113]. Whether HuNoV infects tissue-resident cell subsets is still unknown.

B cells have, however, recently been identified as a target of HuNoV, leading to the development of a HuNoV cell culture system using the BJAB B cell line [11]. While the HuNoV strain GII.4 Sydney 2012 replicated in BJAB cells, replication was decreased upon stool filtration. Infectivity was, however, restored by co-culture with enteric bacteria expressing histo-blood group antigens (HBGAs) or through the addition of synthetic H antigen, suggesting that enteric bacteria act as a stimulatory factor for norovirus infection. In vivo infection of B cells by HuNoVs is supported by the detection of HuNoV antigen in the intestine of infected chimpanzees [71], and by a decrease in viral titer in B cells-deficient patients [114]. A role for enteric bacteria is supported in MNV infection where antibiotic pre-treatment of wild-type mice resulted in a decrease in MNV titer [11]. Additionally, it has been hypothesized that interaction with enteric bacteria may provide another mechanism for norovirus transcytosis across the epithelial cell barrier [108].

While the HuNoV titer was demonstrated to be a log lower in severe combined immunodeficiency (SCID) patients lacking B cells compared to SCID patients with B cells, the former can be infected with HuNoV at high viral loads, suggesting that significant replication occurs in other cell types in vivo [114]. Additionally, the BJAB cell culture system achieves only a modest increase in HuNoV genome copies that is unlikely to recapitulate the level of viral replication in vivo [11, 115]. HuNoV replication has recently been demonstrated in human intestinal enteroids (HIEs) [12]. This cell culture develops from stem cells isolated from human intestinal crypts and forms a multicellular, differentiated and physiologically active culture containing enterocytes, enteroendocrine cells, goblet cells and paneth cells [12, 116]. Replication of viruses from the GI.1, GII.3, GII.4 and GII.17 HuNoV genotypes has been demonstrated in HIEs, with enterocytes being the only cell type within the culture to be infected [12]. Interestingly, while multiple strains from the GII.4 genotype replicate in HIEs without the need for additional factors, the GI.1, GII.3 and GII.17 viruses required a non-proteinaceous component of bile to support their replication, with this component likely acting on the cells to support infection, although the mechanism of action is not yet understood [12]. The addition of bile also enhanced GII.4 infection. Both GII.3 and GII.4 viruses replicated in HIEs derived from duodenal, jejunal and ileal segments, suggesting productive infection of enterocytes from each region of the small intestine. In vivo infection of enterocytes by HuNoV is supported by a study in immunocompromised transplant patients, in which the VP1 capsid protein, VPg and RdRp were identified in enterocytes from the duodenum and jejunum [117]. Importantly, the detection of the non-structural proteins suggests productive infection of enterocytes in vivo. Further support comes from studies examining gnotobiotic pigs infected with GII.4 norovirus and calves infected with the bovine GIII.1 norovirus, which also found evidence of enterocyte infection [118, 119]. While VP1 was detected in macrophages, T cells and dendritic cells in the lamina propria in immunocompromised transplant patients, VPg and RdRp were not detected in T cells or dendritic cells and the presence of these non-structural proteins in macrophages was suggested to be due to phagocytosis of a productively infected enterocyte [117]. Therefore, this study did not provide clear evidence of infection of immune cell types in immunocompromised patients by HuNoV.

The currently available evidence therefore suggests that enterocytes are a major site of HuNoV replication in vivo at least in immunocompromised individuals, with additional replication in B cells and the potential for replication in other immune system cell types [11, 12, 114, 117, 120].

2.6 The Norovirus Replication Cycle

Prior to the recent development of in vitro propagation systems for HuNoV [11, 12], much of our understanding of the norovirus lifecycle has come from studies of MNV and related caliciviruses, for which efficient reverse genetics and cell culture systems exist [104, 121–124]. Reverse genetics allows recovery of infectious virions from transfection of an infectious clone or in vitro transcribed and capped RNA into cell lines under the control of a T7 promoter. The resulting virus can then be propagated on cell lines which permit subsequent rounds of replication. Importantly, this allows recovery of genetically defined viruses and introduction of specific mutations that allows manipulation of the genotype to study the resulting phenotype [59, 121–123].

Viral entry into a target cell is typically a multistep process that involves binding of the viral particle to one or more attachment factors to concentrate the viral particles on the cell surface followed by interaction with one or more receptors that promote uptake into the cell [125]. While norovirus entry is incompletely understood, a similar multistep entry process has been hypothesized [126]. As previously discussed, HBGAs expressed on the surface of intestinal epithelial cells act as attachment factors for HuNoVs [4], possibly with additional contributions from HBGAs expressed by enteric bacteria (Fig. 2.6) [11]. HBGAs can be found on a variety of N- and O-linked glycoproteins and in several groups of glycosphingolipids (GSLs) [127]. GSLs, sphingosine-containing glycolipids, are major components of cellular membranes and are comprised of hydrophobic ceramides that anchor the compound to the cellular membrane, and hydrophilic carbohydrates, such as HBGAs, that extend into the extracellular space. GSLs have been demonstrated to act as attachment factors for multiple non-enveloped viruses and HuNoVs recognize HBGAs on type 1, 2 and 3 GSLs (Fig. 2.6) [127]. In addition to binding to HBGA-containing GSLs, there is evidence that at least some HuNoV strains can bind to the galactose-containing GSL galactosylceramide [103] or the sialic acid-containing gangliosides [102], the latter of which have also been implicated as attachment factors for MNV [128]. The interaction between the HuNoV VP1 P2 domain and HBGAs results in clustering of the GSLs to form a lipid microdomain leading to local invagination of the membrane [126]. As mentioned above, the possibility of alternative attachment factors has been suggested due to the identification of strains with weak affinity for all tested HBGAs [60, 61].

The HBGA-binding site has been mapped in several norovirus genotypes and consists of different residues and a different binding mode in GI and GII viruses [7, 53, 129, 130]. Dimerisation of VP1 is essential for HBGA-binding with the binding site being formed of residues from both monomers [7]. Five residues within the GII.4 VP1 are essential for binding to each HBGA and interact with the fucose moiety in ABH and Lewis HBGAs: T344, R345, D374, G443 and Y444 [7, 53]. Additionally, residues S343, Y390, C441, S442 and the 391–395 loop are important for interactions with specific HBGAs [53, 131]. Repositioning of the 391–395 loop enables binding to HBGA type B, Le^B and Le^Y [53]. VP1 residue T338 is also thought to be essential for HBGA binding due to hydrogen bonds

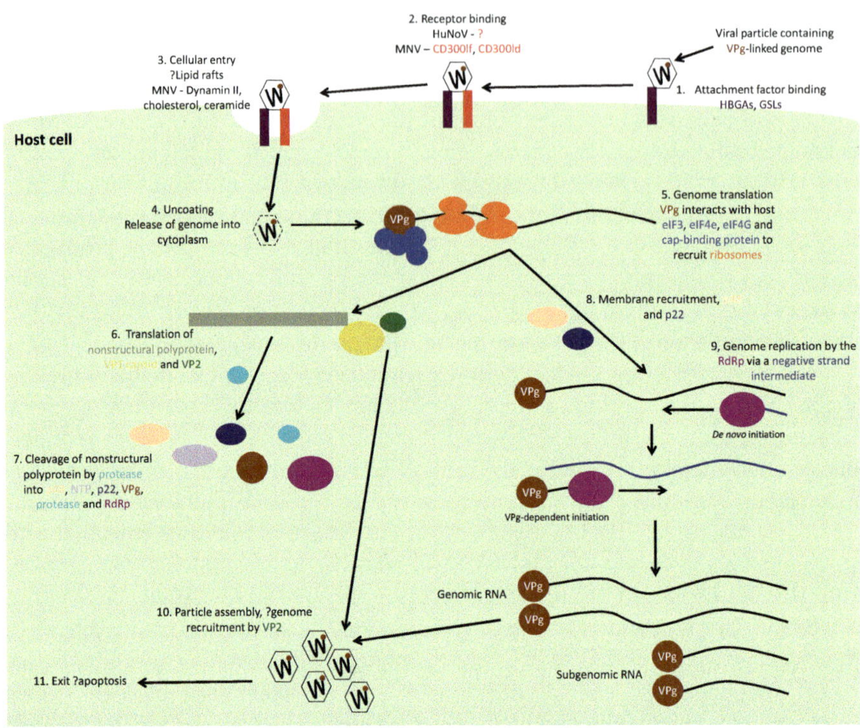

Fig. 2.6 The norovirus replication cycle. Briefly, the virion binds to HBGA cellular attachment factors which enables interaction with proteinaceous cellular receptors. While these proteinaceous receptors are currently unknown for HuNoVs, MNV binds to CD300lf and CD300ld. These interactions likely occur in lipid rafts and MNV cell entry occurs in a process that requires ceramide, cholesterol and dynamin II. After release of the viral genome into the cell cytoplasm, it acts as a template for the 'pioneer round' of translation. The non-structural polyprotein is cleaved by the virally encoded protease. Genome replication occurs in association with cellular membranes derived from the ER, Golgi and endosomes, which are recruited by the non-structural proteins p48 and p22. Genome replication occurs via a negative-sense RNA, which is likely synthesized through de novo initiation to produce a double stranded RNA intermediate. VPg-dependent initiation then occurs to produce the positive-sense genomic RNA. Subgenomic RNA is also synthesized and is translated into the structural proteins, VP1 and VP2. VP1 and VP2 assemble into viral particles that incorporate the VPg-linked genome, possibly through an interaction between the basic domain in VP2 and the viral genome. HuNoV replication is lytic, releasing newly synthesized viruses, although the mechanisms are not fully understood

formed between this residue and R345 [7, 131]. In GI.1, residues D327, H329, S377 and S380 interact with the *N*-acetylglucosamine of HBGA type A [132].

Binding of HuNoV to HBGA attachment factors is not sufficient to mediate entry into target cells, suggesting a further proteinaceous receptor is required. To date, a receptor for HuNoV has not been identified. Two groups recently independently identified CD300lf as a receptor for MNV (Fig. 2.6) [133, 134]. CD300lf was identified as a critical protein for MNV pathogenesis through a CRISPR-cas9 screen in either murine microglial BV2 [133] or murine macrophage RAW264.7 [134] cell lines. Pre-incubation of MNV with soluble CD300lf protected mice from MNV-induced lethality in a dose-dependent manner [133]. Interestingly, the expression of murine CD300lf in the usually non-permissive human HEK293T [134] or HeLa [133] cell lines enabled productive infection of these cells, indicating that the restriction to viral replication in these cases occurs at the level of cell entry. Additionally, stable expression of CD300lf also allowed replication in cat, Chinese hamster, African green monkey and other murine cell lines that are otherwise non-permissive for MNV infection [134]. These results indicate that changes in the CD300lf receptor between species could be an important determinant of MNV species tropism. Substitution of the CC' loop or CDR3 from human CD300lf into the murine protein diminished and abolished MNV infection, respectively, indicating that these regions are important for MNV-binding [133]. However, the reciprocal substitutions from the murine CD300lf into the human protein were not sufficient to enable murine norovirus infection, indicating other regions of the receptor may be important in this interaction. Additionally, the expression of murine CD300ld was also sufficient to enable MNV infection of usually non-permissive human cells (Fig. 2.6) [133]. However, CD300lf-knockout mice were resistant to MNV infection, suggesting CD300lf is likely the predominant MNV receptor [133], although whether CD300lf knockout also alters expression of CD300ld is unknown. Whether CD300lf and/or CD300ld act as functional receptors for HuNoVs is yet to be reported.

Examination of cell lines and in vivo immune system cells permissive to MNV infection demonstrated that these cells express CD300lf to varying degrees [105]. However, IECs from B6 mice exhibit only minimal CD300lf expression [105], despite being permissive to persistent infection with the MNV strain CR6 [110]. Together these results suggest that other as yet unidentified proteins could serve as cell type-specific receptors for MNV. Identification of these receptors and HuNoV receptors on immune cells and IECs could facilitate the infection of other cell lines in vitro and the development of cell culture models to complement the B cell and HIE models for HuNoV, as demonstrated for MNV [133].

Interestingly, clustering of CD300 molecules in lipid rafts is essential for protein function [126], while the GSL attachment factors are known to drive formation of lipid microdomains [127]. A recently proposed model of entry suggests that binding to one or more GSLs results in attachment of the viral particle to the cell surface [126]. The GSLs may subsequently be recruited into lipid rafts or may already be located within lipid rafts, providing a platform for interaction of the MNV particle with CD300 receptors. The interactions between cellular receptors and the viral particle then result in membrane invagination, followed by scission of the invagination by dynamin II [126] (Fig. 2.6). MNV entry occurs via endocytosis, and is known to be dependent on dynamin II, cholesterol and ceramide [4] (Fig. 2.6). Additionally, cell surface proteins CD36, CD44, CD98 and the transferrin receptor have been suggested to be cell type-specific modulators of MNV entry into macrophages and dendritic cells, although the mechanism by which these molecules act is currently unknown [135].

After entry into the cells, the viral capsid disassembles in a process known as uncoating, although the molecular details of the timing and trigger for this are not known. Uncoating results in release of the VPg-linked genome into the cytoplasm where it acts as the template for the 'pioneer round' of protein translation (Fig. 2.6) [4]. Noroviruses do not encode their own translational machinery and translation is therefore dependent on the recruitment of host cell factors and ribosomes. The incoming viral RNA is typically found in very small numbers in comparison with abundant cellular messenger RNAs and viruses have therefore evolved novel mechanisms to enable preferential translation of viral RNA by host cell machinery [136]. In the case of norovirus, translation initiation is driven by the VPg molecule covalently bound to the 5′ end of the viral genome, which acts as a substitute for the 5′ cap found on cellular mRNAs and recruits translation initiation factors, including eIF3, eIF4E, eIF4G and cap binding protein (Fig. 2.6) [4, 5, 137, 138]. These interactions have been demonstrated for HuNoV and MNV and linkage of VPg to the viral genome is essential for infectivity [139]. Binding of these factors by VPg results in recruitment of the 43S ribosomal pre-initiation complex [4]. Other cellular factors are required for translation [139], including eIF4A, the RNA helicase component of eIF4F that has been proposed to unwind RNA structures at the 5′ end of the norovirus genome [4, 139]. Additionally, highly conserved RNA structures at the 5′ and 3′ ends of the Norwalk virus and/or MNV genome bind to cellular proteins including lupus autoantigen (La), poly(rC)-binding protein (PCBP), poly(A)-binding protein (PABP), polypyrimidine tract-binding protein (PTB), heterogeneous nuclear ribonucleoprotein L (hnRNPL) and nucleolin [138], several of which are known to regulate translation of other RNA viral genomes, for example La and PTB stimulate internal ribosome entry site (IRES)-mediated translation of picornavirus [4]. The role of each of these proteins in replication and translation of the norovirus genome, which does not contain an IRES, is not yet fully understood. Binding of the cellular proteins PCPB2 and hnRNPA1 to the genome extremities has been shown to promote HuNoV and MNV genome circularization, known to enhance cellular mRNA translation. Similarly, mutations that disrupt interaction between the 5′ and 3′ ends of the MNV genome resulted in a reduction in viral titer [140, 141]. Recruitment of PTB has been suggested to enable a gradual switch from translation in the early phase of infection to replication in the late phase of infection [138]. Preventing recruitment of PTB to the MNV genome, by mutation of its binding site in the 3′UTR, resulted in viral attenuation in vivo, demonstrating the essential role of host cell proteins in infection [142]. Many other cellular proteins have also been identified to interact at the 5′ and 3′ termini of MNV genomic RNA, but the role of most of these proteins in the MNV lifecycle is not yet known [143].

Similar to other RNA viruses [144–146], noroviruses not only hijack components of the cellular machinery for translation but also selectively switch off host cell protein translation during infection [147]. This effect has recently been shown to be in part mediated by the MNV protease, which cleaves the host protein PABP, a factor essential for cellular cap-dependent mRNA translation [147]. Induction of apoptosis by MNV is also thought to contribute to shut-down of host mRNA translation during infection [147]. Reducing translation of host mRNAs has the additional benefit of reducing synthesis of interferon-stimulated genes and may therefore act as a mechanism to enable antagonism of innate immunity.

Following translation of the ORF1 mRNA, the non-structural polyprotein is cleaved co- and post-translationally by the virally-encoded protease (Fig. 2.6) [148–150]. This results in the mature form of each non-structural protein as well as stable precursor forms [148, 151]. Some of the protein precursors are thought to be functional, for example the protease-polymerase intermediate is capable of catalyzing RNA replication [151]. The norovirus protease has been well characterized compared with most other non-structural proteins and the active site consists of a catalytic triad of residues H30, E54 and C139, with E54 likely to determine substrate specificity rather than being involved in catalysis [152–155]. Translation of VP1 and VP2 occurs primarily from the subgenomic RNA, which is likely a strategy to enable efficient production of the high levels of these proteins required to form new viral particles [4]. The subgenomic RNA is polycistronic and translation of VP2 occurs through a termination-reinitiation mechanism. Upon termination of VP1 translation, ribosomes remain associated with the RNA and reinitiate at the start codon of ORF3 [156].

Like other positive-sense RNA viruses, replication of the norovirus genome occurs in close association with cellular membranes that help form a replication complex (RC), which concentrates the viral proteins alongside host cell proteins required for replication [4]. In MNV-infected cells, the RC forms in the peri-nuclear region in close proximity to the microtubule organizing center (MTOC) [157]. Norovirus recruits membranes from the host cell secretory pathway, including from the endoplasmic reticulum (ER), Golgi apparatus and endosomes (Fig. 2.6) [4, 158]. The mechanisms of membrane recruitment are not completely understood, however the non-structural proteins p48 (MNV NS1/2) and p22 (NS4) have been implicated in this process. Both of these proteins localize to membranes of the secretory pathway and are able to interact with each other [159–161]. The ER-export signal in p22 is thought to promote uptake of p22 into the COPII (conserved coat protein complex II)-coated vesicles that are involved in transport from the ER to the Golgi [4]. This uptake causes mislocalization of COPII-coated vesicles resulting in disassembly of the Golgi. As MNV NS4 does not contain this ER-export signal, it likely acts via a different mechanism. The HuNoV p48 protein is thought to act through the SNARE (soluble *N*-ethylmaleimide-sensitive fusion protein attachment protein receptors) regulator vesicle associated membrane protein-associated protein A (VAP-A) and VAP-B, proteins that regulate cellular vesicle transport [162]. Interaction with these proteins is essential for efficient replication of MNV and results in Golgi disassembly. Inhibition of protein secretion by RNA viruses is thought to be an immune evasion strategy by preventing cytokine secretion and cell surface MHC expression. The 2B and 3A proteins expressed in the same gene order as p48 and p22 in the closely related picornaviruses block host secretory pathways and 3A reduces cytokine secretion and cell surface expression of MHC class I and tumour necrosis factor receptor (TNFR) on infected cells [163–166]. While p48 and p22 are thought to have the same effect [167], this has not been explicitly demonstrated.

Replication of the positive-sense single-stranded norovirus genome occurs via a negative-sense RNA intermediate and is carried out by the RdRp (Fig. 2.6). Structural studies have indicated that the RdRp active site consists of residues E182, D242, Y243, D247, S300, N309, D343 and D344 [168, 169] and the RdRp is thought to function as a homodimer [170]. There are two mechanisms by which the RdRp can initiate synthesis of a new RNA strand: de novo and VPg-dependent [171, 172]. Both of these mechanisms have been demonstrated in vitro and in cells [171]. It has been proposed that synthesis of the negative-sense genomic and subgenomic RNAs occurs by de novo initiation [173]. Interestingly, de novo initiation by the RdRp is enhanced by species-specific interaction with loops in the S domain of VP1 [173]. This has been proposed to drive the early rounds of genome replication when VP1 levels are low [173]. As VP1 levels increase during subsequent rounds of translation, it forms higher level capsid structures, which is proposed to prevent interactions with the RdRp [173]. Therefore, interactions with VP1 may regulate RdRp activity and could influence whether genome replication or translation occurs. Additionally, species-specific interaction between p48 and RdRp enhances RdRp activity, while interaction with VP2 is inhibitory [173], suggesting that other viral proteins may modulate RdRp activity and timing within the viral replication cycle.

Synthesis of the negative-sense RNA results in the formation of a double-stranded replicative form (RF). The synthesis of the positive-sense genomic RNA is then catalyzed by the RdRp through VPg-dependent initiation (Fig. 2.6) [4, 139]. This involves nucleotidylation (also called guanylation) of VPg through formation of a new phosphodiester bond between a conserved tyrosine residue (Y27 in HuNoV and Y26 in MNV) and the first nucleotide of the genome, which is guanine throughout the norovirus genus [172]. The process of nucleotidylation was suggested to be enhanced by an element at the 3′ end of the viral genome, at least in vitro [151]. Following formation of the RF intermediate, synthesis of the positive sense sub-genomic RNA (sgRNA) occurs [4]. Two potential, and non-mutually exclusive, mechanisms have been proposed for sgRNA synthesis based on the identification of both positive-sense and negative-sense genomic RNAs and sgRNAs in infected cells [174]. In the first mechanism, the RdRp reaches an unknown termination signal in the negative sense genomic RNA at the 5′ end of ORF2, resulting in negative-sense subgenomic RNA molecules from which positive-sense subgenomic RNA can be transcribed. More experimental evidence, however, has been provided for the second mechanism, in which VPg-dependent initiation occurs on a promoter present in the genomic negative-sense RNA, downstream of the start of ORF2. This leads to the synthesis of positive sense sgRNA, which could then serve as a template for subsequent rounds of sgRNA replication [4]. The sgRNA promoter has been identified as a conserved RNA stem-loop structure located six nucleotides before the start of the sgRNA, present in all calicivirus genomes [175]. This stem-loop structure has been shown to be essential for MNV infectivity and is sufficient to drive initiation of viral RNA synthesis by the MNV RdRp in vitro, confirming it forms the core of the sgRNA promoter [176]. Furthermore, the specific interactions between the HuNoV and MNV RdRps with their cognate sgRNA promoter have been mapped and mutation of MNV RdRp (NS7) residues that form contacts with the sgRNA promoter also reduce replication and virus production in cells [177]. These interactions with the sgRNA promoter most likely allow the RdRp to specifically recognize the viral RNA in a complex cellular RNA environment.

Newly synthesized viral genomes and capsid proteins, from multiple rounds of genome RNA and sgRNA replication, are packaged together to form nascent virions (Fig. 2.6). As the VP1 capsid protein self-assembles into VLPs in vitro [56], it has been suggested that VP1 may be capable of forming viral particles without assistance from cellular factors (Fig. 2.6). VP2 is incorporated into the interior face of the viral particle through specific interactions with the VP1 shell domain [9]. VP2 contains a basic domain, and as such it has been suggested that VP2 may recruit the genome into the particle [9]; such a domain is not present in VP1. However, specific binding of VP2 to viral RNA has not been demonstrated. Alternatively, it is possible that an interaction between VPg and either VP1 or VP2 recruits the genome into the particle and a possible interaction between VPg and VP1 has been noted in the related FCV [178]. While studies on the mechanisms employed by noroviruses to exit the host cell are largely lacking, apoptosis is induced by MNV and apoptotic cells have been identified in intestinal biopsies from infected patients [117, 179, 180]. Importantly, damaged sites in these biopsies correlate with expression of viral antigens [117]. The induction of apoptosis by MNV is associated with downregulation of the pro-survival factor survivin followed by activation of caspase and cathepsin B and release of cytochrome c [179, 180]. The mechanism of survivin downregulation is currently unknown. The NS1/2 protein is cleaved by caspases (and another currently unidentified cellular protease) in the late phase of the infection cycle [149]. The inhibition of apoptosis reduces the production of MNV particles, suggesting its importance in the replication cycle [180].

2.7 Summary

To date, much of our understanding of norovirus replication and host-pathogen interactions has come from studies with MNV and by making comparisons between conserved mechanisms within other caliciviruses and positive-strand RNA viruses as a whole. Whilst details of the complex interactions between HuNoV and attachment factors, and how this determines host susceptibility, are well studied, our understanding of the molecular details of HuNoV replication are still limited compared to that of other RNA viruses. As we have highlighted here, fundamental questions still remain, such as the identity of the HuNoV proteinaceous receptor and how this influences tropism. However, recent advances in our understanding of norovirus tropism, leading to the development of in vitro propagation systems for HuNoVs make this an exciting time for the norovirus research community. Combined with the available HuNoV reverse genetics system [59] to manipulate and study and produce genetically defined viruses, these systems will allow molecular characterization of norovirus replication and interactions with the host cell. It will enable us to determine whether the details of MNV replication are conserved within HuNoV; MNV has and will continue to provide a tractable surrogate model. Furthermore, these systems will now allow development of specific antiviral and vaccine strategies, and testing of those that have been developed using MNV.

References

1. Green KY. Caliciviridae: the noroviruses. Fields Virol. 2007;2(1):3177.
2. Farkas T, Sestak K, Wei C, Jiang X. Characterization of a rhesus monkey calicivirus representing a new genus of caliciviridae. J Virol. 2008;82(11):5408–16.
3. Smits SL, Rahman M, Schapendonk CME, van Leeuwen M, Faruque ASG, Haagmans BL, et al. Calicivirus from novel recovirus genogroup in human diarrhea, Bangladesh. Emerg Infect Dis. 2012;18(7):1192–5.
4. Thorne LG, Goodfellow IG. Norovirus gene expression and replication. J Gen Virol. 2014;95:278–91.
5. Leen EN, Sorgeloos F, Correia S, Chaudhry Y, Cannac F, Pastore C, et al. A conserved interaction between a C-terminal Motif in norovirus VPg and the HEAT-1 domain of eIF4G is essential for translation initiation. PLoS Pathog. 2016;12(1):e1005379.
6. Prasad BV, Hardy ME, Dokland T, Bella J, Rossmann MG, Estes MK. X-ray crystallographic structure of the Norwalk virus capsid. Science. 1999;286:287–90.
7. Cao S, Lou Z, Tan M, Chen Y, Liu Y, Zhang Z, et al. Structural basis for the recognition of blood group trisaccharides by norovirus. J Virol. 2007;81(11):5949–57.
8. Hardy ME. Norovirus protein structure and function. FEMS Microbiol Lett. 2005;253(1):1–8.
9. Vongpunsawad S, Venkataram Prasad BV, Estes MK. Norwalk virus minor capsid protein VP2 associates within the VP1 shell domain. J Virol. 2013;87(9):4818–25.
10. McFadden N, Bailey D, Carrara G, Benson A, Chaudhry Y, Shortland A, et al. Norovirus regulation of the innate immune response and apoptosis occurs via the product of the alternative open reading frame 4. PLoS Pathog. 2011;7(12):e1002413.
11. Jones MK, Watanabe M, Zhu S, Graves CL, Keyes LR, Grau KR, et al. Enteric bacteria promote human and mouse norovirus infection of B cells. Science. 2014;346(6210):755–9.
12. Ettayebi K, Crawford SE, Murakami K, Broughman JR, Karandikar U, Tenge VR, et al. Replication of human noroviruses in stem cell – derived human enteroids. Science. 2016;5211:1–12.
13. Zheng D-P, Ando T, Fankhauser RL, Beard RS, Glass RI, Monroe SS. Norovirus classification and proposed strain nomenclature. Virology. 2006;346(2):312–23.
14. Kroneman A, Vega E, Vennema H, Vinjé J, White PA, Hansman G, et al. Proposal for a unified norovirus nomenclature and genotyping. Arch Virol. 2013;158(10):2059–68.
15. Vinje J. Advances in laboratory methods for detection and typing of norovirus. J Clin Microbiol. 2015;53(2):373–81.
16. Bull RA, Hansman GS, Clancy LE, Tanaka MM, Rawlinson WD, White PA. Norovirus recombination in ORF1/ORF2 overlap. Emerg Infect Dis. 2005;11(7):1079–85.
17. Bull RA, Tanaka MM, White PA. Norovirus recombination. J Gen Virol. 2007;88 (12):3347–59.
18. Eden J-S, Tanaka MM, Boni MF, Rawlinson WD, White PA. Recombination within the pandemic norovirus GII.4 lineage. J Virol. 2013;87(11):6270–82.
19. Chhabra P, de Graaf M, Parra GI, Chan MC, Green K, Martella V, Wang Q, White PA, Katayama K, Vennema H, Koopmans MPG, Vinjé J. Updated classification of norovirus genogroups and genotypes. J Gen Virol. 2019;100(10):1393–406.
20. Siebenga JJ, Vennema H, Zheng D-P, Vinjé J, Lee BE, Pang X-L, et al. Norovirus illness is a global problem: emergence and spread of norovirus GII.4 variants, 2001–2007. J Infect Dis. 2009;200(5):802–12.

21. Parra GI, Squires RB, Karangwa CK, Johnson JA, Lepore CJ, Sosnovtsev SV, et al. Static and evolving norovirus genotypes: implications for epidemiology and immunity. PLoS Pathog. 2017;13(1):e1006136.
22. van Beek J, Ambert-Balay K, Botteldoorn N, Eden JS, Fonager J, Hewitt J, et al. Indications for worldwide increased norovirus activity associated with emergence of a new variant of genotype II.4, late 2012. Euro Surveill. 2013;18(1):8–9.
23. Sugieda M, Nagaoka H, Kakishima Y, Ohshita T, Nakamura S, Nakajima S. Detection of Norwalk-like virus genes in the caecum contents of pigs. Brief Rep Arch Virol. 1998;143 (6):1215–21.
24. Wang QH, Myung GH, Cheetham S, Souza M, Funk JA, Saif LJ. Porcine noroviruses related to human noroviruses. Emerg Infect Dis. 2005;11(12):1874–81.
25. Wolf S, Williamson W, Hewitt J, Lin S, Rivera-Aban M, Ball A, et al. Molecular detection of norovirus in sheep and pigs in New Zealand farms. Vet Microbiol. 2009;133(1–2):184–9.
26. Liu BL, Lambden PR, Günther H, Otto P, Elschner M, Clarke IN. Molecular characterization of a bovine enteric calicivirus: relationship to the Norwalk-like viruses. J Virol. 1999;73 (1):819–25.
27. Ando T, Noel JS, Fankhauser RL. Genetic classification of "Norwalk-like viruses". J Infect Dis. 2000;181(Suppl 2):S336–S348.
28. Oliver SL, Dastjerdi AM, Wong S, El-Attar L, Gallimore C, Brown DWG, et al. Molecular characterization of bovine enteric caliciviruses: a distinct third genogroup of noroviruses (Norwalk-like viruses) unlikely to be of risk to humans. J Virol. 2003;77(4):2789–98.
29. Park SI, Jeong C, Kim HH, Park SH, Park SJ, Hyun BH, et al. Molecular epidemiology of bovine noroviruses in South Korea. Vet Microbiol. 2007;124(1–2):125–33.
30. Martella V, Campolo M, Lorusso E, Cavicchio P, Camero M, Bellacicco AL, et al. Norovirus in captive lion cub (*Panthera leo*). Emerg Infect Dis. 2007;13(7):1071–3.
31. Bodnar L, Lorusso E, Di Martino B, Catella C, Lanave G, Elia G, et al. Identification of a novel canine norovirus. Infect Genet Evol. 2017;52:75–81.
32. Karst SM. STAT1-dependent innate immunity to a Norwalk-like virus. Science. 2003;299 (5612):1575–8.
33. Wobus CE, Thackray LB, Virgin HW. Murine norovirus: a model system to study norovirus biology and pathogenesis. J Virol. 2006;80(11):5104–12.
34. Smith DB, McFadden N, Blundell RJ, Meredith A, Simmonds P. Diversity of murine norovirus in wild-rodent populations: species-specific associations suggest an ancient divergence. J Gen Virol. 2012;93(2):259–66.
35. Tse H, Lau SKP, Chan W-M, Choi GKY, Woo PCY, Yuen K-Y. Complete genome sequences of novel canine noroviruses in Hong Kong. J Virol. 2012;86(17):9531–2.
36. Martella V, Lorusso E, Decaro N, Elia G, Radogna A, D'Abramo M, et al. Detection and molecular characterization of a canine norovirus. Emerg Infect Dis. 2008;14(8):1306–8.
37. Martella V, Decaro N, Lorusso E, Radogna A, Moschidou P, Amorisco F, et al. Genetic heterogeneity and recombination in canine noroviruses. J Virol. 2009;83(21):11391–6.
38. Mesquita JR, Barclay L, Nascimento MSJ, Vinjé J. Novel norovirus in dogs with diarrhea. Emerg Infect Dis. 2010;16(6):980–2.
39. Caddy SL, de Rougemont A, Emmott E, El-Attar L, Mitchell JA, Hollinshead M, et al. Evidence for human norovirus infection of dogs in the UK. J Clin Microbiol. 2015;53:1873–83.
40. Takano T, Kusuhara H, Kuroishi A, Takashina M, Doki T, Nishinaka T, et al. Molecular characterization and pathogenicity of a genogroup GVI feline norovirus. Vet Microbiol. 2015;178(3–4):201–7.
41. Di Martino B, Di Profio F, Melegari I, Sarchese V, Cafiero MA, Robetto S, et al. A novel feline norovirus in diarrheic cats. Infect Genet Evol. 2016;38:132–7.
42. de Graaf M, Bodewes R, van Elk CE, van de Bildt M, Getu S, Aron GI, et al. Norovirus infection in harbor porpoises. Emerg Infect Dis. 2017;23(1):87–91.

43. Wu Z, Yang L, Ren X, He G, Zhang J, Yang J, et al. Deciphering the bat virome catalog to better understand the ecological diversity of bat viruses and the bat origin of emerging infectious diseases. ISME J. 2016;10(3):609–20.
44. Hu D, Zhu C, Wang Y, Ai L, Yang L, Ye F, et al. Virome analysis for identification of novel mammalian viruses in bats from Southeast China. Sci Rep. 2017;7(1):10917.
45. Yang L, Wang Q, Xu L, Tu C, Huang X, He B. Detection and characterization of a novel norovirus in bats, China. Virol Sin. 2018;33:100–3.
46. Li L, Shan T, Wang C, Cote C, Kolman J, Onions D, et al. The fecal viral flora of California Sea Lions. J Virol. 2011;85(19):9909–17.
47. Siebenga JJ, Lemey P, Kosakovsky Pond SL, Rambaut A, Vennema H, Koopmans M. Phylodynamic reconstruction reveals norovirus GII.4 epidemic expansions and their molecular determinants. PLoS Pathog. 2010;6(5):e1000884.
48. Eden J-S, Hewitt J, Lim KL, Boni MF, Merif J, Greening G, et al. The emergence and evolution of the novel epidemic norovirus GII.4 variant Sydney 2012. Virology. 2014;450–451:106–13.
49. Bok K, Abente EJ, Realpe-Quintero M, Mitra T, Sosnovtsev SV, Kapikian AZ, et al. Evolutionary dynamics of GII.4 noroviruses over a 34-year period. J Virol. 2009;83(22):11890–901.
50. Mori K, Chu PY, Motomura K, Somura Y, Nagano M, Kimoto K, et al. Genomic analysis of the evolutionary lineage of Norovirus GII.4 from archival specimens during 1975–1987 in Tokyo. J Med Virol. 2017;89(2):363–7.
51. Donaldson EF, Lindesmith LC, Lobue AD, Baric RS. Norovirus pathogenesis: mechanisms of persistence and immune evasion in human populations. Immunol Rev. 2008;225:190–211.
52. Bertolotti-Ciarlet A, White LJ, Chen R, Prasad V, Estes MK, Prasad BVV. Structural requirements for the assembly of Norwalk virus-like particles. J Virol. 2002;76(8):4044–55.
53. Singh BK, Leuthold MM, Hansman GS. Human noroviruses' fondness for histo-blood group antigens. J Virol. 2015;89(4):2024–40.
54. Singh BK, Leuthold MM, Hansman GS. Structural constraints on human norovirus binding to histo-blood group antigens. mSphere. 2016;1(2):1–7.
55. Green KY, Lew JF, Jiang X, Kapikian AZ, Estes MK. Comparison of the reactivities of baculovirus-expressed recombinant Norwalk virus capsid antigen with those of the native Norwalk virus antigen in serologic assays and some epidemiologic observations. J Clin Microbiol. 1993;31(8):2185–91.
56. Baric R, Yount B, Lindesmith L. Expression and self-assembly of Norwalk virus capsid protein from Venezuelan equine encephalitis virus replicons. J Virol. 2002;76(6):3023–30.
57. Bertolotti-Ciarlet A, Crawford SE, Hutson AM, Estes MK. The 3′ end of Norwalk virus mRNA contains determinants that regulate the expression and stability of the viral capsid protein VP1: a novel function for the VP2 protein. J Virol. 2003;77(21):11603–15.
58. Jiang X, Wang M, Graham DY, Estes MK. Expression, self-assembly, and antigenicity of the Norwalk virus capsid protein. J Virol. 1992;66(11):6527–32.
59. Katayama K, Murakami K, Sharp TM, Guix S, Oka T, Takai-Todaka R, et al. Plasmid-based human norovirus reverse genetics system produces reporter-tagged progeny virus containing infectious genomic RNA. Proc Natl Acad Sci. 2014;111(38):E4043–52.
60. Lindesmith LC, Donaldson EF, LoBue AD, Cannon JL, Zheng DP, Vinje J, et al. Mechanisms of GII.4 norovirus persistence in human populations. PLoS Med. 2008;5(2):0269–90.
61. Lindesmith LC, Beltramello M, Donaldson EF, Corti D, Swanstrom J, Debbink K, et al. Immunogenetic mechanisms driving norovirus GII.4 antigenic variation. PLoS Pathog. 2012;8 (5):e1002705.
62. Cortes-Penfield NW, Ramani S, Estes MK, Atmar RL. Prospects and challenges in the development of a norovirus vaccine. Clin Ther. 2017;39:1537–49.
63. Siebenga JJ, Vennema H, Renckens B, de Bruin E, van der Veer B, Siezen RJ, et al. Epochal evolution of GGII.4 norovirus capsid proteins from 1995 to 2006. J Virol. 2007;81 (18):9932–41.

64. Allen DJ, Gray JJ, Gallimore CI, Xerry J, Iturriza-Gómara M. Analysis of amino acid variation in the P2 domain of the GII-4 Norovirus VP1 protein reveals putative variant-specific epitopes. PLoS One. 2008;3(1):e1485.
65. Bull RA, Eden J-S, Rawlinson WD, White PA. Rapid evolution of pandemic noroviruses of the GII.4 lineage. PLoS Pathog. 2010;6(3):e1000831.
66. Lindesmith LC, Donaldson EF, Baric RS. Norovirus GII.4 strain antigenic variation. J Virol. 2011;85(1):231–42.
67. Lindesmith LC, Donaldson EF, Beltramello M, Pintus S, Corti D, Swanstrom J, et al. Particle conformation regulates antibody access to a conserved GII.4 norovirus blockade epitope. J Virol. 2014;88(16):8826–42.
68. Lindesmith LC, Costantini V, Swanstrom J, Debbink K, Donaldson EF, Vinjé J, et al. Emergence of a norovirus GII.4 strain correlates with changes in evolving blockade epitopes. J Virol. 2013;87(5):2803–13.
69. Debbink K, Lindesmith LC, Donaldson EF, Costantini V, Beltramello M, Corti D, et al. Emergence of new pandemic gII.4 sydney norovirus strain correlateswith escape from herd immunity. J Infect Dis. 2013;208(11):1877–87.
70. Reeck A, Kavanagh O, Estes MK, Opekun AR, Gilger MA, Graham DY, et al. Serological correlate of protection against norovirus-induced gastroenteritis. J Infect Dis. 2010;202 (8):1212–8.
71. Bok K, Parra GI, Mitra T, Abente E, Shaver CK, Boon D, et al. Chimpanzees as an animal model for human norovirus infection and vaccine development. Proc Natl Acad Sci USA. 2011;108(1):325–30.
72. Debbink K, Donaldson EF, Lindesmith LC, Baric RS. Genetic mapping of a highly variable norovirus GII.4 blockade epitope: potential role in escape from human herd immunity. J Virol. 2012;86(2):1214–26.
73. Lindesmith LC, Debbink K, Swanstrom J, Vinjé J, Costantini V, Baric RS, et al. Monoclonal antibody-based antigenic mapping of norovirus GII.4-2002. J Virol. 2012;86(2):873–83.
74. Parrino TA, Schreiber DS, Trier JS, Kapikian AZ, Blacklow NR. Clinical immunity in acute gastroenteritis caused by Norwalk agent. N Engl J Med. 1977;297(2):86–9.
75. Lindesmith L, Moe C, Marionneau S, Ruvoen N, Jiang X, Lindblad L, et al. Human susceptibility and resistance to Norwalk virus infection. Nat Med. 2003;9(5):548–53.
76. Kindberg E, Åkerlind B, Johnsen C, Knudsen JD, Heltberg O, Larson G, et al. Host genetic resistance to symptomatic norovirus (GGII.4) infections in Denmark. J Clin Microbiol. 2007;45(8):2720–2.
77. Lindesmith LC, Donaldson EF, Lobue AD, Cannon JL, Zheng D-P, Vinje J, et al. Mechanisms of GII.4 norovirus persistence in human populations. PLoS Med. 2008;5(2):e31.
78. Bucardo F, Kindberg E, Paniagua M, Vildevall M, Svensson L. Genetic susceptibility to symptomatic norovirus infection in nicaragua. J Med Virol. 2009;81(4):728–35.
79. Kindberg E, Svensson L. Genetic basis of host resistance to norovirus infection. Future Virol. 2009;4(4):369–82.
80. Nordgren J, Nitiema LW, Ouermi D, Simpore J, Svensson L. Host genetic factors affect susceptibility to norovirus infections in Burkina Faso. PLoS One. 2013;8(7):e69557.
81. Ruvoën-Clouet N, Belliot G, Le Pendu J. Noroviruses and histo-blood groups: the impact of common host genetic polymorphisms on virus transmission and evolution. Rev Med Virol. 2013;23:355–66.
82. Kambhampati A, Payne DC, Costantini V, Lopman BA. Host genetic susceptibility to enteric viruses: a systematic review and metaanalysis. Clin Infect Dis. 2015;62:11–8.
83. Storry JR, Olsson ML. Genetic basis of blood group diversity. Br J Haematol. 2004;126:759–71.
84. Marionneau S, Cailleau-Thomas A, Rocher J, Le Moullac-Vaidye B, Ruvoën N, Clément M, et al. ABH and Lewis histo-blood group antigens, a model for the meaning of oligosaccharide diversity in the face of a changing world. Biochimie. 2001;83:565–73.

85. Le Pendu J, Ruvoën-Clouet N, Kindberg E, Svensson L. Mendelian resistance to human norovirus infections. Semin Immunol. 2006;18:375–86.
86. Nordgren J, Sharma S, Kambhampati A, Lopman B, Svensson L, Lynfield R. Innate resistance and susceptibility to norovirus infection. PLoS Pathog. 2016;12(4):e1005385.
87. Carlsson B, Kindberg E, Buesa J, Rydell GE, Lidón MF, Montava R, et al. The G428A nonsense mutation in FUT2 provides strong but not absolute protection against symptomatic GII.4 norovirus infection. PLoS One. 2009;4(5):e5593.
88. Huang P, Farkas T, Marionneau S, Zhong W, Ruvoën-Clouet N, Morrow AL, et al. Noroviruses bind to human ABO, Lewis, and secretor histo-blood group antigens: identification of 4 distinct strain-specific patterns. J Infect Dis. 2003;188(1):19–31.
89. Marionneau S, Ruvoën N, Le MoullacVaidye B, Clement M, CailleauThomas A, RuizPalacois G, et al. Norwalk virus binds to histo-blood group antigens present on gastroduodenal epithelial cells of secretor individuals. Gastroenterology. 2002;122(7):1967–77.
90. Liu P, Wang X, Lee J-C, Teunis P, Hu S, Paradise HT, et al. Genetic susceptibility to norovirus GII.3 and GII.4 infections in Chinese pediatric diarrheal disease. Pediatr Infect Dis J. 2014;33(11):e305–9.
91. Lopman BA, Trivedi T, Vicuña Y, Costantini V, Collins N, Gregoricus N, et al. Norovirus infection and disease in an ecuadorian birth cohort: association of certain norovirus genotypes with host FUT2 secretor status. J Infect Dis. 2015;211(11):1813–21.
92. Currier RL, Payne DC, Staat MA, Selvarangan R, Shirley SH, Halasa N, et al. Innate susceptibility to norovirus infections influenced by FUT2 genotype in a United States pediatric population. Clin Infect Dis. 2015;60(11):1631–8.
93. Donaldson EF, Lindesmith LC, Lobue AD, Baric RS. Viral shape-shifting: norovirus evasion of the human immune system. Nat Rev Microbiol. 2010;8(3):231–41.
94. Ruvoën-Clouet N, Magalhaes A, Marcos-Silva L, Breiman A, Figueiredo C, David L, et al. Increase in genogroup II.4 norovirus host spectrum by CagA-positive helicobacter pylori infection. J Infect Dis. 2014;210(2):183–91.
95. Frenck R, Bernstein DI, Xia M, Huang P, Zhong W, Parker S, et al. Predicting susceptibility to norovirus GII.4 by use of a challenge model involving humans. J Infect Dis. 2012;206(9):1386–93.
96. BHG R, Vennema H, CJPA H, Duizer E, MPG K. Association of histo-blood group antigens and susceptibility to norovirus infections. J Infect Dis. 2005;191:749–54.
97. Nordgren J, Kindberg E, Lindgren PE, Matussek A, Svensson L. Norovirus gastroenteritis outbreak with a secretor-independent susceptibility pattern, Sweden. Emerg Infect Dis. 2010;16(1):81–7.
98. Lindesmith L, Moe C, Le Pendu J, Frelinger JA, Treanor J, Baric RS. Cellular and humoral immunity following Snow Mountain virus challenge. J Virol. 2005;79(5):2900–9.
99. Shirato H, Ogawa S, Ito H, Sato T, Kameyama A, Narimatsu H, et al. Noroviruses distinguish between type 1 and type 2 histo-blood group antigens for binding. J Virol. 2008;82(21):10756–67.
100. Ayouni S, Estienney M, Sdiri-Loulizi K, Ambert-Balay K, de Rougemont A, Aho S, et al. Relationship between GII.3 norovirus infections and blood group antigens in young children in Tunisia. Clin Microbiol Infect. 2015;21(9):874.e1-874.e8.
101. Tamura M, Natori K, Kobayashi M, Miyamura T, Takeda N. Genogroup II noroviruses efficiently bind to heparan sulfate proteoglycan associated with the cellular membrane. J Virol. 2004;78(8):3817–26.
102. Han L, Tan M, Xia M, Kitova EN, Jiang X, Klassen JS. Gangliosides are ligands for human noroviruses. J Am Chem Soc. 2014;136(36):12631–7.
103. Bally M, Rydell GE, Zahn R, Nasir W, Eggeling C, Breimer ME, et al. Norovirus GII.4 virus-like particles recognize galactosylceramides in domains of planar supported lipid bilayers. Angew Chemie Int Ed. 2012;51(48):12020–4.

104. Wobus CE, Karst SM, Thackray LB, Chang KO, Sosnovtsev SV, Belliot G, et al. Replication of norovirus in cell culture reveals a tropism for dendritic cells and macrophages. PLoS Biol. 2004;2(12):e432.
105. Grau KR, Roth AN, Zhu S, Hernandez A, Colliou N, DiVita BB, et al. The major targets of acute norovirus infection are immune cells in the gut-associated lymphoid tissue. Nat Microbiol. 2017;2:1586–91.
106. Ward JM, Wobus CE, Thackray LB, Erexson CR, Faucette LJ, Belliot G, et al. Pathology of immunodeficient mice with naturally occurring murine norovirus infection. Toxicol Pathol. 2006;34(6):708–15.
107. Gonzalez-Hernandez MB, Liu T, Blanco LP, Auble H, Payne HC, Wobus CE. Murine norovirus transcytosis across an in vitro polarized murine intestinal epithelial monolayer is mediated by M-like cells. J Virol. 2013;87:12685–93.
108. Karst SM, Wobus CE. A working model of how noroviruses infect the intestine. PLoS Pathog. 2015;11(2):e1004626.
109. Gonzalez-Hernandez MB, Liu T, Payne HC, Stencel-Baerenwald JE, Ikizler M, Yagita H, et al. Efficient norovirus and reovirus replication in the mouse intestine requires microfold (M) cells. J Virol. 2014;88(12):6934–43.
110. Lee S, Wilen CB, Orvedahl A, McCune BT, Kim KW, Orchard RC, et al. Norovirus cell tropism is determined by combinatorial action of a viral non-structural protein and host cytokine. Cell Host Microbe. 2017. doi: https://doi.org/10.1016/j.chom.2017.08.021
111. Wilen CB, Lee S, Hsieh LL, Orchard RC, Desai C, Hykes BL, et al. Tropism for tuft cells determines immune promotion of norovirus pathogenesis. Science. 2018;360(6385):204–8.
112. Taube S, Kolawole AO, Höhne M, Wilkinson JE, Handley SA, Perry JW, et al. A mouse model for human norovirus. MBio. 2013;4(4) https://doi.org/10.1128/mBio.00450-13.
113. Lay MK, Atmar RL, Guix S, Bharadwaj U, He H, Neill FH, et al. Norwalk virus does not replicate in human macrophages or dendritic cells derived from the peripheral blood of susceptible humans. Virology. 2010;406(1):1–11.
114. Brown JR, Gilmour K, Breuer J. Norovirus infections occur in B-cell-deficient patients. Clin Infect Dis. 2016;62(9):1136–8.
115. Jones MK, Grau KR, Costantini V, Kolawole AO, de Graaf M, Freiden P, et al. Human norovirus culture in B cells. Nat Protoc. 2015;10(12):1939–47.
116. Sato T, Stange DE, Ferrante M, Vries RGJ, Van Es JH, Van Den Brink S, et al. Long-term expansion of epithelial organoids from human colon, adenoma, adenocarcinoma, and Barrett's epithelium. Gastroenterology. 2011;141(5):1762–72.
117. Karandikar UC, Crawford SE, Ajami NJ, Murakami K, Kou B, Ettayebi K, et al. Detection of human norovirus in intestinal biopsies from immunocompromised transplant patients. J Gen Virol. 2016;97(9):2291–300.
118. Cheetham S, Souza M, Meulia T, Grimes S, Han MG, Saif LJ. Pathogenesis of a genogroup II human norovirus in gnotobiotic pigs. J Virol. 2006;80(21):10372–81.
119. Otto PH, Clarke IN, Lambden PR, Salim O, Reetz J, Liebler-Tenorio EM. Infection of calves with bovine norovirus GIII.1 strain Jena Virus: an experimental model to study the pathogenesis of norovirus infection. J Virol. 2011;85(22):12013–21.
120. Green KY. Editorial commentary: noroviruses and B cells. Clin Infect Dis. 2016;62 (9):1139–40.
121. Chaudhry Y, Skinner MA, Goodfellow IG. Recovery of genetically defined murine norovirus in tissue culture by using a fowlpox virus expressing T7 RNA polymerase. J Gen Virol. 2007;88(8):2091–100.
122. Yunus MA, Chung LMW, Chaudhry Y, Bailey D, Goodfellow I. Development of an optimized RNA-based murine norovirus reverse genetics system. J Virol Methods. 2010;169 (1):112–8.
123. Arias A, Bailey D, Chaudhry Y, Goodfellow I. Development of a reverse-genetics system for murine norovirus 3: long-term persistence occurs in the caecum and colon. J Gen Virol. 2012;93(Part 7):1432–41.

124. Hwang S, Alhatlani B, Arias A, Caddy SL, Christodoulou C, Cunha JB, et al. Murine norovirus: propagation, quantification, and genetic manipulation. Curr Protoc Microbiol. 2014;33 https://doi.org/10.1002/9780471729259.mc15k02s33.
125. Mercer J, Schelhaas M, Helenius A. Virus entry by endocytosis. Annu Rev Biochem. 2010;79 (1):803–33.
126. Bartnicki E, Cunha JB, Kolawole AO, Wobus CE. Recent advances in understanding noroviruses. F1000Research. 2017;6:79.
127. Taube S, Jiang M, Wobus CE. Glycosphingolipids as receptors for non-enveloped viruses. Viruses. 2010;2:1011–49.
128. Taube S, Perry JW, Yetming K, Patel SP, Auble H, Shu L, et al. Ganglioside-linked terminal sialic acid moieties on murine macrophages function as attachment receptors for murine noroviruses. J Virol. 2009;83:4092–101.
129. Koromyslova AD, Leuthold MM, Bowler MW, Hansman GS. The sweet quartet: binding of fucose to the norovirus capsid. Virology. 2015;483:203–8.
130. Singh BK, Koromyslova A, Hefele L, Gurth C, Hansman GS. Structural evolution of the emerging 2014-2015 GII.17 noroviruses. J Virol. 2016;90(5):2710–5.
131. Donaldson EF, Lindesmith LC, Lobue AD, Baric RS. Norovirus pathogenesis: mechanisms of persistence and immune evasion in human populations. Immunol Rev. 2008 Oct;225:190–211.
132. Bu W, Mamedova A, Tan M, Xia M, Jiang X, Hegde RS. Structural basis for the receptor binding specificity of Norwalk virus. J Virol. 2008;82(11):5340–7.
133. Orchard RC, Wilen CB, Doench JG, Baldridge MT, McCune BT, Lee Y-CJ, et al. Discovery of a proteinaceous cellular receptor for a norovirus. Science. 2016;353(6302):933–6.
134. Haga K, Fujimoto A, Takai-Todaka R, Miki M, Doan YH, Murakami K, et al. Functional receptor molecules CD300lf and CD300ld within the CD300 family enable murine noroviruses to infect cells. Proc Natl Acad Sci USA. 2016;113:E6248.
135. Bragazzi Cunha J, Wobus CE. Select membrane proteins modulate MNV-1 infection of macrophages and dendritic cells in a cell type-specific manner. Virus Res. 2016;222:64–70.
136. Firth AE, Brierley I. Non-canonical translation in RNA viruses. J Gen Virol. 2012;93:1385–409.
137. Chung L, Bailey D, Leen EN, Emmott EP, Chaudhry Y, Roberts LO, et al. Norovirus translation requires an interaction between the C terminus of the genome-linked viral protein VPg and eukaryotic translation initiation factor 4G. J Biol Chem. 2014;289(31):21738–50.
138. Alhatlani B, Vashist S, Goodfellow I. Functions of the 5′ and 3′ ends of calicivirus genomes. Virus Res. 2015;206:134–43.
139. Chaudhry Y, Nayak A, Bordeleau ME, Tanaka J, Pelletier J, Belsham GJ, et al. Caliciviruses differ in their functional requirements for eIF4F components. J Biol Chem. 2006;281 (35):25315–25.
140. López-Manríquez E, Vashist S, Ureña L, Goodfellow I, Chavez P, Mora-Heredia JE, et al. Norovirus genome circularization and efficient replication are facilitated by binding of PCBP2 and hnRNP A1. J Virol. 2013;87(21):11371–87.
141. Sandoval-Jaime C, Gutiérrez-Escolano AL. Cellular proteins mediate 5′-3′ end contacts of Norwalk virus genomic RNA. Virology. 2009;387(2):322–30.
142. Bailey D, Karakasiliotis I, Vashist S, Chung LMW, Reese J, McFadden N, et al. Functional analysis of rna structures present at the 3′ extremity of the murine norovirus genome: the variable polypyrimidine tract plays a role in viral virulence. J Virol. 2010;84(6):2859–70.
143. Vashist S, Urena L, Chaudhry Y, Goodfellow I. Identification of RNA-protein interaction networks involved in the norovirus life cycle. J Virol. 2012;86(22):11977–90.
144. Kuyumcu-Martinez NM, Van Eden ME, Younan P, Lloyd RE. Cleavage of poly(A)-binding protein by poliovirus 3C protease inhibits host cell translation: a novel mechanism for host translation shutoff. Mol Cell Biol. 2004;24(4):1779–90.

145. Kuyumcu-Martinez M, Belliot G, Sosnovtsev SV, Chang K-O, Green KY, Lloyd RE. Calicivirus 3C-like proteinase inhibits cellular translation by cleavage of poly(A)-binding protein. J Virol. 2004;78(15):8172–82.
146. Smith RWP, Gray NK. Poly(A)-binding protein (PABP): a common viral target. Biochem J. 2010;426(1):1–12.
147. Emmott E, Sorgeloos F, Caddy SL, Heesom K. Norovirus-mediated modification of the translational landscape via virus and host-induced cleavage of translation initiation factors. Mol Cell Proteomics. 2017;16:1–32.
148. Belliot G, Sosnovtsev SV, Mitra T, Hammer C, Garfield M, Green KY. In vitro proteolytic processing of the MD145 norovirus ORF1 nonstructural polyprotein yields stable precursors and products similar to those detected in calicivirus-infected cells. J Virol. 2003;77(20):10957–74.
149. Sosnovtsev SV, Belliot G, Chang K-O, Prikhodko VG, Thackray LB, Wobus CE, et al. Cleavage map and proteolytic processing of the murine norovirus nonstructural polyprotein in infected cells. J Virol. 2006;80(16):7816–31.
150. Emmott E, de Rougemont A, Haas J, Goodfellow I. Spatial and temporal control of norovirus protease activity is determined by polyprotein processing and intermolecular interactions within the viral replication complex. bioRxiv. 2017; https://doi.org/10.1101/175463.
151. Belliot G, Sosnovtsev SV, Chang K-O, Babu V, Uche U, Arnold JJ, et al. Norovirus proteinase-polymerase and polymerase are both active forms of RNA-dependent RNA polymerase. J Virol. 2005;79(4):2393–403.
152. Someya Y, Takeda N, Miyamura T. Identification of active-site amino acid residues in the Chiba virus 3C-like protease. J Virol. 2002;76(12):5949–58.
153. Someya Y, Takeda N, Wakita T. Saturation mutagenesis reveals that GLU54 of norovirus 3C-like protease is not essential for the proteolytic activity. J Biochem. 2008;144(6):771–80.
154. Zeitler CE, Estes MK, Venkataram Prasad BV. X-ray crystallographic structure of the Norwalk virus protease at 1.5-A resolution. J Virol. 2006;80(10):5050–8.
155. Leen EN, Baeza G, Curry S. Structure of a murine norovirus NS6 protease-product complex revealed by adventitious crystallisation. PLoS One. 2012;7(6):e38723.
156. Napthine S, Lever RA, Powell ML, Jackson RJ, Brown TDK, Brierley I. Expression of the VP2 protein of murine norovirus by a translation termination-reinitiation strategy. PLoS One. 2009;4(12):e8390.
157. Hyde JL, Gillespie LK, Mackenzie JM. Mouse norovirus 1 utilizes the cytoskeleton network to establish localization of the replication complex proximal to the microtubule organizing center. J Virol. 2012;86(8):4110–22.
158. Hyde JL, Sosnovtsev SV, Green KY, Wobus C, Virgin HW, Mackenzie JM. Mouse norovirus replication is associated with virus-induced vesicle clusters originating from membranes derived from the secretory pathway. J Virol. 2009;83(19):9709–19.
159. Bailey D, Kaiser WJ, Hollinshead M, Moffat K, Chaudhry Y, Wileman T, et al. Feline calicivirus p32, p39 and p30 proteins localize to the endoplasmic reticulum to initiate replication complex formation. J Gen Virol. 2010;91(3):739–49.
160. Hyde JL, Mackenzie JM. Subcellular localization of the MNV-1 ORF1 proteins and their potential roles in the formation of the MNV-1 replication complex. Virology. 2010;406(1):138–48.
161. Thorne L, Bailey D, Goodfellow I. High-resolution functional profiling of the norovirus genome. J Virol. 2012;86(21):11441–56.
162. McCune BT, Tang W, Lu J, Eaglesham JB, Thorne L, Mayer AE, et al. Noroviruses co-opt the function of host proteins VAPA and VAPB for replication via a phenylalanine-phenylalanine-acidic-tract-motif mimic in nonstructural viral protein NS1/2. MBio. 2017;8(4):1–17.
163. Doedens JR, Kirkegaard K. Inhibition of cellular protein secretion by poliovirus proteins 2B and 3A. EMBO J. 1995;14(5):894–907.
164. Deitz SB, Dodd DA, Cooper S, Parham P, Kirkegaard K. MHC I-dependent antigen presentation is inhibited by poliovirus protein 3A. Proc Natl Acad Sci USA. 2000;97(25):13790–5.

165. Dodd DA, Giddings TH, Kirkegaard K. Poliovirus 3A protein limits interleukin-6 (IL-6), IL-8, and beta interferon secretion during viral infection. J Virol. 2001;75(17):8158–65.
166. Neznanov N, Kondratova A, Chumakov KM, Angres B, Zhumabayeva B, Agol VI, et al. Poliovirus protein 3A inhibits tumor necrosis factor (TNF)-induced apoptosis by eliminating the TNF receptor from the cell surface. J Virol. 2001;75(21):10409–20.
167. Roth AN, Karst SM. Norovirus mechanisms of immune antagonism. Curr Opin Virol. 2016;16:24–30.
168. Ng KK-S, Pendás-Franco N, Rojo J, Boga JA, Machín A, Alonso JMM, et al. Crystal structure of norwalk virus polymerase reveals the carboxyl terminus in the active site cleft. J Biol Chem. 2004;279(16):16638–45.
169. Zamyatkin DF, Parra F, Alonso JMM, Harki DA, Peterson BR, Grochulski P, et al. Structural insights into mechanisms of catalysis and inhibition in Norwalk virus polymerase. J Biol Chem. 2008;283(12):7705–12.
170. Högbom M, Jäger K, Robel I, Unge T, Rohayem J. The active form of the norovirus RNA-dependent RNA polymerase is a homodimer with cooperative activity. J Gen Virol. 2009;90(2):281–91.
171. Rohayem J, Robel I, Jager K, Scheffler U, Rudolph W. Protein-primed and de novo initiation of RNA synthesis by norovirus 3Dpol. J Virol. 2006;80(14):7060–9.
172. Subba-Reddy CV, Goodfellow I, Kao CC. VPg-primed RNA synthesis of norovirus RNA-dependent RNA polymerases by using a novel cell-based assay. J Virol. 2011;85 (24):13027–37.
173. Subba-Reddy CV, Yunus MA, Goodfellow IG, Kao CC. Norovirus RNA synthesis is modulated by an interaction between the Viral RNA-dependent RNA polymerase and the major capsid protein, VP1. J Virol. 2012;86(18):10138–49.
174. Green KY, Mory A, Fogg MH, Weisberg A, Belliot G, Wagner M, et al. Isolation of enzymatically active replication complexes from feline calicivirus-infected cells. J Virol. 2002;76(17):8582–95.
175. Simmonds P, Karakasiliotis I, Bailey D, Chaudhry Y, Evans DJ, Goodfellow IG. Bioinformatic and functional analysis of RNA secondary structure elements among different genera of human and animal caliciviruses. Nucleic Acids Res. 2008;36(8):2530–46.
176. Yunus MA, Lin X, Bailey D, Karakasiliotis I, Chaudhry Y, Vashist S, et al. The murine norovirus core subgenomic RNA promoter consists of a stable stem-loop that can direct accurate initiation of RNA synthesis. J Virol. 2015;89(2):1218–29.
177. Lin X, Thorne L, Jin Z, Hammad LA, Li S, Deval J, et al. Subgenomic promoter recognition by the norovirus RNA-dependent RNA polymerases. Nucleic Acids Res. 2015;43(1):446–60.
178. Kaiser WJ, Chaudhry Y, Sosnovtsev SV, Goodfellow IG. Analysis of protein-protein interactions in the feline calicivirus replication complex. J Gen Virol. 2006;87(Pt 2):363–8.
179. Bok K, Prikhodko VG, Green KY, Sosnovtsev SV. Apoptosis in murine norovirus-infected RAW264.7 cells is associated with downregulation of survivin. J Virol. 2009;83(8):3647–56.
180. Furman LM, Maaty WS, Petersen LK, Ettayebi K, Hardy ME, Bothner B. Cysteine protease activation and apoptosis in murine norovirus infection. Virol J. 2009;6:139.

Viral Diversity, Evolution, and Selective Pressure

3

Angelique Ealy and Kari Debbink

3.1 Norovirus Viral Diversity

Noroviruses are a highly diverse group of viruses. They are divided into ten genogroups based on capsid sequence, which differ by >60% [1–4]. Those genogroups are further divided into over 30 genotypes, which vary by about 40% [1]. Three of these genogroups (GI, GII and GIV) are known to infect humans, with GI and GII noroviruses causing the vast majority of human infections. The GI genogroup is divided into nine genotypes, while GII noroviruses are divided into at least 25 genotypes based on VP1 sequence [5]. Noroviruses can also be classified according to RdRp sequence, with at least 14 GI.P and 27 GII.P types [6]. Noroviruses antigenic evolution is typically studied by examining genetic changes in ORF2, which codes for the structural gene, VP1 (the major capsid protein). VP1 is divided into two domains, the shell domain (S) and the protruding domain (P). The P domain is further divided into P1 and P2, in which P1 forms a stalk protruding from the surface of the virion, while P2 forms a cap on top of P1, making it the most surface exposed region of the capsid. The P2 domain contains the known antibody neutralization and attachment/receptor binding sites for norovirus. Genotype GII.4 is typically responsible for the majority of outbreaks [7], although other genotypes fluctuate in prevalence depending on year and geographic region.

When looking at the origin of GII.4 strains predominantly circulating during the past 3 decades, Siebenga et al. [8] reported that the most recent common ancestor of these highly successful variants originated in the early 1980s, although it is known that other GII.4 variants existed before that time, as Bok et al. described the emergence of an ancestral GII.4 genotype in the 1960s [9]. Two studies addressed the divergence of GI and GII genogroups [10, 11]. The GI norovirus study suggests that GI noroviruses diverged from the other norovirus genogroups around

A. Ealy · K. Debbink (✉)
Department of Natural Sciences, Bowie State University, Bowie, MD, USA
e-mail: kdebbink@bowiestate.edu

N. M. Melhem (ed.), *Norovirus*,
https://doi.org/10.1007/978-3-030-27209-8_3

2800 years ago. GI noroviruses then broke into two lineages around 750 years ago: Lineage 1 contains genotypes 1, 2, 4, 5, and 6, with genotype 1 evolving independently of the others who share a common ancestor. Lineage 2 contains 3, 7, 8, and 9, and genogroup 3 evolved separately from 7, 8, and 9 [10]. A similar analysis on GII genotypes estimated three lineages that diverged from GIV noroviruses around 380 years ago [11]. Since then, these GII noroviruses have branched off into over 25 genotypes [5, 12].

Despite assumptions that the greater norovirus prevalence of GII.4 noroviruses was likely linked to higher mutation rates, studies on norovirus evolution rates indicate that this is not necessarily the case. Bok et al. evaluated the evolutionary rate of GII.4 noroviruses from 1974 through 2007 and reported an evolutionary rate of 4.3×10^{-3} substitutions/site/year [9]. Similar results were reported by Siebenga et al. using sequences from 1987 through 2008 (rate of 5.3×10^{-3} substitutions/site/year) [8]. In an attempt to determine potential differences among genotypes, Bull et al. evaluated evolutionary rates for GII.4, GII.3, GII.b/GII.3, and GII.7 genotypes [13]. GII.4 noroviruses evolved at a rate of 3.9×10^{-3} substitutions/site/year, similar to what was reported by Bok et al., while GII.3 evolved at 1.9×10^{-3}, GII.b/GII.3 at 2.4×10^{-3}, and GII.7 at 2.3×10^{-3} substitutions/site/year. It was concluded that GII.4 norovirus capsid evolution rates are significantly higher compared to other noroviruses [13]. However, later work by Boon et al. studied evolution rates in GII.3 noroviruses and estimated that they evolve at a rate of 4.16×10^{-3} substitutions/site/year [14], two-fold higher than the previously estimated rate [13], and in line with the rate of evolution previously reported for GII.4 noroviruses [9]. As measuring evolutionary rates relies heavily on an accurate representative sampling of sequences over time, bias can occur when the sequences used for analysis are not representative of the sequences present in the population at a given time point. Bias is more likely to happen in cases where few sequences are available, so reported evolutionary rates should always be considered in terms of the number of sequences used for the calculation and in the context of the broader literature. Likewise, care should be taken in extrapolating differences in evolutionary rate to overall evolutionary patterns in cases where few sequences were used in the analysis. Evolutionary rates in GII studies are in line with data for GI genotypes. Rackoff et al. estimated that GI.1 and GI.3 noroviruses evolve at a rate of $1.25–3.52 \times 10^{-3}$ substitutions/site/year [15]. Another study of GI noroviruses attempted to evaluate evolutionary rates in all nine genotypes, and were able to do so for genotypes 2–6 [10]. The mean evolutionary rate ranged from 1.66×10^{-3} to 3.28×10^{-3} substitutions/site/year for these genotypes, supporting earlier work that substitution rates are similar among norovirus genotypes. Discordant with this, recent work on the emergent GII.17 strain suggests an increase in evolutionary rate (1.2×10^{-2}–2.1×10^{-2} substitutions/site/year) that corresponds with epidemic emergence of this genotype in 2014–2015 [16]. This rate is an order of magnitude different than other norovirus groups, and may have been affected by the low number of early strain sequences available for analysis. Another study that examined the GII.17 evolutionary rate and estimated an evolutionary rate between 4.83×10^{-3} and 5.6×10^{-3} substitutions/site/year, which is similar to the other norovirus genotypes and supports the idea that the low number of early sequences in the Chan et al. study biased the results [17]. Together, the data demonstrate similar evolutionary rates among norovirus genotypes, suggesting that observed differences in epidemiological prevalence of different norovirus genotypes are likely due to factors other than evolutionary rate. Further work should be conducted in order to establish precise evolutionary rates among norovirus genogroups and genotypes. Ideally, as cell culture methods and small animal models become established for human noroviruses, carefully controlled in vitro studies can be conducted that will allow for discernment of even small differences in evolutionary rate for different norovirus genotypes.

While the evolutionary rate does not generally appear to account for differences in prevalence of different norovirus genotypes, the fixation rate of amino acid changes in the capsid may help explain why GII.4 noroviruses are more epidemiologically important than other genotypes. Studies suggest that over time, mutations in many norovirus genotypes, including GI.3, GI.5, GI.6, GII.3, GII.6, GII.7, and GII.17, are recycled and revert to amino acids found in earlier variants [14, 15]. Specifically, Boon et al. demonstrated that while substitution rates were similar between GII.3 and GII.4 noroviruses, fixation of amino acid changes over time was not. GII.3 noroviruses accumulate diversity in the capsid region more slowly than GII.4 noroviruses. GII.3 strains tended to cycle the same genetic variants over time and displayed around 4% diversity change from 1974 to 2008. GII.4 noroviruses, on the other hand, generated greater diversity and were able to accumulate 10% amino acid variation between 1974 and 2008 [14]. This suggests that genotype-specific limits on structural flexibility in the capsid region may account for differences in mutation fixation rates over time. This also suggests that GII.4 noroviruses may encode a greater ability to tolerate mutations and retain a functional capsid compared to other genotypes.

Several groups have conducted bioinformatics studies linking epidemiologically important GII.4 strains with amino acid replacements at key sites in the capsid gene. Informative sites and/or those under positive selection have been found to predominate in the surface-exposed P2 domain of the capsid, a region hypothesized to drive antigenic evolution over time [9, 18, 19]. Analysis of GII.4 capsid sequences from 1996 through 2006 identified 48 informative sites in the capsid, 30 of which were in the P2 domain [18]. Hypervariable sites were restricted to P domain positions, with 13 occurring in P2 while only two were located in P1. When overall changes from major epidemic strains 1996, 2002, 2004, and 2006b were considered, varying residues were found to change with each strain at five P2 domain amino acid positions: 193, 255, 340, 407, and 534 [18]. Lindesmith et al. also reported that amino acid replacements were most common in the surface-exposed regions of the capsid, suggesting that these substitutions in the P2 domain may be the result of positive selection in response to escaping protective antibodies [19]. Sequences representing major epidemic strains from 1987 through 2006 were used and ten sites under positive selection were identified: 6, 9, 355, 372, 393, 394 (GII.4.1995 Grimsby numbering), 412, 505, and 534 [19]. Work by Bok et al. on GII.4 strains between 1974 and 2007 identified 73 variable sites within the GII.4 capsid, 36 of which resided in P2. When variable sites were evaluated for positive selection, four

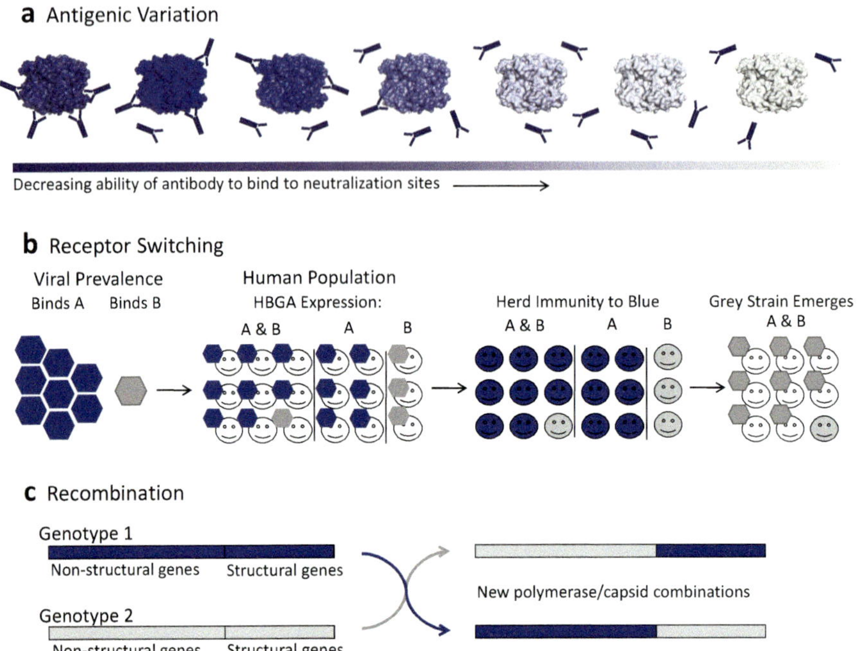

Fig. 3.1 Noroviruses evolve by antigenic variation, receptor switching, and recombination. (**a**) Antigenic variation: changing color from dark blue to grey represents accumulating changes in the capsid P2 domain over time. As changes occur, neutralizing antibodies against the blue virus are less and less potent. Eventually they no longer recognize, bind, or neutralize the virus. (**b**) Receptor Switching: noroviruses exhibit differential HBGA recognition and human populations vary in their HBGA expression patterns. As herd immunity to the dominant strain is established, novel strains that can access new susceptible populations may emerge. (**c**) Recombination: Novel combinations of polymerase and capsid genes may give rise to noroviruses with new characteristics

were found in the shell domain—6, 9, 15, and 47, one in P1 (534) and one in P2 (395) [9]. While some of these positively-selected sites have been experimentally tested for their roles in antigenic variation, it is still unclear what roles, if any, many of these play in GII.4 norovirus evolution.

3.2 General Mechanisms of Evolution and Experimental Systems

The main evolutionary mechanisms described and hypothesized for norovirus include antigenic variation, receptor switching, and recombination (Fig. 3.1). Antigenic variation results when antibodies, once able to neutralize a virus, can no longer bind in a way that neutralizes the virus to stop infection (Fig. 3.1a). Amino acid changes in the antibody neutralization sites of the capsid P2 domain result in structural alterations, preventing antibody binding, thus facilitating immune escape.

This recurring pattern has been observed in GII.4 noroviruses whereby neutralizing antibodies against one pandemic strain are unable to neutralize the next pandemic strain.

Receptor switching and antigenic drift are intertwined. Random mutations in the viral genome result in a pool of viral variants that may have slightly different abilities to bind different receptors or different antibodies (genetic drift). Through application of selective pressures (bottlenecks, receptor availability, antibody, etc.. . .), this can lead to receptor switching through which a virus gains the ability to bind a new receptor or to escape from antibody neutralization (antigenic drift). Noroviruses utilize a diverse family of carbohydrates, the histo-blood group antigens (HBGAs), as receptors or co-receptors. Noroviruses demonstrate genotype, and sometimes strain-specific, patterns of HBGAs receptor usage (Fig. 3.1b). Humans differ in their expression of these HBGAs, providing unique, strain-restricted pools of susceptible hosts. In GII.4 noroviruses, mutations near the receptor binding domains (RBD) can restrict or expand the HBGA repertoire a particular virus can bind, modulating the pool of susceptible hosts. When herd immunity is established for a dominant norovirus strain in the pool of susceptible hosts, it leaves an opening for other strains with different or more broad HBGA binding profiles to establish infection. This can result in the emergence of a new strain utilizing a different set of HBGAs for attachment.

Recombination results when specific sections of DNA from two or more norovirus genotypes combine to form a new strain. Recombination of human noroviruses typically occurs at the ORF1 (codes for the non-structural genes) and ORF2 (codes for the structural protein, VP1) junction. This often results in novel polymerase/capsid gene combinations that may confer a selective advantage to the recombinant virus (Fig. 3.1c).

Until recently and due to the lack of a cell culture model to study infectious human norovirus evolution, these processes have been evaluated using alternative systems. Antigenic variation and receptor switching are typically studied using virus-like particles (VLPs) or P domain particles (P particles) in ELISA-based assays. VLPs form spontaneously when VP1 is expressed in cell culture, creating a particle that is conformationally equivalent to native virus, allowing for meaningful studies of HBGA binding, antibody reactivity/cross-reactivity, and blockade (surrogate neutralization). Several expression systems have been used to produce VLPs including baculovirus [20], Venezuelan equine encephalitis virus [21], and plant-based systems [22]. Blockade assays are surrogate neutralization assays that measure how much antibody is needed to prevent a VLP from binding to an HBGA ligand [23]. Older studies of blockade utilized individual synthetic biotinylated carbohydrates. However, due to variation in binding efficiency between carbohydrate batches, current studies are often conducted using animal mucins, which contain multiple HBGAs and demonstrate consistent binding efficiency from batch to batch [24]. Human saliva samples have also been used to characterize norovirus blockade as they contain HBGAs and can be used as a binding ligand in assays [25]. While viral recombination of viruses with stable tissue culture models, like poliovirus, can be studied experimentally in vitro with live virus, these systems have been unavailable in the human norovirus field [26]. Therefore, norovirus recombination studies have typically been done using bioinformatics analysis of publicly available viral sequences from epidemiological and patient samples, although the development of new cell culture systems for human noroviruses will make in vitro recombination studies feasible [27].

3.3 Human Norovirus Evolutionary Mechanisms: Antigenic Variation

Noroviruses display complex patterns of inter- and intra-genogroup and genotype cross-reactivity and cross-blockade. Antibodies raised against GI noroviruses are typically cross-reactive and cross-blocking among other GI genotypes [23, 28, 29]. Human subjects experimentally infected with GI.1 norovirus displayed stronger blockade responses against heterologous GI VLPs [29]. This suggests that GI noroviruses contain conserved neutralization sites among different genotypes. GII noroviruses do not exhibit the same ability to induce cross-blockade responses indicating that there are not genogroup-level conserved neutralization sites [30, 31]. However, there is some evidence of limited cross-reactivity between different genotypes. Cannon et al. demonstrated that GII.3 antibodies cross-react with GII.4 VLPs and vice versa, but do not exhibit cross-blockade responses in mice not previously exposed to norovirus antigen [31]. Even within some genotypes, there are temporal differences in cross-reactivity and cross-neutralization. Both GII.4 and GII.17 noroviruses demonstrate this. A recent study of GII.17 noroviruses found that GII.17-2015-blocking mouse sera did not block GII.17-1987, GII.17-2002, or GII.17-2005 VLPs [32]. In addition, a large body of research demonstrates that GII.4 noroviruses undergo antigenic variation over time, and antibodies against one strain do not necessarily block VLP-HBGA interactions to other strains [24, 33–35].

Norovirus antigenic variation is driven by genetic changes in the capsid gene, likely as a result of immune pressure to escape neutralization. GII.4 antigenic variation in major outbreak strains has been characterized most completely, although other norovirus genotypes have also been evaluated to a lesser degree and are discussed at the end of this section. GII.4 noroviruses appear to follow an epochal pattern of evolution. This is hypothesized to occur when the dominant strain undergoes a period of antigenic stasis as mutations accumulate but do not impact antibody neutralization. After several years of accumulation, these mutations allow antibody escape from herd immunity and a new strain emerges and replaces the previous strain. Between 1997 and 2017, there have been several major GII.4 replacements—GII.4-1997-Grimbsy, GII.4-2002-Farmington Hills, GII.4-2006-Minerva, GII.4-2009-New Orleans, and GII.4-2012-Sydney. Interestingly, there has not been a GII.4 replacement since Sydney; however, both GII.17 and GII.2 noroviruses have increased in prevalence and even replaced GII.4 as the dominant genotype in some areas of the world [36, 37].

Studies of GII.4 norovirus evolution have demonstrated a clear relationship between mutations in specific blockade sites and emergence of novel strains.

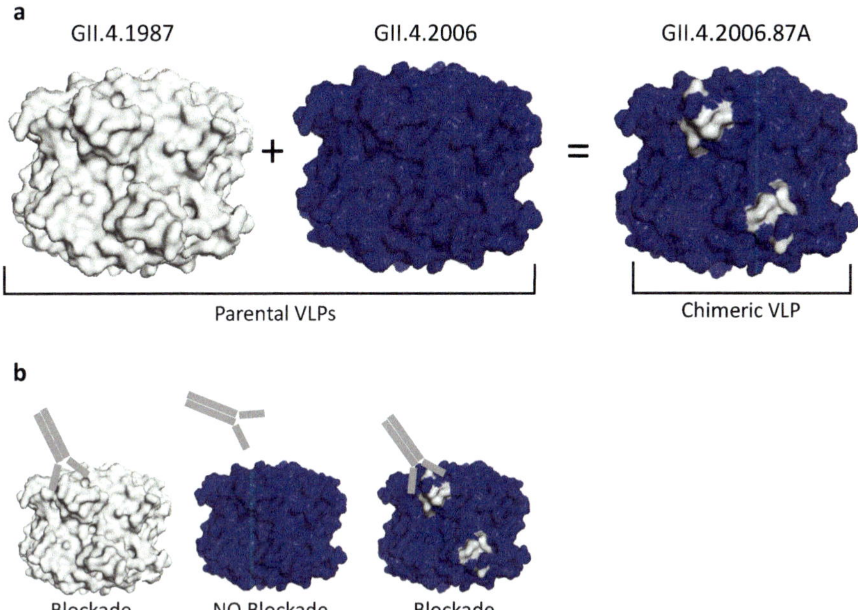

Fig. 3.2 Antigenic evolution can be evaluated using chimeric VLPs. Chimeric VLPs were created for each of the five predicted antigenic sites. An example using site A is shown. (**a**) A site A chimeric VLP was created from antigenically distinct strains GII.4-1987 and GII.4-2006, where the site A amino acids from GII.4-1987 were inserted into the GII.4-2006 backbone. (**b**) Antibody blockade characteristics of the chimeric VLP were compared to each parental strain

Computational work by Allen et al. predicted that P2 residues 296–298 and 393–395 were likely important for emergence of new GII.4 strains [38]. Work by Debbink et al. predicted five antigenic sites (Sites A-E) that are important for new strain emergence, which included the residues predicted by Allen et al. [33]. The importance of these antigenic sites was tested by creating chimeric VLPs representing antigenically distinct parental strains of GII.4-1987-Camberwell and GII.4-2006-Minerva in addition to chimeric VLPs between the two strains. To generate the latter, the amino acids in the predicted antigenic sites of the GII.4-1987 Camberwell strain were inserted into the GII.4-2006-Minerva background (Fig. 3.2a). Then, using a panel of strain-specific mouse and human monoclonal antibodies and polyclonal sera against GII.4-1987-Camberwell, GII.4-2002-Farmington Hills, and GII.4-2006-Minerva in reactivity and blockade assays, they were able to demonstrate that sites A (positions 294, 296–298, 368, and 372) (Fig. 3.2b) and D (positions 393–395) are important antigenic blockade sites for GII.4 noroviruses [33] (Fig. 3.3). Specifically, mouse mAbs anti-GII.4-1987-G1, anti-GII.4-1987-G4, anti-GII.4-1987-G5 mapped site A, while human mAb NVB 97 mapped site D. Using a similar approach with mouse mAb anti-GII.4-2002-G6, Lindesmith et al. demonstrated that site E (positions 407, 412–413) is also a GII.4 blockade site [35] (Fig. 3.3).

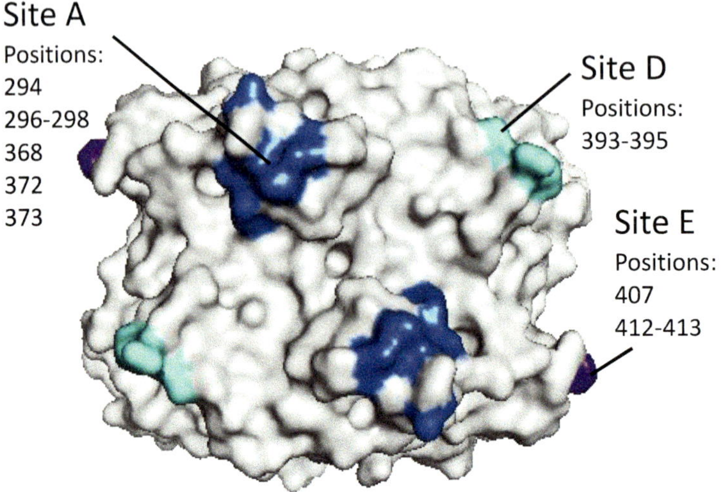

Fig. 3.3 GII.4 confirmed blockade sites. A grey GII.4 norovirus capsid P2 dimer shows the location of confirmed GII.4 blockade sites A (blue), D (cyan), and E (purple)

To evaluate the relationship between blockade sites and the emergence of new GII.4 strains, additional chimeric VLPs were generated to represent the time-ordered changes in these sites in epidemic strains GII.4-2009-New Orleans and GII.4-2012-Sydney. Blockade mouse and human monoclonal antibodies and polyclonal sera against GII.4-2009-New Orleans demonstrated that these sites, in particular, site A, was antigenically distinct with each new epidemic strain [24, 34]. Furthermore, site A was expanded to include nearby residue 373 [34]. Collectively, work with chimeric VLPs and antibody demonstrated that antigenic evolution occurred with each emerging GII.4 epidemic from 1997 through 2012.

Epitope mapping studies by Lindesmith et al. revealed an antibody that targets a non-surface exposed site on the GII.4 norovirus capsid that is conserved among GII.4 strains but whose blockade potency is variable among strains [39] (Fig. 3.4). The conserved antibody blockade site was impacted by temperature and altered the residues at positions 310, 316, 484, and 493, which are outside of the antibody binding site. This suggests that structural flexibility and conformational particle dynamics can impact the antigenic characteristics of the virus. Since this conserved site is under limited immune selection, it could provide a stable target for effective, universal GII.4 vaccines [39].

Some GII.4 outbreak strains have been shown to circulate at very low levels a year or more prior to emerging more widely [40]. It is presently unclear how GII.4 genetic diversity is generated on a global scale or why certain variants emerge and are more epidemiologically successful. Some have suggested that immunocompromised people including HIV patients, those with genetic immunodeficiencies, those on immunosuppressive therapies, and those undergoing cancer chemotherapy could serve as reservoirs for norovirus genetic diversity [41]. In healthy individuals,

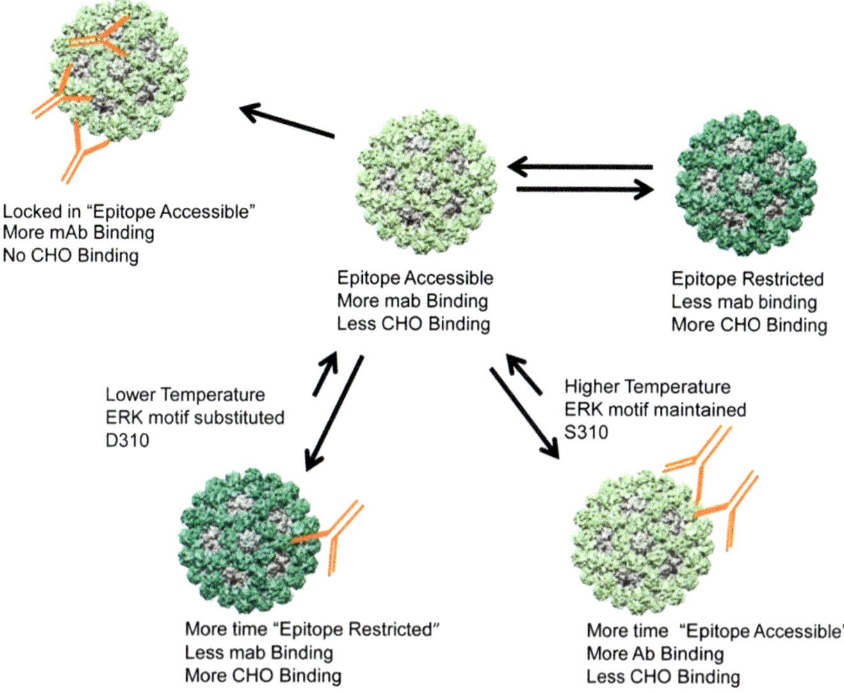

Fig. 3.4 Particle conformation impacts antibody binding. Viral particles have some structural flexibility, which likely varies by virus type and strain. Factors such as temperature and mutations within an antibody binding site can impact how exposed a specific antibody epitope is for antibody binding and neutralization. (Reproduced from [39])

norovirus is an acute disease that resolves within days, although shedding of virus can persist for weeks. Consensus sequencing of virus from healthy patients has determined that few mutations arise within individuals [42]. People with compromised immune systems, however, can develop chronic norovirus infections that can last a year or more [43]. Sequencing of virus taken at different time points during chronic infection yields an accumulation of mutations over time, many of which occur in known antigenic sites [43–45]. A study by Debbink et al. evaluated the antigenic evolution of VLPs representing temporally sequenced samples from an immunocompromised kidney transplant patient and found that over the course of 9 months, the virus had escaped the antibody response generated against the initial virus [46]. This demonstrates that immunocompromised patients are able to give rise to viral variants that are capable of antibody escape and suggests that they could contribute to global disease emergence. However, two major recent studies have challenged this idea. A recent modeling study by Eden et al. reported that while immunocompromised hosts do generate high levels of viral variants, when considering the rarity of immunocompromised people with norovirus and their tendency to be isolated from the general population, immunocompromised hosts are less likely to be a source of norovirus genetic diversity on a population level than those who are acutely infected [47]. Moreover, work analyzing virus sequences from acutely infected individuals using next-generation sequencing identified viral variants that had not been detected using consensus sequencing, demonstrating that there is more underlying diversity present in acutely infected patients than was previously appreciated [48]. In this study, shed virus was collected from one particular child infected with GII.4 norovirus within 2–3 weeks of infection and well after resolution of symptoms. Next generation sequencing analysis found minority variants containing mutations in known antigenic sites. This underlying diversity suggests that minority variants with potential antigenic differences can be transmitted among acutely infected individuals and that acutely infected patients are the likely sources of norovirus diversity after all [48].

Norovirus evolution has been evaluated in some non-GII.4 noroviruses as well. During the 2014–2015 season, GII.17 cluster IIIb norovirus emerged as the dominant genotype in Hong Kong, China, and Japan [36]. Lindesmith et al. conducted a study evaluating the antigenic evolution of cluster I, II, and IIIb GII.17 strains to determine if changes in antigenic sites correlated with predominance of the GII.17-2015 strain [32]. By creating chimeric VLPs between GII.17-1978 and GII.17-2015 in residues 393–396 of VP1, they demonstrated that this is an important antigenic site for GII.17 viruses. Interestingly, these residues also comprise an important antigenic site in GII.4 noroviruses (site D), suggesting that although each genotype is antigenically distinct with different amino acids at these positions, the specific positions that are antigenically important may be similar. This work suggests that antigenically important changes in the capsid may contribute to the epidemiological success of GII.17-2015.

GII.2 typically causes around 2% of norovirus outbreaks [49]; however, this genotype dominated in China, Japan, and Germany during the 2016–2017 winter season [50–52]. Tohma et al. conducted a study comparing VP1 of GII.2 noroviruses over the past four decades and observed that amino acid substitution rates were constant over time with very few changes [37]. It was concluded that factors other than changes at the capsid protein could have impacted the wider circulation of GII.2 noroviruses. This is in line with previous work by Swanstrom et al., who evaluated the blockade response of several GII.2 monoclonal antibodies against a panel of VLPs representing strains GII.2-1976, GII.2-2002, GII.2-2008, and GII.2-2010. Results demonstrated no major differences in blockade responses among these strains, suggesting that GII.2 viruses undergo limited antigenic evolution [25].

GII.3 noroviruses are more prevalent in children than in adults, suggesting that most people are exposed to GII.3 viruses early in life and maintain protection via herd immunity into adulthood. Mahar et al. analyzed sequence data from GII.3 strains spanning 1975 through 2010 [53]. Variable, lineage-defining residues were detected in the capsid P2 domain that are analogous to sites A and E in GII.4 noroviruses [33, 35] and to HBGA binding sites [54]. In further work, Mahar et al. conducted an antibody mapping study of GII.3 noroviruses in order to evaluate cross-reactivity among different GII.3 strains [55]. VLPs against seven GII.3 strains circulating between 1975 and 2008 were generated, and six antibody epitopes were identified using mass spectrometry of peptide-linked protease-digested VLPs. Two epitopes were variable among GII.3 strains and located in the capsid P2 domain. Despite this variability, they did not detect major differences in human or rabbit serum reactivity among the strains, possibly due to other more immunodominant antigenic sites that were conserved among strains [55]. Coupled with the observation that GII.3 are primarily acquired during childhood, this supports the hypothesis that antigenic evolution is not the primary driver of GII.3 diversity.

Studies on GI norovirus, a genogroup containing nine genotypes, suggest that they are antigenically similar and do not undergo significant antigenic variation over time. A study by Noel et al. reported that antigenic profiles of several GI genotypes were similar [30]. This was supported by work reported by Lindesmith et al. whereby GI.1-1968 and GI.1-2001, temporally separated by over 30 years, encoded five changes in the P2 domain, only one of which is surface exposed. GI.1-1968 serum reactivity against heterotypic GI VLPs representing GI.2, GI.3, and GI.4 strains was not significantly different, demonstrating that GI intra- and inter-genotype antigenic profiles are similar and that antigenic variation is not a major driver of GI norovirus outbreaks.

Further research evaluating antigenic differences within and between norovirus genotypes is warranted. This is particularly true for non-GII.4 noroviruses, as the lower disease incidence and fewer available sequences to use to evaluate antigenic variation over time have limited the information on evolution in these groups. Work by Parra et al. evaluated the evolutionary mechanisms of GI and GII noroviruses with more than ten sequences available in Genbank [48]. They were able to include 16 norovirus genotypes in their analysis. Their results showed two patterns of diversification: evolving and static. GII.4 noroviruses fit into the evolving group, while the other 15 genotypes studied were classified as static. Importantly, this study also introduced the concept of norovirus immunotypes, which are norovirus genotypes that cluster into groups based on antigenic profile [48]. This is vitally important to vaccine design since protection against multiple norovirus genotypes may be induced by careful choice of genotypes that represent the major immunotypes in a multivalent vaccine. Clearly further work with non-GII.4 strains may uncover vital patterns of antigenic diversity among genotypes that will help in the development of novel norovirus vaccines and epidemiological surveillance tools.

3.4 Human Norovirus Evolutionary Mechanisms: Receptor Switching

Norovirus evolution is likely impacted by strain-specific differences in the ability to utilize carbohydrate receptors expressed differentially in human hosts. This phenomenon leads to variation in the availability of susceptible human populations. Histo-blood group antigens (HBGAs) are a diverse family of carbohydrates, which are expressed on mucosal surfaces such as the saliva and the gut mucosal linings. These carbohydrates serve as binding ligands and putative receptors for norovirus attachment and entry, and noroviruses bind different HBGAs in a strain-dependent manner. Likewise, humans display individual patterns of HBGA expression, which is governed both by genetic variation at the locus coding for these genes and the presence of a functional *FUT2* gene. The *FUT2* gene encodes fucosyltransferase, which adds side chains to a precursor carbohydrate molecule. This allows for the construction of different ABH HBGAs, which are then expressed on mucosal surfaces. Individuals without a functional *FUT2* gene do not express fucosyltransferase and are referred to as "non-secretors" as they do not express most HBGAs on their mucosal surfaces. Non-secretors make up about 20% of the population and are resistant to certain norovirus strains. For instance, non-secretors have been shown to be resistant to infection by GI.1 noroviruses [56]. Other norovirus strains, however, can infect non-secretors because they are able to bind to HBGAs, such as Lewis antigens, that do not require fucosyltransferase for their assembly [57–60].

Studies mapping the location of norovirus HBGA binding sites have identified distinct genogroup differences [54, 61]. This genogroup-based difference suggests that HBGA binding preferences are evolutionarily important. Interestingly, recent work identified two GII genotypes that do not share the conserved GII HBGA binding site. Both GII.21 and GII.13 appear to have a novel HBGA binding interface, while closely related GII.17 retains the conserved GII interface [62]. Despite a general GII conserved HBGA binding site, GII.4 noroviruses, predominantly circulating for over the past three decades, sometimes demonstrate changes in HBGA specificity between epidemic strains. This is modulated by mutations in and around the HBGA binding pocket, suggesting that HBGA binding may play a role in evolution within the GII.4 genogroup. Crystal structures of Hu/NoV/GII.4/VA387/1998/US resolved the P domain in complex with HBGAs, thus first revealing the HBGA binding site for GII.4 noroviruses [54]. This site contains both conserved and variable residues, which modulate HBGA binding affinity and/or avidity for each norovirus strain. The GII.4 ancestral strain, GII.4–1974, and GII.4-1987-Camberwell bind H type carbohydrates (H3 and Lewis-y) [9]. The pandemic strain GII.4-1997-Grimsby binds to A, B, and H HBGAs, while GII.4-2004-Farmington Hills binds B and H HBGAs [19]. GII.4-2006-Minerva binds A, B, and H HBGAs [31], as do both GII.4-2009-New Orleans and GII.4-2012-Sydney [63]. This variability in HBGA binding in different strains is hypothesized to lead to infection in different susceptible populations [64].

Several groups have conducted studies that identify amino acid positions that can modulate HBGA binding for both ABH and Lewis HBGAs. In the absence of a tissue culture model for norovirus infection, alternative in vitro systems were developed to probe carbohydrate receptor usage for different noroviruses. In these functional binding assays, VLPs or P particles are surrogates for viral capsid, and synthetic carbohydrates, saliva, or mucins are HBGA sources. Using synthetic carbohydrates and GII.4 VLPs, representing capsids from major epidemic strains, the emergence of new GII.4 strains was linked to alterations in HBGA binding [19]. Using the VA387 sequence (GII.4-1997-Grimsby-like strain) as a reference for creating mutant P particles using alanine scanning, Tan et al. showed the importance of amino acids at positions 346 and 441 to ABH HBGA binding by demonstrating that changes to these residues could ablate receptor-ligand binding. Furthermore, they demonstrated that several other amino acids in and around the carbohydrate binding pocket are involved in HBGA specificity [65].

Evidence that receptor switching of HBGA ligands impacts norovirus evolution was assessed by Donaldson et al. by studying the emergence of pandemic GII.4-1997-Grimbsy [64]. While this strain was antigenically similar to GII.4-1987-Camberwell, the two strains displayed very different HBGA binding characteristics, with GII.4-1997-Grimsby able to bind A and B HBGAs in addition to the H carbohydrates that GII.4-1987 Camberwell could bind. This difference in HBGA binding potentially allows infection of new host populations previously unavailable to GII.4-1987-Camberwell. GII.4-1997-Grimsby contains a mutation within the HBGA interaction region at position 393 relative to GII.4-1987-Camberwell (D393G). Functional studies demonstrated that introducing the D393G mutation into the GII.4-1987-Camberwell capsid conferred the ability of GII.4-1987-Camberwell to bind HBGA B. This altered HBGA binding profile and subsequent increase in susceptible hosts is hypothesized to be the reason behind the ability of GII.4-1997-Grimsby's to spread worldwide [64]. This was supported by crystal structure studies by Shanker et al. showing that GII.4 strains had temporal differences in HBGA affinity between the Hu/NoV/GII.4/VA387/1998/US capsid and the GII.4-Farmington Hills-2004 capsid. Specifically, they determined that residues 393–395 directly impact Lewis HBGA binding [66] and may indirectly impact ABH HBGA binding affinity through conformational and electrostatic effects. Together, this work supports a model whereby individual or clusters of mutations can alter HBGA binding profiles over time.

Furthermore, as Donaldson et al. proposed, when HBGA binding differences are considered in conjunction with antigenic variation and population-wide herd immunity, the true impact of receptor switching may be appreciated [19, 64]. As susceptible populations are infected with an epidemic norovirus strain utilizing a specific HBGA, the population of susceptible individuals is naturally reduced as antibody-induced protection prevents re-infection of individuals expressing that specific HBGA, resulting ultimately in herd immunity to the circulating strain. This allows new strains with either different HBGA binding or antibody binding characteristics to emerge. New strains that bind novel HBGAs are likely to be successful because they can infect susceptible populations that were not available to the old strain and thus are not protected by herd immunity.

Recent studies have also identified sialylated carbohydrates, like gangliosides, as binding factors and potential receptors for norovirus infection [67]; however, the role of these carbohydrates in evolution has not yet been established. Further work to determine the extent that noroviruses use these molecules to invade host cells is warranted in order to better understand how receptor switching impacts norovirus evolutionary trends.

3.5 Human Norovirus Evolutionary Mechanisms: Recombination

Recombination is an important driver of evolution for many RNA viruses and has been hypothesized to drive norovirus evolution as well [66, 68]. Recombination shuffles existing genetic diversity, allowing for novel genetic combinations that may confer new or altered abilities to the virus, such as increased transmission, virulence, or evolutionary rate. A role for recombination in norovirus evolution, especially with different polymerase/capsid combinations, is supported by a study that found that the norovirus major capsid protein modulates RdRp activity and the degree to which this occurs varies by capsid/RdRp combination [69].

Several studies using available database sequences and various computational tools have frequently identified norovirus recombinants in clinical and surveillance samples from both local and epidemic outbreaks worldwide [70, 71]. Recombinant viruses have been detected worldwide in countries and regions where there are concerted surveillance efforts including Japan, USA, South Korea, China, Brazil, Europe, Africa, and Australia. Historically, noroviruses were classified based only on the capsid gene; however, as more groups launched into investigating recombinant noroviruses, it was clear that recombination is a common phenomenon. As such, more recently isolates are often typed based on both the polymerase and capsid genes: GII.P17-GII.12, which indicates the polymerase from GII.17 and the capsid from GII.12 [72]. Norovirus recombination occurs most often at the ORF1-ORF2 junction, but also occurs less frequently at the ORF2-ORF3 junction and within an ORF [73]. To date, at least 14 GI and 27 GII different polymerase types and 9 GI, 22 GII, and 2 GIV capsid genotypes have been described for human noroviruses [72]. Clearly, this diversity has enormous potential to create genetic combinations with novel phenotypes. Unfortunately, due mainly to lack of appropriate model systems, there is an absence of direct experimental evidence for norovirus recombination to generate phenotypic diversity. However, the available phylogenetic and epidemiological evidence suggests that recombination is likely an important mechanism of norovirus evolution on a global scale. With recent developments in in vitro human norovirus culture systems, we will be able to study recombination more precisely than is currently done with epidemiological samples. For example, measuring recombination rates for human norovirus will be possible using microfluidics to analyze virus at the single cell level [74]. Similar studies have already been done with murine norovirus [75].

The first evidence for recombination as a mechanism to generate norovirus diversity was recognized in 1997 when Hardy et al. first identified a recombinant GII.2 Snow Mountain virus [76]. Jiang et al. also identified naturally occurring homologous recombination between GII.3 and GII.4 noroviruses [77]. Subsequent work by Katayama et al. identified recombination events at the ORF1/ORF2 junction for both a novel GI strain and a GII strain [78], identifying a specific breakpoint (BP) and demonstrating that intra-genogroup recombination occurs for both GI and GII noroviruses. This is important in that it shows that recombination is probably a common method for genetic variation within different norovirus genogroups. This was confirmed by Bull et al. who used sequence databases and lab sample sequences to identify a total of 23 GII recombinants that could be clustered into nine groups, as well as one G1 recombinant that had previously been identified [79]. The ORF1/ORF2 junction was the recombination point for all of them. Rohayem et al. evaluated recombination within ORF2 in several genotypes from the GI and GII genogroups and found several examples of intra-genotype recombination within the capsid gene [80]. Break points within the capsid tended to occur at the P1-P2 interface or, more rarely, within P2 itself [80]. In addition, recombinants have been identified in animal noroviruses [81]. Since these earlier studies, numerous novel norovirus recombinants have been identified.

Recombination appears to be a driving force specifically in the evolution of GII.3 noroviruses. A study by Mahar et al. found that GII.3 capsids with a GIIb polymerase appear to have higher substitution rates, suggesting that recombination is a way to increase evolutionary rate and subsequent adaptability [53]. To what extent recombination impacts norovirus evolution as a whole is still unclear, although a 30-year study of norovirus sequences in Brazil indicates that their role may be significant, as 47.6% of the sequences were recombinants [82]. Research shows that some recombinants become widely distributed, epidemiologically important strains [83]. During the 1999–2000 season in Japan, recombinant Arg320-like virus was detected in an Osaka City study of pediatric gastroenteritis [84]. This virus was also the causative agent of a norovirus outbreak during the same year at an infant home in Sapporo, Japan, that sickened 23 children under the age of two [85]. In 2001–2002, a recombinant GII.b/GII.3 norovirus caused high rates of disease in Europe, Australia, and Asia [79, 86, 87]. A later study evaluated virus that circulated from 2001–2004 in Hungary and found that the GIIb polymerase formed recombinants with several capsids including Mexico, Hawaii, Snow Mountain, and Lordsdale. These recombinant viruses were the second most prevalent noroviruses circulating during that time [88]. More recently, during the 2016–2017 season, GII.2 noroviruses predominated in several countries including Germany, Japan, and China, primarily in pediatric populations [37, 50, 51]. While GII.2 noroviruses have been paired with many different polymerases, a recombinant GII.P16-GII.2 with minor changes in the RdRp was responsible for these epidemics [37]. In addition, recent work has suggested that the continued success of the GII.4 Sydney capsid since 2012 is the result of recombination with GII.16 polymerase. This recombination event led to the emergence of the GII.P16-GII.4 Sydney variant emerging in 2015 to overtake the previous GII.4 Sydney variant [83].

3.6 Norovirus Evolution Concluding Remarks

Norovirus evolution is complex, owing in part to the diversity of genotypes and genogroups. GII.4 noroviruses appear to be unique in their ability to undergo rapid antigenic evolution and generate new variants every few years. While several factors including antigenic variation, receptor switching, and recombination appear to impact norovirus evolution, the respective impact of these mechanisms likely varies based on the unique characteristics of each genotype or strain. There is no doubt that further work studying each of these mechanisms as well as the relationship among these mechanisms is warranted to better appreciate and characterize norovirus evolution and identify unifying evolutionary themes among different types.

References

1. Zheng D-P, Ando T, Fankhauser RL, Beard RS, Glass RI, Monroe SS. Norovirus classification and proposed strain nomenclature. Virology. 2006;346(2):312–23.
2. Tse H, Lau SKP, Chan W-M, Choi GKY, Woo PCY, Yuen K-Y. Complete genome sequences of novel canine noroviruses in Hong Kong. J Virol. 2012;86(17):9531–2.
3. Wu Z, Yang L, Ren X, He G, Zhang J, Yang J, et al. Deciphering the bat virome catalog to better understand the ecological diversity of bat viruses and the bat origin of emerging infectious diseases. ISME J. 2016;10(3):609–20.
4. Chhabra P, de Graaf M, Parra GI, Chan MC, Green K, Martella V, Wang Q, White PA, Katayama K, Vennema H, Koopmans MPG, Vinjé J. Updated classification of norovirus genogroups and genotypes. J Gen Virol. 2019;100(10):1393–406.
5. Chhabra P, Aswath K, Collins N, Ahmed T, Olórtegui MP, Kosek M, et al. Near-complete genome sequences of several new norovirus Genogroup II genotypes. Genome Announc. 2018;6(6):e00007-18.
6. Cannon JL, Barclay L, Collins NR, Wikswo ME, Castro CJ, Magaña LC, et al. Genetic and epidemiologic trends of norovirus outbreaks in the United States from 2013 to 2016 demonstrated emergence of novel GII.4 recombinant viruses. J Clin Microbiol. 2017;55 (7):2208–21.
7. Scallan E, Hoekstra RM, Angulo FJ, Tauxe RV, Widdowson M-A, Roy SL, et al. Foodborne illness acquired in the United States—major pathogens. Emerg Infect Dis. 2011;17(1):7–15.
8. Siebenga JJ, Lemey P, Kosakovsky Pond SL, Rambaut A, Vennema H, Koopmans M. Phylodynamic reconstruction reveals norovirus GII.4 epidemic expansions and their molecular determinants. PLoS Pathog. 2010;6(5):e1000884.
9. Bok K, Abente EJ, Realpe-Quintero M, Mitra T, Sosnovtsev SV, Kapikian AZ, et al. Evolutionary dynamics of GII.4 noroviruses over a 34-year period. J Virol. 2009;83(22):11890–901.
10. Kobayashi M, Yoshizumi S, Kogawa S, Takahashi T, Ueki Y, Shinohara M, et al. Molecular evolution of the capsid gene in norovirus Genogroup I. Sci Rep. 2015;5:13806.
11. Kobayashi M, Matsushima Y, Motoya T, Sakon N, Shigemoto N, Okamoto-Nakagawa R, et al. Molecular evolution of the capsid gene in human norovirus genogroup II. Sci Rep. 2016;6:29400.
12. Kroneman A, Vega E, Vennema H, Vinjé J, White PA, Hansman G, et al. Proposal for a unified norovirus nomenclature and genotyping. Arch Virol. 2013;158(10):2059–68.
13. Bull RA, Eden J-S, Rawlinson WD, White PA. Rapid evolution of pandemic noroviruses of the GII.4 lineage. PLoS Pathog. 2010;6(3):e1000831.
14. Boon D, Mahar JE, Abente EJ, Kirkwood CD, Purcell RH, Kapikian AZ, et al. Comparative evolution of GII.3 and GII.4 norovirus over a 31-year period. J Virol. 2011;85(17):8656–66.
15. Rackoff LA, Bok K, Green KY, Kapikian AZ. Epidemiology and evolution of rotaviruses and noroviruses from an archival WHO global study in children (1976-79) with implications for vaccine design. PLoS One. 2013;8(3):e59394.
16. Chan MCW, Lee N, Hung T-N, Kwok K, Cheung K, Tin EKY, et al. Rapid emergence and predominance of a broadly recognizing and fast-evolving norovirus GII.17 variant in late 2014. Nat Commun. 2015;6:10061.
17. Lu J, Fang L, Zheng H, Lao J, Yang F, Sun L, et al. The evolution and transmission of epidemic GII.17 noroviruses. J Infect Dis. 2016;214(4):556–64.

18. Siebenga JJ, Vennema H, Renckens B, de Bruin E, van der Veer B, Siezen RJ, et al. Epochal evolution of GGII.4 norovirus capsid proteins from 1995 to 2006. J Virol. 2007;81 (18):9932–41.
19. Lindesmith LC, Donaldson EF, Lobue AD, Cannon JL, Zheng D-P, Vinje J, et al. Mechanisms of GII.4 norovirus persistence in human populations. PLoS Med. 2008;5(2):e31.
20. Jiang X, Matson DO, Ruiz-Palacios GM, Hu J, Treanor J, Pickering LK. Expression, self-assembly, and antigenicity of a snow mountain agent-like calicivirus capsid protein. J Clin Microbiol. 1995;33(6):1452–5.
21. Harrington PR, Yount B, Johnston RE, Davis N, Moe C, Baric RS. Systemic, mucosal, and heterotypic immune induction in mice inoculated with Venezuelan equine encephalitis replicons expressing Norwalk virus-like particles. J Virol. 2002;76(2):730–42.
22. Huang Z, Elkin G, Maloney BJ, Beuhner N, Arntzen CJ, Thanavala Y, et al. Virus-like particle expression and assembly in plants: hepatitis B and Norwalk viruses. Vaccine. 2005;23 (15):1851–8.
23. LoBue AD, Lindesmith L, Yount B, Harrington PR, Thompson JM, Johnston RE, et al. Multivalent norovirus vaccines induce strong mucosal and systemic blocking antibodies against multiple strains. Vaccine. 2006;24(24):5220–34.
24. Lindesmith LC, Costantini V, Swanstrom J, Debbink K, Donaldson EF, Vinjé J, et al. Emergence of a norovirus GII.4 strain correlates with changes in evolving blockade epitopes. J Virol. 2013;87(5):2803–13.
25. Swanstrom J, Lindesmith LC, Donaldson EF, Yount B, Baric RS. Characterization of blockade antibody responses in GII.2.1976 snow mountain virus-infected subjects. J Virol. 2014;88 (2):829–37.
26. Muslin C, Joffret M-L, Pelletier I, Blondel B, Delpeyroux F. Evolution and emergence of enteroviruses through intra- and inter-species recombination: plasticity and phenotypic impact of modular genetic exchanges in the 5' Untranslated region. PLoS Pathog. 2015;11(11): e1005266.
27. Ludwig-Begall LF, Mauroy A, Thiry E. Norovirus recombinants: recurrent in the field, recalcitrant in the lab – a scoping review of recombination and recombinant types of noroviruses. J Gen Virol. 2018;99(8):970–88.
28. LoBue AD, Thompson JM, Lindesmith L, Johnston RE, Baric RS. Alphavirus-adjuvanted norovirus-like particle vaccines: heterologous, humoral, and mucosal immune responses protect against murine norovirus challenge. J Virol. 2009;83(7):3212–27.
29. Lindesmith LC, Donaldson E, Leon J, Moe CL, Frelinger JA, Johnston RE, et al. Heterotypic humoral and cellular immune responses following Norwalk virus infection. J Virol. 2010;84 (4):1800–15.
30. Noel JS, Ando T, Leite JP, Green KY, Dingle KE, Estes MK, et al. Correlation of patient immune responses with genetically characterized small round-structured viruses involved in outbreaks of nonbacterial acute gastroenteritis in the United States, 1990 to 1995. J Med Virol. 1997;53(4):372–83.
31. Cannon JL, Lindesmith LC, Donaldson EF, Saxe L, Baric RS, Vinjé J. Herd immunity to GII.4 noroviruses is supported by outbreak patient sera. J Virol. 2009;83(11):5363–74.
32. Lindesmith LC, Kocher JF, Donaldson EF, Debbink K, Mallory ML, Swann EW, et al. Emergence of novel human norovirus GII.17 strains correlates with changes in blockade antibody epitopes. J Infect Dis. 2017;216(10):1227–34.
33. Debbink K, Donaldson EF, Lindesmith LC, Baric RS. Genetic mapping of a highly variable norovirus GII.4 blockade epitope: potential role in escape from human herd immunity. J Virol. 2012;86(2):1214–26.
34. Debbink K, Lindesmith LC, Donaldson EF, Costantini V, Beltramello M, Corti D, et al. Emergence of new pandemic GII.4 Sydney norovirus strain correlates with escape from herd immunity. J Infect Dis. 2013;208(11):1877–87.
35. Lindesmith LC, Debbink K, Swanstrom J, Vinjé J, Costantini V, Baric RS, et al. Monoclonal antibody-based antigenic mapping of norovirus GII.4-2002. J Virol. 2012;86(2):873–83.

36. Chan MCW, Hu Y, Chen H, Podkolzin AT, Zaytseva EV, Komano J, et al. Global spread of Norovirus GII.17 Kawasaki 308, 2014-2016. Emerg Infect Dis. 2017;23(8):1359–4.
37. Tohma K, Lepore CJ, Ford-Siltz LA, Parra GI. Phylogenetic analyses suggest that factors other than the capsid protein play a role in the epidemic potential of GII.2 norovirus. mSphere. 2017;2 (3):e00187–17. https://doi.org/10.1128/mSphereDirect.00187-17.
38. Allen DJ, Gray JJ, Gallimore CI, Xerry J, Iturriza-Gómara M. Analysis of amino acid variation in the P2 domain of the GII-4 norovirus VP1 protein reveals putative variant-specific epitopes. PLoS One. 2008;3(1):e1485.
39. Lindesmith LC, Donaldson EF, Beltramello M, Pintus S, Corti D, Swanstrom J, et al. Particle conformation regulates antibody access to a conserved GII.4 norovirus blockade epitope. J Virol. 2014;88(16):8826–42.
40. Eden J-S, Hewitt J, Lim KL, Boni MF, Merif J, Greening G, et al. The emergence and evolution of the novel epidemic norovirus GII.4 variant Sydney 2012. Virology. 2014;450–451:106–13.
41. Bok K, Green KY. Norovirus gastroenteritis in immunocompromised patients. N Engl J Med. 2013;368(10):971.
42. Bull RA, Eden J-S, Luciani F, McElroy K, Rawlinson WD, White PA. Contribution of intra- and interhost dynamics to norovirus evolution. J Virol. 2012;86(6):3219–29.
43. Siebenga JJ, Beersma MFC, Vennema H, van Biezen P, Hartwig NJ, Koopmans M. High prevalence of prolonged norovirus shedding and illness among hospitalized patients: a model for in vivo molecular evolution. J Infect Dis. 2008;198(7):994–1001.
44. Schorn R, Höhne M, Meerbach A, Bossart W, Wüthrich RP, Schreier E, et al. Chronic norovirus infection after kidney transplantation: molecular evidence for immune-driven viral evolution. Clin Infect Dis Off Publ Infect Dis Soc Am. 2010;51(3):307–14.
45. van Beek J, de Graaf M, Smits SL, Schapendonk CME, Verjans GMGM, Vennema H, et al. Whole genome next generation sequencing to study within-host evolution of chronic norovirus infection among immunocompromised patients. J Infect Dis. 2017;216(12):1513–24. https://doi.org/10.1093/infdis/jix520.
46. Debbink K, Lindesmith LC, Ferris MT, Swanstrom J, Beltramello M, Corti D, et al. Within-host evolution results in antigenically distinct GII.4 noroviruses. J Virol. 2014;88(13):7244–55.
47. Eden J-S, Chisholm RH, Bull RA, White PA, Holmes EC, Tanaka MM. Persistent infections in immunocompromised hosts are rarely sources of new pathogen variants. Virus Evol. 2017;3(2): vex018.
48. Parra GI, Squires RB, Karangwa CK, Johnson JA, Lepore CJ, Sosnovtsev SV, et al. Static and evolving norovirus genotypes: implications for epidemiology and immunity. PLoS Pathog. 2017;13(1):e1006136.
49. Iritani N, Vennema H, Siebenga JJ, Siezen RJ, Renckens B, Seto Y, et al. Genetic analysis of the capsid gene of genotype GII.2 noroviruses. J Virol. 2008;82(15):7336–45.
50. Ao Y, Wang J, Ling H, He Y, Dong X, Wang X, et al. Norovirus GII.P16/GII.2-associated gastroenteritis, China, 2016. Emerg Infect Dis. 2017;23(7):1172–5.
51. Lu J, Fang L, Sun L, Zeng H, Li Y, Zheng H, et al. Association of GII.P16-GII.2 recombinant Norovirus strain with increased norovirus outbreaks, Guangdong, China, 2016. Emerg Infect Dis. 2017;23(7):1188–90.
52. Thongprachum A, Okitsu S, Khamrin P, Maneekarn N, Hayakawa S, Ushijima H. Emergence of norovirus GII.2 and its novel recombination during the gastroenteritis outbreak in Japanese children in mid-2016. Infect Genet Evol J Mol Epidemiol Evol Genet Infect Dis. 2017;51:86–8.
53. Mahar JE, Bok K, Green KY, Kirkwood CD. The importance of intergenic recombination in norovirus GII.3 evolution. J Virol. 2013;87(7):3687–98.
54. Cao S, Lou Z, Tan M, Chen Y, Liu Y, Zhang Z, et al. Structural basis for the recognition of blood group trisaccharides by norovirus. J Virol. 2007;81(11):5949–57.
55. Mahar JE, Donker NC, Bok K, Talbo GH, Green KY, Kirkwood CD. Identification and characterization of antibody-binding epitopes on the norovirus GII.3 capsid. J Virol. 2014;88 (4):1942–52.

56. Lindesmith L, Moe C, Marionneau S, Ruvoen N, Jiang X, Lindblad L, et al. Human susceptibility and resistance to Norwalk virus infection. Nat Med. 2003;9(5):548–53.
57. Huang P, Farkas T, Marionneau S, Zhong W, Ruvoën-Clouet N, Morrow AL, et al. Noroviruses bind to human ABO, Lewis, and secretor histo-blood group antigens: identification of 4 distinct strain-specific patterns. J Infect Dis. 2003;188(1):19–31.
58. Huang P, Farkas T, Zhong W, Tan M, Thornton S, Morrow AL, et al. Norovirus and histo-blood group antigens: demonstration of a wide spectrum of strain specificities and classification of two major binding groups among multiple binding patterns. J Virol. 2005;79(11):6714–22.
59. Lindesmith L, Moe C, Lependu J, Frelinger JA, Treanor J, Baric RS. Cellular and humoral immunity following Snow Mountain virus challenge. J Virol. 2005;79(5):2900–9.
60. de Rougemont A, Ruvoen-Clouet N, Simon B, Estienney M, Elie-Caille C, Aho S, et al. Qualitative and quantitative analysis of the binding of GII.4 norovirus variants onto human blood group antigens. J Virol. 2011;85(9):4057–70.
61. Tan M, Xia M, Chen Y, Bu W, Hegde RS, Meller J, et al. Conservation of carbohydrate binding interfaces: evidence of human HBGA selection in norovirus evolution. PLoS One. 2009;4(4): e5058.
62. Liu W, Chen Y, Jiang X, Xia M, Yang Y, Tan M, et al. A unique human norovirus lineage with a distinct HBGA binding Interface. PLoS Pathog. 2015;11(7):e1005025.
63. Zhang J, Shen Z, Zhu Z, Zhang W, Chen H, Qian F, et al. Genotype distribution of norovirus around the emergence of Sydney_2012 and the antigenic drift of contemporary GII.4 epidemic strains. J Clin Virol. 2015;72:95–101.
64. Donaldson EF, Lindesmith LC, Lobue AD, Baric RS. Norovirus pathogenesis: mechanisms of persistence and immune evasion in human populations. Immunol Rev. 2008;225:190–211.
65. Tan M, Xia M, Cao S, Huang P, Farkas T, Meller J, et al. Elucidation of strain-specific interaction of a GII-4 norovirus with HBGA receptors by site-directed mutagenesis study. Virology. 2008;379(2):324–34.
66. Shanker S, Choi J-M, Sankaran B, Atmar RL, Estes MK, Prasad BVV. Structural analysis of histo-blood group antigen binding specificity in a norovirus GII.4 epidemic variant: implications for epochal evolution. J Virol. 2011;85(17):8635–45.
67. Wegener H, Mallagaray Á, Schöne T, Peters T, Lockhauserbäumer J, Yan H, et al. Human norovirus GII.4(MI001) P dimer binds fucosylated and sialylated carbohydrates. Glycobiology. 2017;27(11):1027–37.
68. Bull RA, Tanaka MM, White PA. Norovirus recombination. J Gen Virol. 2007;88 (Pt 12):3347–59.
69. Subba-Reddy CV, Yunus MA, Goodfellow IG, Kao CC. Norovirus RNA synthesis is modulated by an interaction between the viral RNA-dependent RNA polymerase and the major capsid protein, VP1. J Virol. 2012;86(18):10138–49.
70. Lochridge VP, Hardy ME. Snow mountain virus genome sequence and virus-like particle assembly. Virus Genes. 2003;26(1):71–82.
71. Etherington GJ, Dicks J, Roberts IN. High throughput sequence analysis reveals hitherto unreported recombination in the genus Norovirus. Virology. 2006;345(1):88–95.
72. Vinjé J. Advances in laboratory methods for detection and typing of norovirus. J Clin Microbiol. 2015;53(2):373–81.
73. Eden J-S, Tanaka MM, Boni MF, Rawlinson WD, White PA. Recombination within the pandemic norovirus GII.4 lineage. J Virol. 2013;87(11):6270–82.
74. Bartnicki E, Cunha JB, Kolawole AO, Wobus CE. Recent advances in understanding noroviruses. F1000Research. 2017;6:79.
75. Tao Y, Rotem A, Zhang H, Cockrell SK, Koehler SA, Chang CB, et al. Artifact-free quantification and sequencing of rare recombinant viruses by using drop-based microfluidics. Chembiochem Eur J Chem Biol. 2015;16(15):2167–71.
76. Hardy ME, Kramer SF, Treanor JJ, Estes MK. Human calicivirus genogroup II capsid sequence diversity revealed by analyses of the prototype Snow Mountain agent. Arch Virol. 1997;142 (7):1469–79.

77. Jiang X, Espul C, Zhong WM, Cuello H, Matson DO. Characterization of a novel human calicivirus that may be a naturally occurring recombinant. Arch Virol. 1999;144(12):2377–87.
78. Katayama K, Shirato-Horikoshi H, Kojima S, Kageyama T, Oka T, Hoshino F, et al. Phylogenetic analysis of the complete genome of 18 Norwalk-like viruses. Virology. 2002;299 (2):225–39.
79. Bull RA, Hansman GS, Clancy LE, Tanaka MM, Rawlinson WD, White PA. Norovirus recombination in ORF1/ORF2 overlap. Emerg Infect Dis. 2005;11(7):1079–85.
80. Rohayem J, Münch J, Rethwilm A. Evidence of recombination in the norovirus capsid gene. J Virol. 2005;79(8):4977–90.
81. Han MG, Smiley JR, Thomas C, Saif LJ. Genetic recombination between two genotypes of genogroup III bovine noroviruses (BoNVs) and capsid sequence diversity among BoNVs and Nebraska-like bovine enteric caliciviruses. J Clin Microbiol. 2004;42(11):5214–24.
82. Siqueira JAM, Bandeira R da S, Oliveira D de S, Dos Santos LFP, Gabbay YB. Genotype diversity and molecular evolution of noroviruses: a 30-year (1982-2011) comprehensive study with children from northern Brazil. PLoS One. 2017;12(6):e0178909.
83. Ambert-Balay K, Bon F, Le Guyader F, Pothier P, Kohli E. Characterization of new recombinant noroviruses. J Clin Microbiol. 2005;43(10):5179–86.
84. Iritani N, Seto Y, Kubo H, Murakami T, Haruki K, Ayata M, et al. Prevalence of Norwalk-like virus infections in cases of viral gastroenteritis among children in Osaka City, Japan. J Clin Microbiol. 2003;41(4):1756–9.
85. Tsugawa T, Numata-Kinoshita K, Honma S, Nakata S, Tatsumi M, Sakai Y, et al. Virological, serological, and clinical features of an outbreak of acute gastroenteritis due to recombinant genogroup II norovirus in an infant home. J Clin Microbiol. 2006;44(1):177–82.
86. Phan TG, Takanashi S, Kaneshi K, Ueda Y, Nakaya S, Nishimura S, et al. Detection and genetic characterization of norovirus strains circulating among infants and children with acute gastroenteritis in Japan during 2004-2005. Clin Lab. 2006;52(9–10):519–25.
87. Phan TG, Kuroiwa T, Kaneshi K, Ueda Y, Nakaya S, Nishimura S, et al. Changing distribution of norovirus genotypes and genetic analysis of recombinant GIIb among infants and children with diarrhea in Japan. J Med Virol. 2006;78(7):971–8.
88. Reuter G, Vennema H, Koopmans M, Szücs G. Epidemic spread of recombinant noroviruses with four capsid types in Hungary. J Clin Virol. 2006;35(1):84–8.

Pathogenesis and Clinical Features 4

Molly Steele and Ben Lopman

4.1 Introduction

Noroviruses are non-enveloped, single-stranded RNA viruses of the family *Caliciviridae* [1]. These viruses are highly infectious, with an infectious dose estimated to be between 18 and 2800 genome equivalent copies [2, 3]. Individuals become infected after ingesting the virus and replication takes place in intestinal enterocytes within the small intestines [4–12]. Infected individuals will shed copious amounts of virus both in stool (10^5–10^{11} viral copies per gram of feces) and vomitus (10^3–10^6 gene equivalent copies per mL of vomitus) for 2–4 weeks, with peaks in shedding occurring 2–5 days following infection [13–15].

Noroviruses readily transmit through person-to-person, foodborne, waterborne and environmental pathways, thus humans can be exposed to the virus in many ways. Globally, noroviruses are estimated to cause approximately 677 million AGE cases and 213,000 AGE related deaths annually [16–20]. In this chapter, we describe in detail the natural history of infection, infectivity, pathogenesis, viral shedding, and transmission of noroviruses.

M. Steele (✉)
Environmental Health Sciences, Rollins School of Public Health, Emory University, Atlanta, GA, USA
e-mail: molly.steele@emory.edu

B. Lopman
Department of Epidemiology, Rollins School of Public Health, Emory University, Atlanta, GA, USA
e-mail: blopman@emory.edu

N. M. Melhem (ed.), *Norovirus*,
https://doi.org/10.1007/978-3-030-27209-8_4

4.2 Incubation, Symptoms, and Duration of Infectiousness and Illness

Noroviruses are highly infectious. The infectious dose sufficient to cause infection in 50% of those exposed (ID_{50}) for Norwalk virus (GI.1) has been estimated to be between 18 and 2800 gene equivalent copies [2, 3]. Following exposure to noroviruses, individuals will enter a relatively short incubation period [21, 22]. A recent meta-analysis of data on 1022 norovirus outbreaks reported the average incubation period to be 32.8 h (95% CI: 30.9–34.6) [22]. Symptoms of norovirus infection include a sudden onset of vomiting, watery, non-bloody diarrhea, abdominal cramps, fever and malaise [23–26]. The average duration of symptoms, also estimated by meta-analysis, is 44.2 h (95% CI: 38.9–50.7) [22].

Children less than 1 year old are more likely to experience diarrhea; vomiting and diarrhea are the predominant symptoms of norovirus in individuals who are older than 1 year [27, 28]. Persons infected with norovirus are more likely to experience vomiting, and less likely to have bloody diarrhea and long-term sequelae compared to other causes of gastroenteritis, especially bacterial pathogens such as *Salmonella* and *Campylobacter* [27–34].

While norovirus illness affects all age groups, the severity of disease outcomes differs between age groups [1, 27]. Children suffer high incidence of norovirus gastroenteritis, outpatient and emergency department visits and hospitalizations [35–38] relative to other age groups. However, severe outcomes are more common among the elderly, with the vast majority of norovirus-associated deaths occurring among those 65 years of age and older [39–41]. Among immunocompetent individuals, noroviruses typically cause acute, self-limiting infections; however, as is described in more detail below, immunocompromised individuals experience more severe disease and, sometimes, chronic norovirus infections [29, 42–45]. Norovirus is often detected in stool of individuals who are not exhibiting symptoms of acute gastroenteritis. In challenge studies and community cohorts, 12–35% of individuals with norovirus detected in stool do not have symptoms of disease [13, 46–48]. Additionally, many individuals will shed norovirus in stool for several weeks after symptoms of illness have resolved, as described below [13, 14, 49, 50].

4.3 Infectivity and Pathogenesis

Until very recently, attempts to culture norovirus *in vitro* using an array of cell systems have been unsuccessful, [51–53] thus limiting progress in our understanding of the biology of human noroviruses. Accordingly, much of our knowledge of the infectivity and pathogenesis of human noroviruses is inferred from studies of surrogate viruses and human intestinal biopsies. In an important breakthrough, human noroviruses have been successfully cultured in stem cell-derived, human intestinal enteroids (HIEs) [4]. This development portends many important applications, though this or a derivative system will take many years to be adopted widely in research labs.

Data from human intestinal biopsies and the HIEs culture system indicate that human noroviruses replicate in intestinal enterocytes and the virus can be found in the duodenal, jejunal, and ileal segments of the small intestines of infected individuals [4–9]. During infection, histological changes of the duodenal and jejunal regions of the small intestine can be observed including broadening and blunting of villi, shortening of the microvilli, disorganization of epithelial cells, crypt cell hyperplasia, cytoplasmic vacuolization, and invasion of inflammatory cells in the lamina propria [9–12]. Severe epithelial barrier dysfunction has been reported as a result of norovirus infection, and is thought to be the result of epithelial apoptosis and down regulation of tight junction proteins [9]. Some have hypothesized that the release of cytotoxic T-cells in response to norovirus infection leads to increases in epithelial apoptosis [9]. Moreover, down regulation of tight junction proteins may increase the permeability of the small intestines, allowing water and ions to diffuse from subepithelial capillaries back into the intestinal lumen, resulting in diarrhea [9].

Human challenge studies conducted in the 1970s and 1980s found that not everyone challenged with norovirus became infected, indicating variable susceptibility to norovirus infection [25, 46, 54]. Current knowledge suggests that certain host genetic factors are associated with the ability of norovirus to establish an infection within a human host [48, 55–57]. Secretor status is known to be associated with norovirus infections. Secretor positive individuals have a functional *fucosyltransferase-2* (*FUT2*) gene, which encodes for α(1,2)fucosyltransferases that produce H type-1 antigens, which serve as precursors to A and B antigens [58]. Secretor negative individuals have non-functional *FUT2* genes and do not produce H type-1, resulting in the absence of histo-blood group antigens (HBGAs) in saliva and mucosa [58]. Certain norovirus genotypes are strongly associated with HBGAs; secretors (i.e., those who express HGBAs in mucosa) are susceptible to genogroup I genotype 1 (GI.1) and genogroup II genotype 4 (GII.4) norovirus infections, however non-secretors are almost completely resistant to GI.1 and GII.4 norovirus infection [48, 56, 57, 59–64]. Non-secretors, however, can be infected by other norovirus genotypes, thus HBGAs are not the only receptors for norovirus infections [65–68]. A recent study reported that a GII.4 norovirus strain (Sydney) successfully replicated in a B-cell line culture suggesting that B-cells may also act as a receptor for norovirus [69].

The prevalence of secretor phenotypes (presence of a functional *FUT2* gene) varies by ethnicity. In Mesoamerican populations, up to 95% of individuals are secretors, [70] and secretors comprise 70–80% of European, North American, Asian and some African populations [66, 71–73]. There is also a weak secretor phenotype; the presence of a homozygous mutation at nucleotide 385 of the *FUT2* gene (A385T mutation) results in low expression of secretor HGBAs [74]. 15–20% of Asian populations are weak secretors [75]. In China, children with weak secretor phenotypes have exhibited partial protection against GII.4 norovirus infections [76].

The severity of symptoms may vary by norovirus genotype. GII.4 norovirus infections tend to have longer durations of vomiting and diarrhea and are more often associated with severe outcomes, such as hospitalizations and death, than non-GII.4 infections [77]. However, it is difficult to discern what determines the severity of infections as age, setting, and genotype effects may be highly correlated. For example, GII.4 norovirus outbreaks often occur in long-term care facilities where individuals are older, and the vast majority of norovirus deaths occur among the elderly [37, 78]. While GII.4 noroviruses are highly diverse, with new strains emerging every 2–5 years, [79] it is unclear whether there are differences in the severity of infection between strains [80].

4.4 Shedding

4.4.1 Stool

Infected individuals will shed virus in their stool, prior to symptomatic illness, during the symptomatic period, and after symptoms resolve. The viral shedding period is highly variable between individuals and may begin as early as 18 h after exposure and extend as long as 8 weeks post-exposure among immunocompetent hosts [13–15, 49, 81]. Several human challenge studies have documented the duration of norovirus shedding to be on the order of weeks [13, 14]. In one Norwalk virus (GI.1) challenge study, the median duration of shedding was 28 days (range: 13–56 days) [13]. Another Norwalk challenge study in 2011 showed similar results, with a median duration of shedding of 21 days (range: 8–35 days) [49]. The duration of viral shedding also varies depending on the genogroup causing infection. One recent challenge study found that among participants challenged with Norwalk virus, the median duration of shedding was 17 days (range: 5–27 days), while participants challenged with Snow Mountain (GII.2) virus tended to have a shorter duration of shedding, with a median of 5 days (range 2–25 days) [14].

Infected individuals can shed up to 10^5–10^{11} viral copies per gram of feces; [2, 13, 14, 81] however, the amount of viral shedding varies during the shedding period. Viral shedding often peaks 2–5 days following infection, thus the peak of viral shedding may occur after symptoms have resolved. 69% of participants in one challenge study had the highest levels of shedding after their symptoms resolved [13]. One recent challenge study found that among participants challenged with Norwalk virus, peak viral shedding occurred between 3 and 15 days post-infection, while participants challenged with Snow Mountain virus had peak viral shedding 2–6 days post-infection [14].

The duration of viral shedding in stool does not differ substantially between symptomatically and asymptomatically infected individuals. In one Norwalk virus challenge study, the median durations of shedding for symptomatic and asymptomatic infections was 15.5 days (range: 5–27) and 18 days (range: 5–26), respectively [14]. Results from a longitudinal analysis of serial stool samples from a GII.4 norovirus outbreak showed that the duration of shedding is similar during symptomatic and asymptomatic infections [50]. Little is known about the degree to which asymptomatically shedding individuals contribute to the transmission of norovirus;

however, studies have shown that while asymptomatic individuals are able to cause subsequent infections, they are less infectious than symptomatically infected individuals [50, 82–84].

4.4.2 Vomit

Virus is also shed in vomitus, though few studies have analyzed the shedding dynamics of vomitus. In one Norwalk challenge study, among participants with vomiting, the median concentration of virus was 4.1×10^5 genome equivalent copies per mL (range <2200 to 1.2×10^7) [2]. In another challenge study, the average shedding concentration among participants with vomiting was 8.0×10^5 and 3.9×10^4 gene equivalent copies per mL in vomitus for GI and GII viruses, respectively [85]. Generally, the concentration of norovirus shed in vomitus is lower than the concentration of virus shed in stool (Fig. 4.1).

While concentrations of norovirus are lower in vomitus than in stool, vomiting events can pose a significant risk of transmission [86–93]. From norovirus outbreak investigations, it was shown that vomiting events can contaminate public surfaces,

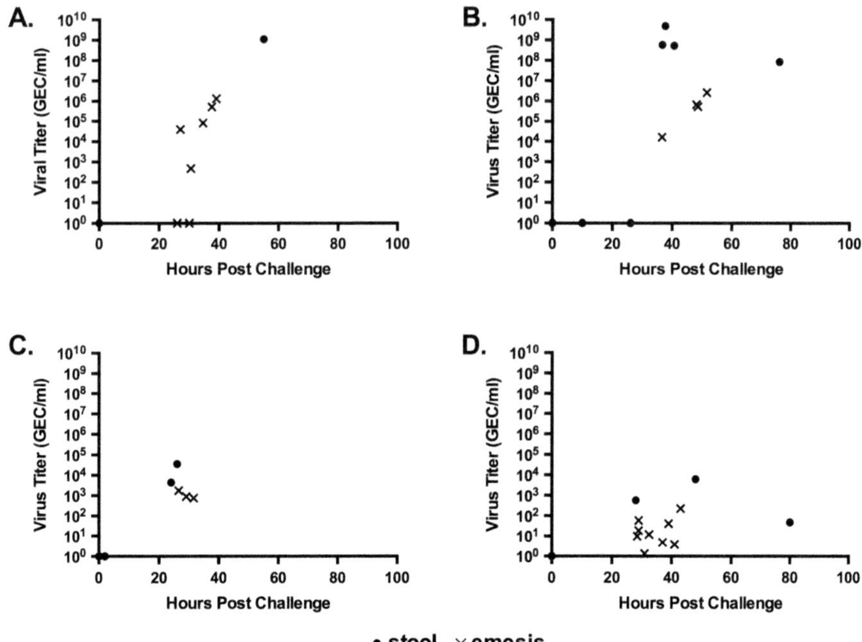

Fig. 4.1 Viral titers in stool and vomitus (emesis) samples over time from participants of a norovirus challenge study. (**a**, **b**) Virus titers from participants challenged with Norwalk virus (GI.1). (**c**) Virus titers from a participant challenged with Snow Mountain virus (GII.2). (**d**) Virus titers from a participant challenged with Hawaii virus (GII.1). (Figure reproduced from [85])

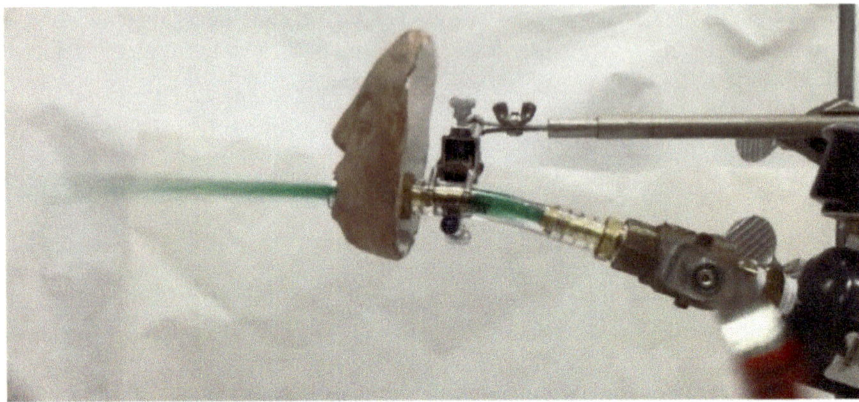

Fig. 4.2 A machine constructed by Tung-Thompson et al. simulates a vomiting event. The "vomitus" contains Bacteriophage MS2, a surrogate for norovirus, which allows researchers to measure virus aerosolized during a vomiting event. (Image from [95])

[88–93] and unlike diarrhea, vomiting produces aerosols that can be inhaled [86, 87]. To this end, vomiting simulation machines are being used to characterize the potential spread of viral contamination [94] and the amount of norovirus that can be aerosolized during a vomiting event (Fig. 4.2) [95]. These simulations show that vomiting events have the potential to contaminate surfaces >3 m longitudinally and >2.2 m laterally [94]. Additionally, between 36 and 13,350 Plaque Forming Units (PFU) of Bacteriophage MS2, a surrogate for norovirus, can be aerosolized during simulated vomiting events. These results suggest that noroviruses can be aerosolized during vomiting events and pose a risk for disease transmission [95].

Currently, it is not clear whether infection from aerosolized vomitus droplets varies in severity or pathogenesis relative to infections from contact with fecal material. However, the probability of norovirus infection increases as viral dose increases (ID_{50} between 18 and 2800 gene equivalent copies), [2, 3] therefore the probability of norovirus infection could vary if the dose that individuals are exposed to differs between contact with stool or vomitus. More studies are needed to assess whether there are differences in norovirus infections from exposure to vomitus versus feces.

4.5 Shedding Among Immunocompromised

While norovirus infections are acute and self-limiting among immunocompetent individuals, immunocompromised individuals may experience long term or chronic infection as well as more severe illness from norovirus [29, 42–45]. Immunocompromised individuals tend to shed norovirus for longer periods [82, 96, 97]. One prospective cohort study of children under the age of 18 with inherited immune deficiencies reported chronic shedding and symptoms after a median follow-up period of 9.5 months among four children with norovirus infections [42]. A study of pediatric cancer patients reported prolonged norovirus shedding from 22–433 days; [97] and, chronic shedding has been reported to last for 15 months in one HIV patient [44]. In addition to longer shedding periods, norovirus infections among immunocompromised individuals have high viral diversity and evolution rates [98–100]. Several studies have indicated that viral capsid mutations occur in immunocompromised patients which potentially lead to the emergence of new norovirus strains [96, 98, 99]. One such study reported a higher mutation rate of norovirus in an immunocompetent child (0.13 amino acid mutations per day) compared to immunocompromised children (0.03–0.07 amino acid mutations per day); however, due to prolonged infection, some of the immunocompromised children had higher total amino acid substitutions (5–11 amino acid substitutions) relative to the immunocompetent child (4 amino acid substitutions) [96]. Given their longer duration of shedding and harboring of viral diversity, some have suggested that immunocompromised individuals may serve as a heightened source of transmission as well as a reservoir for emerging norovirus variants [96, 98–100]. Counter to this idea, a recent mathematical modeling study found that transmissible viruses most often occur among infections in immunocompetent hosts, as immunocompromised hosts are rare relative to immunocompetent hosts, and are generally isolated, limiting their transmission potential [101].

4.6 Transmission

Due to its low infectious dose combined with capacity for extended environmental persistence, noroviruses can transmit through person-to-person, foodborne, waterborne, and environmental pathways [88, 92, 102, 103].

4.6.1 Person-to-Person

Data from surveillance systems indicate that the predominant mode of norovirus transmission is person-to-person (Fig. 4.3), and the genotype most often associated with person-to-person transmission is GII.4 [78, 104–106]. Currently, there are three surveillance systems with recent publications of norovirus case and/or outbreak data predominately reported from developed countries: NoroNET Surveillance data on norovirus outbreaks and sporadic cases from 19 countries across Europe, Asia, Oceania, and Africa are available from 2005 to 2016; the Center for Disease Control's National Outbreak Reporting System (NORS) surveillance data from norovirus outbreaks reported in the U.S. between 2009 and 2016; and the National Epidemiological Surveillance of Infectious Diseases' (NESID) data on norovirus outbreak reports in Japan between 2000 and 2016. There is a lack of norovirus surveillance data that captures the burden of norovirus among older children and adults, as well as the burden of norovirus in low- and middle-income countries [16, 107]. Both NORS and NESID are national surveillance systems, to which local

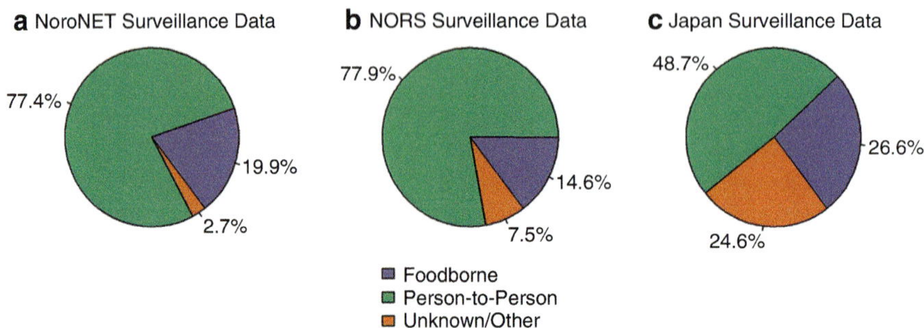

Fig. 4.3 Proportion of outbreaks reported to surveillance systems by transmission mode. (**a**) Proportion of norovirus outbreaks and cases reported to NoroNET from Europe, Asia, Oceania, and Africa between 2005 and 2016 [104]. (**b**) Proportion of norovirus outbreaks reported to NORS from the U.S. between 2009 and 2016 [78]. (**c**) Proportion of norovirus outbreaks reported to Japan's National Epidemiological Surveillance of Infectious Disease [105]

health departments will report norovirus outbreaks; [78, 105] NoroNET, however, is an informal, international collaboration where participants contribute molecular and epidemiological data from norovirus outbreaks and cases [104]. Of 6446 norovirus outbreaks and cases reported to NoroNET, 77% had person-to-person transmission, 20% had foodborne, 2% had waterborne and 0.7% had other modes of transmission [104]. Of 15,148 norovirus outbreaks reported to NORS, 78% had person-to-person transmission, 15% had foodborne, 0.17% had waterborne, 0.30% had environmental and 7% had other/unknown modes of transmission [78]. Of 9264 norovirus outbreaks reported to NESID, 49% had person-to-person transmission, 26% had foodborne and 25% had other/unknown modes of transmission [105]. Outbreaks that spread by person-to-person transmission can occur in many different settings; however in the U.S. the top settings where outbreaks occur are long term care facilities (78%), schools (6%), hospitals (4%), and child care facilities (2%) [108].

While noroviruses affect persons of all ages, young children are particularly important in the transmission of noroviruses. Observational studies have identified contact with a young child with norovirus illness is a significant risk factor for diarrhea in older children and adults [38]. Mathematical transmission modeling studies have estimated higher transmission parameter values for young children, relative to older children, adults and the elderly; estimates of R_0 (i.e., the average number of infections that arise from a primary infection in a fully susceptible population) for children under 5 years old are between 3.98 and 4.84, whereas estimates of R_0 for those 5 years and older are between 0.4 and 1.88 [109, 110].

4.6.2 Foodborne

Foodborne transmission is the second most commonly reported mode of norovirus transmission (Fig. 4.3). Foodborne norovirus outbreaks are more often caused by non-GII.4 genotypes [106]. Noroviruses are the leading cause of foodborne disease in the U.S., with an estimated annual burden of 5.5 million cases [111, 112]. Approximately half of foodborne outbreaks with identified etiology are caused by noroviruses [23, 113–116]. More than half (62%) of reported foodborne outbreaks occurred in restaurants. In outbreaks where specific foods were identified as the source, 92% were caused by contamination of food during preparation or service. Foods most commonly involved in norovirus outbreaks were leafy vegetables (33%), fruits and nuts (16%), and mollusks (13%) [116]. The results of a recent literature review of foodborne outbreaks associated with produce reported between 2004 and 2012 showed foods implicated in norovirus outbreaks in the U.S. were: salads (43%), leafy greens (28%), melons (4%), tomatoes (2%), berries (2%) and juices (1%). Foods implicated in norovirus outbreaks in the European Union were: berries (51%), leafy greens (24%), salads (14%), and tomatoes (1%) [117]. Contamination of food products can occur prior to harvesting, during harvesting, and during processing and handling prior to consumption [118–122].

Prior to harvesting, contamination of produce, such as leafy greens and berries, can occur when pathogens are present in the water used for irrigation or in manure-based fertilizers, [118, 123, 124] and when infected field workers transfer pathogens when handling produce [119]. Few studies have tested for norovirus contamination of produce prior to harvest. In one study, real-time polymerase chain reaction (RT-PCR) was used to detect norovirus on produce samples collected at the point of sale from Belgium, Canada, and France. Norovirus was detected on 28.2% (181/641) to 50% (3/6) of leafy greens, 6.7% (10/150) to 34.5% (10/29) of soft red fruits (e.g., raspberries and strawberries), and 55.5% (10/18) of other types of produce [125]. A second study analyzed produce originating from farms in Poland, Serbia and Spain as well as fruit salads prepared in Belgium. Norovirus was detected on 40% (4/10) of raspberries, 23% (7/30) of cherry tomatoes, 30% (6/20) of strawberries and about 7% (1/15) of fruit salads tested [126]. A more recent study analyzed frozen raspberries across four processing facilities in Poland. Approximately 8.6% (6/70) of samples tested positive for norovirus among frozen raspberries that were to be processed for puree, while 8.3% (1/12) of samples tested positive for norovirus among frozen raspberries that were bulk individually quick-frozen [127].

Bivalve mollusks, such as oysters, that are harvested from water contaminated with norovirus are often implicated as a source of foodborne norovirus outbreaks [116, 128, 129]. As filter feeders, bivalve mollusks are readily contaminated with noroviruses and can concentrate virus from their environment [130]. Additionally, many bivalves have carbohydrate structures that noroviruses bind to, which leads to bioaccumulation of the virus in shellfish tissue [131–133]. Several outbreak investigations associated with oysters have indicated environmental contamination occurred as a result of prior heavy rainfall or flooding events [128, 134]. One investigation of an international norovirus outbreak linked to contaminated oysters revealed that several days of heavy rainfall occurred 1 week prior to the first reported cases of norovirus. The environmental data collected for this investigation showed that heavy rainfall caused local rivers to flood; shortly after the flooding, authorities documented that there were failures in sewage treatment, which ultimately led to the contamination of oyster production beds [128]. An outbreak investigation with multiple enteric viruses (including norovirus) from oyster consumption linked heavy rainfall to flooding and contamination of a shellfish production lagoon in southern France [134].

Contamination of food products can occur during processing and preparation prior to consumption, and food workers (or individuals preparing food for consumption) infected with norovirus are often the source of this contamination [102, 116, 135]. In the U.S., food workers were implicated as the source of 70% of the 1008 foodborne norovirus outbreaks reported to NORS between 2009 and 2012 [116]. Infected food workers can spread norovirus when touching ready-to-eat foods, such as raw fruits and vegetables, with their bare hands before serving them. An observational study of food worker hand washing practices found that proper hand washing only occurred in 27% of activities where hand washing is recommended, and decreased to 16% when food workers were using gloves [136]. Moreover, approximately 20% of food workers report having worked at least one shift while experiencing diarrhea or vomiting, often citing fear of losing their job or leaving coworkers short-staffed as significant influences on their decision to work while ill [137].

4.6.3 Waterborne

GI noroviruses are most often associated with waterborne transmission [106, 138]. Noroviruses have been documented to persist in water for up to 3 years and contaminated water can remain infectious for at least 2 months [139]. Waterborne outbreaks tend to have more primary cases and larger populations at risk than foodborne outbreaks, [138] as noroviruses can contaminate drinking water within distribution systems and affect communities [140–151]. For example, in Greece, thousands of residents developed gastroenteritis during two outbreaks that occurred within a year of each other due to contaminated well and municipal water after heavy rainfall events and flooding [150, 151]. Failures in water treatment systems [142, 144] and contamination of the water source have been identified as causes of contamination of drinking water [140, 141, 143, 144, 146, 152–154]. Waterborne outbreaks of norovirus have also been linked to recreational activities. In one literature review, it was found that there were 25 reported norovirus outbreaks, primarily in the U.S., linked to recreational waters between 1977 and 2006. The majority of these outbreaks were a result of exposure to water in either lakes or swimming pools [155]. Investigations of outbreaks linked to recreational pools and fountains have documented maintenance system failures, inadequate maintenance checks, or poorly trained staff as the likely causes of contamination [156–160]. Outbreaks of norovirus have also occurred from swimming in natural bodies of water, such as lakes or rivers [152, 161].

4.6.4 Environmental

Noroviruses are highly infectious, hardy viruses that can persist on surfaces for up to 2 weeks [162]. These viruses are infectious from 0 to 60°C and resistant to many common household chemical cleaners [1, 163–165]. A low infectious dose combined with environmental persistence allow noroviruses to transmit through environmental pathways [3, 4, 103]. Noroviruses readily transfer from contaminated surfaces to hands and vice versa [166]. Outbreaks from environmental transmission of norovirus are less likely to be recognized and, therefore, more likely to be underreported than outbreaks spread by other transmission modes [88, 92, 103].

Studies that have documented individuals who are subsequently infected with no known direct contact have given the best evidence of environmental transmission. One such study involved a norovirus outbreak after a vomiting event on an aircraft. Over a period of 6 days, 29 flight attendants who worked in the aircraft developed gastroenteritis. Additionally, two flight attendants who became ill were found to be infected with the same, rare genotype (GI.6); these two individuals were determined to have no direct contact with each other, thus investigators concluded they acquired the infection from working in the same aircraft [88]. A second example involved a norovirus outbreak that affected six consecutive sailings on a single cruise ship. An outbreak investigation of the first sailing determined that the source of the outbreak was foodborne. However, after the ship was sanitized, passengers and crew members continued to develop gastroenteritis. The investigators concluded that environmental transmission likely sustained the outbreak between sailings, and/or that infected crew members may have continued to introduce the virus between sailings [167]. A third example of environmental transmission of norovirus comes from an outbreak following a vomiting event in a concert hall. More than 300 individuals developed gastroenteritis in the 5 days following the initial vomiting event. Outbreak investigators determined that individuals who sat in seats near where the initial vomiting event occurred days earlier had the highest risk of developing gastroenteritis [92].

4.7 Controlling Transmission

Currently, there is no licensed vaccine available for norovirus, thus control measures focus on preventing or interrupting transmission [168]. The CDC recommends that hands should be washed with soap and water for at least 20 s, as studies using surrogate viruses have shown that this practice reduces the amount of virus by 0.7–1.2 $\log_{10}$ [169, 170]. It is not clear whether alcohol-based sanitizers are effective against norovirus, therefore alcohol-based sanitizers should be used in addition to proper hand washing [171–174]. In healthcare settings or other settings where there is close person-to-person contact, such as long-term care facilities, isolation of symptomatic individuals can limit transmission [168, 175].

To prevent foodborne transmission of norovirus, produce should be properly washed and shellfish should be properly cooked and not consumed raw [168]. Any contaminated food should be removed, and all contaminated utensils and surfaces should be cleaned with a 1000 ppm sodium hypochlorite bleach solution [168, 176, 177]. To prevent contamination of food during preparation, food workers should not directly contact ready-to-eat foods; rather, food workers should wear gloves and use proper hand hygiene [168, 178, 179]. Food workers should not work shifts while experiencing diarrhea and vomiting, and should return to work 48 h after their symptoms of illness have resolved [168, 178, 180].

To prevent waterborne transmission in recreational waters, the CDC recommends treating water with $\geq$1 ppm concentration of chlorine to inactivate pathogens [181]. When a diarrhea episode occurs in recreational water, the CDC recommends that the venue be closed to all swimmers, the gross contamination be removed, and the concentration of chlorine should be raised to 20 ppm for about 13 h (in the presence of chlorine stabilizers, a higher concentration of chlorine for longer duration is required). Swimmers can return to the venue once the water reaches the state and local operation requirements for pH and chlorine concentration [182].

As noroviruses can persist on environmental surfaces and are resistant to many common household chemical cleaners, contaminated surfaces should be cleaned with a 1000–5000 ppm concentration of chlorine bleach [163, 164, 176]. Steam cleaning and laundering are recommended for decontamination of soft surfaces; however more studies are needed to determine the efficacy of these methods in eliminating norovirus contamination [183].

4.8 Conclusion

Noroviruses are an important cause of acute gastroenteritis globally, accounting for approximately 677 million cases and 213,000 deaths annually [16–20]. The ability of these viruses to circulate widely is in part due to several characteristics of the virus that enable rapid and efficient transmission. As discussed in this chapter, noroviruses are highly infectious, and infected individuals shed copious amounts of virus both in stool and vomitus for long periods of time [2, 3, 13, 14, 42, 50, 96, 97]. Environmental stability allows these viruses to transmit through person-to-person, foodborne, waterborne and environmental pathways, thus humans can be exposed to the virus in many ways [102, 162]. The low infectious dose of noroviruses, their ability to transmit through multiple pathways, and their resistance to common household cleaners pose great challenges to prevention and control measures [2, 3, 102, 163–166].

References

1. Glass RI, Parashar UD, Estes MK. Norovirus gastroenteritis. N Engl J Med. 2009;361(18):1776–85.
2. Atmar RL, Opekun AR, Gilger MA, Estes MK, Crawford SE, Neill FH, et al. Determination of the 50% human infectious dose for Norwalk virus. J Infect Dis. 2013;209(7):1016–22.

3. Teunis PF, Moe CL, Liu P, Miller SE, Lindesmith L, Baric RS, et al. Norwalk virus: how infectious is it? J Med Virol. 2008;80(8):1468–76.
4. Ettayebi K, Crawford SE, Murakami K, Broughman JR, Karandikar U, Tenge VR, et al. Replication of human noroviruses in stem cell-derived human enteroids. Science. 2016;353 (6306):1387–93.
5. Karandikar UC, Crawford SE, Ajami NJ, Murakami K, Kou B, Ettayebi K, et al. Detection of human norovirus in intestinal biopsies from immunocompromised transplant patients. J Gen Virol. 2016;97(9):2291–300.
6. Otto PH, Clarke IN, Lambden PR, Salim O, Reetz J, Liebler-Tenorio EM. Infection of calves with bovine norovirus GIII. 1 strain Jena virus: an experimental model to study the pathogenesis of norovirus infection. J Virol. 2011;85(22):12013–21.
7. Cheetham S, Souza M, Meulia T, Grimes S, Han MG, Saif LJ. Pathogenesis of a genogroup II human norovirus in gnotobiotic pigs. J Virol. 2006;80(21):10372–81.
8. Agus SG, Dolin R, Wyatt RG, Tousimis A, Northrup RS. Acute infectious nonbacterial gastroenteritis: intestinal histopathology: histologic and enzymatic alterations during illness produced by the Norwalk agent in man. Ann Intern Med. 1973;79(1):18–25.
9. Troeger H, Loddenkemper C, Schneider T, Schreier E, Epple H-J, Zeitz M, et al. Structural and functional changes of the duodenum in human norovirus infection. Gut. 2009;58(8):1070–7.
10. Mumphrey SM, Changotra H, Moore TN, Heimann-Nichols ER, Wobus CE, Reilly MJ, et al. Murine norovirus 1 infection is associated with histopathological changes in immunocompetent hosts, but clinical disease is prevented by STAT1-dependent interferon responses. J Virol. 2007;81(7):3251–63.
11. Schreiber DS, Blacklow NR, Trier JS. The mucosal lesion of the proximal small intestine in acute infectious nonbacterial gastroenteritis. N Engl J Med. 1973;288(25):1318–23.
12. Schreiber DS, Blacklow NR, Trier JS. The small intestinal lesion induced by Hawaii agent acute infectious nonbacterial gastroenteritis. J Infect Dis. 1974;129(6):705–8.
13. Atmar RL, Opekun AR, Gilger MA, Estes MK, Crawford SE, Neill FH, et al. Norwalk virus shedding after experimental human infection. Emerg Infect Dis. 2008;14(10):1553.
14. Kirby A, Shi J, Montes J, Lichtenstein M, Moe C. Disease course and viral shedding in experimental Norwalk virus and Snow Mountain virus infection. J Med Virol. 2014;86 (12):2055–64.
15. Bernstein DI, Atmar RL, Lyon GM, Treanor JJ, Chen WH, Jiang X, et al. Norovirus vaccine against experimental human GII.4 virus illness: a challenge study in healthy adults. J Infect Dis. 2015;211(6):870–8.
16. Ahmed SM, Hall AJ, Robinson AE, Verhoef L, Premkumar P, Parashar UD, et al. Global prevalence of norovirus in cases of gastroenteritis: a systematic review and meta-analysis. Lancet Infect Dis. 2014;14(8):725–30.
17. Hall AJ, Lopman BA, Payne DC, Patel MM, Gastañaduy PA, Vinjé J, et al. Norovirus disease in the United States. Emerg Infect Dis. 2013;19(8):1198–205.
18. Bartsch SM, Lopman BA, Ozawa S, Hall AJ, Lee BY. Global economic burden of norovirus gastroenteritis. PLoS One. 2016;11(4):e0151219.
19. Lopman BA, Steele D, Kirkwood CD, Parashar UD. The vast and varied global burden of norovirus: prospects for prevention and control. PLoS Med. 2016;13(4):e1001999.
20. Pires SM, Fischer-Walker CL, Lanata CF, Devleesschauwer B, Hall AJ, Kirk MD, et al. Aetiology-specific estimates of the global and regional incidence and mortality of diarrhoeal diseases commonly transmitted through food. PLoS One. 2015;10(12):e0142927.
21. Lee RM, Lessler J, Lee RA, Rudolph KE, Reich NG, Perl TM, et al. Incubation periods of viral gastroenteritis: a systematic review. BMC Infect Dis. 2013;13(1):446.
22. Devasia T, Lopman B, Leon J, Handel A. Association of host, agent and environment characteristics and the duration of incubation and symptomatic periods of norovirus gastroenteritis. Epidemiol Infect. 2015;143(11):2308–14.

23. Hall AJ, Eisenbart VG, Etingue AL, Gould LH, Lopman BA, Parashar UD. Epidemiology of foodborne norovirus outbreaks, United States, 2001-2008. Emerg Infect Dis. 2012;18 (10):1566–73.
24. Atmar RL, Estes MK. The epidemiologic and clinical importance of norovirus infection. Gastroenterol Clin N Am. 2006;35(2):275–90.
25. Wyatt RG, Dolin R, Blacklow NR, DuPont HL, Buscho RF, Thornhill TS, et al. Comparison of three agents of acute infectious nonbacterial gastroenteritis by cross-challenge in volunteers. J Infect Dis. 1974;129(6):709–14.
26. Dedman D, Laurichesse H, Caul E, Wall P. Surveillance of small round structured virus (SRSV) infection in England and Wales, 1990-5. Epidemiol Infect. 1998;121(01):139–49.
27. Rockx B, de Wit M, Vennema H, Vinjé J, de Bruin E, van Duynhoven Y, et al. Natural history of human calicivirus infection: a prospective cohort study. Clin Infect Dis. 2002;35(3):246–53.
28. Kaplan JE, Gary GW, Baron RC, Singh N, Schonberger LB, Feldman R, et al. Epidemiology of Norwalk gastroenteritis and the role of Norwalk virus in outbreaks of acute nonbacterial gastroenteritis. Ann Intern Med. 1982;96(6 Pt 1):756–61.
29. Lopman BA, Reacher MH, Vipond IB, Sarangi J, Brown DW. Clinical manifestation of norovirus gastroenteritis in health care settings. Clin Infect Dis. 2004;39(3):318–24.
30. Kawada JI, Arai N, Nishimura N, Suzuki M, Ohta R, Ozaki T, et al. Clinical characteristics of norovirus gastroenteritis among hospitalized children in Japan. Microbiol Immunol. 2012;56 (11):756–9.
31. Barreira DMPG, Ferreira MSR, Fumian TM, Checon R, de Sadovsky ADI, Leite JPG, et al. Viral load and genotypes of noroviruses in symptomatic and asymptomatic children in southeastern Brazil. J Clin Virol. 2010;47(1):60–4.
32. de Andrade Jda SR, Rocha MS, Carvalho-Costa FA, Fioretti JM, Xavier Mda TP, Nunes ZMA, et al. Noroviruses associated with outbreaks of acute gastroenteritis in the state of Rio Grande do Sul, Brazil, 2004–2011. J Clin Virol. 2014;61(3):345–52.
33. Wu T-C, Liu H-H, Chen Y-J, Tang R-B, Hwang B-T, Yuan H-C. Comparison of clinical features of childhood norovirus and rotavirus gastroenteritis in Taiwan. J Chin Med Assoc. 2008;71(11):566–70.
34. Ternhag A, Törner A, Svensson Å, Ekdahl K, Giesecke J. Short-and long-term effects of bacterial gastrointestinal infections. Emerg Infect Dis. 2008;14(1):143.
35. Lopman BA, Hall AJ, Curns AT, Parashar UD. Increasing rates of gastroenteritis hospital discharges in US adults and the contribution of norovirus, 1996–2007. Clin Infect Dis. 2011;52(4):466–74.
36. Gastañaduy PA, Hall AJ, Curns AT, Parashar UD, Lopman BA. Burden of norovirus gastroenteritis in the ambulatory setting—United States, 2001–2009. J Infect Dis. 2013;207 (7):1058–65.
37. Kotloff KL, Nataro JP, Blackwelder WC, Nasrin D, Farag TH, Panchalingam S, et al. Burden and aetiology of diarrhoeal disease in infants and young children in developing countries (the Global Enteric Multicenter Study, GEMS): a prospective, case-control study. Lancet. 2013;382(9888):209–22.
38. Phillips G, Tam CC, Conti S, Rodrigues LC, Brown D, Iturriza-Gomara M, et al. Community incidence of norovirus-associated infectious intestinal disease in England: improved estimates using viral load for norovirus diagnosis. Am J Epidemiol. 2010;171:1014–22.
39. Hall AJ, Curns AT, McDonald LC, Parashar UD, Lopman BA. The roles of Clostridium difficile and norovirus among gastroenteritis-associated deaths in the United States, 1999-2007. Clin Infect Dis. 2012;55(2):216–23.
40. van Asten L, Siebenga J, van den Wijngaard C, Verheij R, van Vliet H, Kretzschmar M, et al. Unspecified gastroenteritis illness and deaths in the elderly associated with norovirus epidemics. Epidemiology. 2011;22(3):336–43.
41. Harris JP, Edmunds WJ, Pebody R, Brown DW, Lopman BA. Deaths from norovirus among the elderly, England and Wales. Emerg Infect Dis. 2008;14(10):1546–52.

42. Frange P, Touzot F, Debré M, Héritier S, Leruez-Ville M, Cros G, et al. Prevalence and clinical impact of norovirus fecal shedding in children with inherited immune deficiencies. J Infect Dis. 2012;206(8):1269–74.
43. Munir N, Liu P, Gastañaduy P, Montes J, Shane A, Moe C. Norovirus infection in immunocompromised children and children with hospital-acquired acute gastroenteritis. J Med Virol. 2014;86(7):1203–9.
44. Wingfield T, Gallimore CI, Xerry J, Gray JJ, Klapper P, Guiver M, et al. Chronic norovirus infection in an HIV-positive patient with persistent diarrhoea: a novel cause. J Clin Virol. 2010;49(3):219–22.
45. Mattner F, Sohr D, Heim A, Gastmeier P, Vennema H, Koopmans M. Risk groups for clinical complications of norovirus infections: an outbreak investigation. Clin Microbiol Infect. 2006;12(1):69–74.
46. Graham DY, Jiang X, Tanaka T, Opekun AR, Madore HP, Estes MK. Norwalk virus infection of volunteers: new insights based on improved assays. J Infect Dis. 1994;170(1):34–43.
47. Phillips G, Tam CC, Rodrigues L, Lopman B. Prevalence and characteristics of asymptomatic norovirus infection in the community in England. Epidemiol Infect. 2010;138(10):1454–8.
48. Lindesmith L, Moe C, Marionneau S, Ruvoen N, Jiang X, Lindblad L, et al. Human susceptibility and resistance to Norwalk virus infection. Nat Med. 2003;9(5):548–53.
49. Leon JS, Kingsley DH, Montes JS, Richards GP, Lyon GM, Abdulhafid GM, et al. Randomized, double-blinded clinical trial for human norovirus inactivation in oysters by high hydrostatic pressure processing. Appl Environ Microbiol. 2011;77(15):5476–82.
50. Teunis P, Sukhrie F, Vennema H, Bogerman J, Beersma M, Koopmans M. Shedding of norovirus in symptomatic and asymptomatic infections. Epidemiol Infect. 2015;143 (08):1710–7.
51. Papafragkou E, Hewitt J, Park GW, Greening G, Vinje J. Challenges of culturing human norovirus in three-dimensional organoid intestinal cell culture models. PLoS One. 2013;8(6): e63485.
52. Duizer E, Schwab KJ, Neill FH, Atmar RL, Koopmans MP, Estes MK. Laboratory efforts to cultivate noroviruses. J Gen Virol. 2004;85(1):79–87.
53. Herbst-Kralovetz MM, Radtke AL, Lay MK, Hjelm BE, Bolick AN, Sarker SS, et al. Lack of norovirus replication and histo-blood group antigen expression in 3-dimensional intestinal epithelial cells. Emerg Infect Dis. 2013;19(3):431–8.
54. Parrino TA, Schreiber DS, Trier JS, Kapikian AZ, Blacklow NR. Clinical immunity in acute gastroenteritis caused by Norwalk agent. N Engl J Med. 1977;297(2):86–9.
55. Hutson AM, Airaud F, LePendu J, Estes MK, Atmar RL. Norwalk virus infection associates with secretor status genotyped from sera. J Med Virol. 2005;77(1):116–20.
56. Hutson AM, Atmar RL, Graham DY, Estes MK. Norwalk virus infection and disease is associated with ABO histo-blood group type. J Infect Dis. 2002;185(9):1335–7.
57. Hutson AM, Atmar RL, Marcus DM, Estes MK. Norwalk virus-like particle hemagglutination by binding to h histo-blood group antigens. J Virol. 2003;77(1):405–15.
58. Marionneau S, Cailleau-Thomas A, Rocher J, Le Moullac-Vaidye B, Ruvoën N, Clément M, et al. ABH and Lewis histo-blood group antigens, a model for the meaning of oligosaccharide diversity in the face of a changing world. Biochimie. 2001;83(7):565–73.
59. Marionneau S, Airaud F, Bovin NV, Pendu JL, Ruvoën-Clouet N. Influence of the combined ABO, FUT2 and FUT3 polymorphism on susceptibility to Norwalk virus attachment. J Infect Dis. 2005;192(6):1071–7.
60. Marionneau S, Ruvoën N, Le Moullac-Vaidye B, Clement M, Cailleau-Thomas A, Ruiz-Palacois G, et al. Norwalk virus binds to histo-blood group antigens present on gastroduodenal epithelial cells of secretor individuals. Gastroenterology. 2002;122(7):1967–77.
61. Tan M, Jin M, Xie H, Duan Z, Jiang X, Fang Z. Outbreak studies of a GII-3 and a GII-4 norovirus revealed an association between HBGA phenotypes and viral infection. J Med Virol. 2008;80(7):1296–301.

62. Larsson MM, Rydell GE, Grahn A, Rodríguez-Díaz J, Åkerlind B, Hutson AM, et al. Antibody prevalence and titer to norovirus (genogroup II) correlate with secretor (FUT2) but not with ABO phenotype or Lewis (FUT3) genotype. J Infect Dis. 2006;194:1422–7.
63. Thorven M, Grahn A, Hedlund K-O, Johansson H, Wahlfrid C, Larson G, et al. A homozygous nonsense mutation (428G→A) in the human secretor (FUT2) gene provides resistance to symptomatic norovirus (GGII) infections. J Virol. 2005;79(24):15351–5.
64. Frenck R, Bernstein DI, Xia M, Huang P, Zhong W, Parker S, et al. Predicting susceptibility to norovirus GII.4 by use of a challenge model involving humans. J Infect Dis. 2012;206 (9):1386–93.
65. Van Trang N, Vu HT, Le NT, Huang P, Jiang X, Anh DD. Association between norovirus and rotavirus infection and histo-blood group antigen types in Vietnamese children. J Clin Microbiol. 2014;52:1366–74.
66. Nordgren J, Nitiema LW, Ouermi D, Simpore J, Svensson L. Host genetic factors affect susceptibility to norovirus infections in Burkina Faso. PLoS One. 2013;8(7):e69557.
67. Ayouni S, Estienney M, Sdiri-Loulizi K, Ambert-Balay K, de Rougemont A, Aho S, et al. Relationship between GII.3 norovirus infections and blood group antigens in young children in Tunisia. Clin Microbiol Infect. 2015;21(9):874.e1-8.
68. Lindesmith L, Moe C, LePendu J, Frelinger JA, Treanor J, Baric RS. Cellular and humoral immunity following Snow Mountain virus challenge. J Virol. 2005;79(5):2900–9.
69. Jones MK, Watanabe M, Zhu S, Graves CL, Keyes LR, Grau KR, et al. Enteric bacteria promote human and mouse norovirus infection of B cells. Science. 2014;346(6210):755–9.
70. Bucardo F, Kindberg E, Paniagua M, Grahn A, Larson G, Vildevall M, et al. Genetic susceptibility to symptomatic norovirus infection in Nicaragua. J Med Virol. 2009;81 (4):728–35.
71. Ferrer-Admetlla A, Sikora M, Laayouni H, Esteve A, Roubinet F, Blancher A, et al. A natural history of FUT2 polymorphism in humans. Mol Biol Evol. 2009;26(9):1993–2003.
72. Kindberg E, Hejdeman B, Bratt G, Wahren B, Lindblom B, Hinkula J, et al. A nonsense mutation (428G→A) in the fucosyltransferase FUT2 gene affects the progression of HIV-1 infection. AIDS. 2006;20(5):685–9.
73. Thorne L, Nalwoga A, Mentzer AJ, de Rougemont A, Hosmillo M, Webb E, et al. The first Norovirus longitudinal Seroepidemiological study from sub-Saharan Africa reveals high seroprevalence of diverse genotypes associated with host susceptibility factors. J Infect Dis. 2018;218(5):716–25. https://doi.org/10.1093/infdis/jiy219.
74. Henry S, Mollicone R, Fernandez P, Samuelsson B, Oriol R, Larson G. Molecular basis for erythrocyte Le (a+ b+) and salivary ABH partial-secretor phenotypes: expression of a FUT2 secretor allele with an A→ T mutation at nucleotide 385 correlates with reduced α (1, 2) fucosyltransferase activity. Glycoconj J. 1996;13(6):985–93.
75. Chang JG, Yang TY, Liu TC, Lin TP, Hu CJ, Kao MC, et al. Molecular analysis of secretor type alpha(1,2)-fucosyltransferase gene mutations in the Chinese and Thai populations. Transfusion. 1999;39(9):1013–7.
76. Liu P, Wang X, Lee JC, Teunis P, Hu S, Paradise HT, et al. Genetic susceptibility to Norovirus GII.3 and GII.4 infections in Chinese pediatric diarrheal children. Pediatr Infect Dis J. 2014;33 (11):e305–9.
77. Huhti L, Szakal ED, Puustinen L, Salminen M, Huhtala H, Valve O, et al. Norovirus GII-4 causes a more severe gastroenteritis than other Noroviruses in young children. J Infect Dis. 2011;203(10):1442–4.
78. Centers for Disease Control and Prevention (CDC). National Outbreak Reporting System Dashboard. Atlanta, GA: U.S. Department of Health and Human Services, Centers for Disease Control and Prevention.. wwwn.cdc.gov/norsdashboard
79. Siebenga JJ, Lemey P, SLK P, Rambaut A, Vennema H, Koopmans M. Phylodynamic reconstruction reveals norovirus GII. 4 epidemic expansions and their molecular determinants. PLoS Pathog. 2010;6(5):e1000884.

80. Gustavsson L, Nordén R, Westin J, Lindh M, Andersson LM. Slow clearance of Norovirus following infection with emerging variants of genotype GII.4 strains. J Clin Microbiol. 2017;55(5):1533–9.
81. Aoki Y, Suto A, Mizuta K, Ahiko T, Osaka K, Matsuzaki Y. Duration of norovirus excretion and the longitudinal course of viral load in norovirus-infected elderly patients. J Hosp Infect. 2010;75(1):42–6.
82. Sukhrie FH, Siebenga JJ, Beersma MF, Koopmans M. Chronic shedders as reservoir for nosocomial transmission of norovirus. J Clin Microbiol. 2010;48(11):4303–5.
83. Sukhrie FH, Teunis P, Vennema H, Copra C, Beersma MFT, Bogerman J, et al. Nosocomial transmission of norovirus is mainly caused by symptomatic cases. Clin Infect Dis. 2012;54 (7):931–7.
84. Milbrath M, Spicknall I, Zelner J, Moe C, Eisenberg J. Heterogeneity in norovirus shedding duration affects community risk. Epidemiol Infect. 2013;141(8):1572–84.
85. Kirby AE, Streby A, Moe CL. Vomiting as a symptom and transmission risk in norovirus illness: evidence from human challenge studies. PLoS One. 2016;11(4):e0143759.
86. Marks P, Vipond I, Carlisle D, Deakin D, Fey R, Caul E. Evidence for airborne transmission of Norwalk-like virus (NLV) in a hotel restaurant. Epidemiol Infect. 2000;124(03):481–7.
87. Marks P, Vipond I, Regan F, Wedgwood K, Fey R, Caul E. A school outbreak of Norwalk-like virus: evidence for airborne transmission. Epidemiol Infect. 2003;131(01):727–36.
88. Thornley CN, Emslie NA, Sprott TW, Greening GE, Rapana JP. Recurring norovirus transmission on an airplane. Clin infect Dis. 2011;53:515–20.
89. Patterson W, Haswell P, Fryers P, Green J. Outbreak of small round structured virus gastroenteritis arose after kitchen assistant vomited. Commun Dis Rep CDR Rev. 1997;7(7):R101–3.
90. Petrignani M, van Beek J, Borsboom G, Richardus J, Koopmans M. Norovirus introduction routes into nursing homes and risk factors for spread: a systematic review and meta-analysis of observational studies. J Hosp Infect. 2015;89(3):163–78.
91. Repp KK, Keene WE. A point-source norovirus outbreak caused by exposure to fomites. J Infect Dis. 2012;205(11):1639–41.
92. Evans MR, Meldrum R, Lane W, Gardner D, Ribeiro C, Gallimore C, et al. An outbreak of viral gastroenteritis following environmental contamination at a concert hall. Epidemiol Infect. 2002;129(02):355–60.
93. Wikswo ME, Cortes J, Hall AJ, Vaughan G, Howard C, Gregoricus N, et al. Disease transmission and passenger behaviors during a high morbidity Norovirus outbreak on a cruise ship, January 2009. Clin Infect Dis. 2011;52(9):1116–22.
94. Booth CM. Vomiting Larry: a simulated vomiting system for assessing environmental contamination from projectile vomiting related to norovirus infection. J Infect Prev. 2014;15:176–80.
95. Tung-Thompson G, Libera DA, Koch KL, Francis L III, Jaykus L-A. Aerosolization of a human norovirus surrogate, bacteriophage MS2, during simulated vomiting. PLoS One. 2015;10(8):e0134277.
96. Siebenga JJ, Beersma MF, Vennema H, van Biezen P, Hartwig NJ, Koopmans M. High prevalence of prolonged norovirus shedding and illness among hospitalized patients: a model for in vivo molecular evolution. J Infect Dis. 2008;198(7):994–1001.
97. Ludwig A, Adams O, Laws HJ, Schroten H, Tenenbaum T. Quantitative detection of norovirus excretion in pediatric patients with cancer and prolonged gastroenteritis and shedding of norovirus. J Med Virol. 2008;80(8):1461–7.
98. Debbink K, Lindesmith LC, Ferris MT, Swanstrom J, Beltramello M, Corti D, et al. Within-host evolution results in antigenically distinct GII. 4 noroviruses. J Virol. 2014;88 (13):7244–55.
99. Nilsson M, Hedlund K-O, Thorhagen M, Larson G, Johansen K, Ekspong A, et al. Evolution of human calicivirus RNA in vivo: accumulation of mutations in the protruding P2 domain of the capsid leads to structural changes and possibly a new phenotype. J Virol. 2003;77 (24):13117–24.

100. Vega E, Donaldson E, Huynh J, Barclay L, Lopman B, Baric R, et al. RNA populations in immunocompromised patients as reservoirs for novel norovirus variants. J Virol. 2014;88 (24):14184–96.
101. Eden J-S, Chisholm RH, Bull RA, White PA, Holmes EC, Tanaka MM. Persistent infections in immunocompromised hosts are rarely sources of new pathogen variants. Virus Evol. 2017;3 (2):vex018.
102. Mathijs E, Stals A, Baert L, Botteldoorn N, Denayer S, Mauroy A, et al. A review of known and hypothetical transmission routes for noroviruses. Food Environ Virol. 2012;4(4):131–52.
103. Lopman B, Gastañaduy P, Park GW, Hall AJ, Parashar UD, Vinjé J. Environmental transmission of norovirus gastroenteritis. Curr Opin Virol. 2012;2(1):96–102.
104. van Beek J, de Graaf M, Al-Hello H, Allen DJ, Ambert-Balay K, Botteldoorn N, et al. Molecular surveillance of norovirus, 2005-16: an epidemiological analysis of data collected from the NoroNet network. Lancet Infect Dis. 2018;18(5):545–53.
105. Matsuyama R, Miura F, Nishiura H. The transmissibility of noroviruses: statistical modeling of outbreak events with known route of transmission in Japan. PLoS One. 2017;12(3): e0173996.
106. Verhoef L, Hewitt J, Barclay L, Ahmed SM, Lake R, Hall AJ, et al. Norovirus genotype profiles associated with foodborne transmission, 1999-2012. Emerg Infect Dis. 2015;21 (4):592–9.
107. Mans J, Armah GE, Steele AD, Taylor MB. Norovirus epidemiology in Africa: a review. PLoS One. 2016;11(4):e0146280.
108. Wikswo ME, Kambhampati A, Shioda K, Walsh KA, Bowen AB, Hall AJ. Outbreaks of acute gastroenteritis transmitted by person-to-person contact, environmental contamination, and unknown modes of transmission—United States, 2009-2013. 2015.
109. Simmons K, Gambhir M, Leon J, Lopman B. Duration of immunity to norovirus gastroenteritis. Emerg Infect Dis. 2013;19(8):1260–7.
110. Steele MK, Remais JV, Gambhir M, Glasser JW, Handel A, Parashar UD, et al. Targeting pediatric versus elderly populations for norovirus vaccines: a model-based analysis of mass vaccination options. Epidemics. 2016;17:42–9.
111. Hall AJ, Wikswo ME, Manikonda K, Roberts VA, Yoder JS, Gould LH. Acute gastroenteritis surveillance through the National Outbreak Reporting System, United States. Emerg Infect Dis. 2013;19(8):1305–9.
112. Scallan E, Hoekstra RM, Angulo FJ, Tauxe RV, Widdowson M-A, Roy SL, et al. Foodborne illness acquired in the United States—major pathogens. Emerg Infect Dis. 2011;17(1):7–15.
113. Centers for Disease Control and Prevention. Surveillance for foodborne disease outbreaks—United States, 2006. Ann Emerg Med. 2010;55(1):47–9.
114. Centers for Disease Control and Prevention. Surveillance for foodborne disease outbreaks—United States, 2007. MMWR Morb Mortal Wkly Rep. 2010;59(31):973.
115. Centers for Disease Control and Prevention. Surveillance for foodborne disease outbreaks—United States, 2008. MMWR Morb Mortal Wkly Rep. 2011;60(35):1197.
116. Hall AJ, Wikswo ME, Pringle K, Gould LH, Parashar UD. Vital signs: foodborne norovirus outbreaks—United States, 2009–2012. MMWR Morb Mortal Wkly Rep. 2014;63(22):491–5.
117. Callejón RM, Rodríguez-Naranjo MI, Ubeda C, Hornedo-Ortega R, Garcia-Parrilla MC, Troncoso AM. Reported foodborne outbreaks due to fresh produce in the United States and European Union: trends and causes. Foodborne Pathog Dis. 2015;12(1):32–8.
118. Wei J, Kniel KE. Pre-harvest viral contamination of crops originating from fecal matter. Food Environ Virol. 2010;2(4):195–206.
119. Lynch MF, Tauxe RV, Hedberg CW. The growing burden of foodborne outbreaks due to contaminated fresh produce: risks and opportunities. Epidemiol Infect. 2009;137(3):307–15.
120. Beuchat LR. Pathogenic microorganisms associated with fresh produce. J Food Prot. 1996;59 (2):204–16.
121. Moe CL. Preventing norovirus transmission: how should we handle food handlers? Chicago: The University of Chicago Press; 2009.

122. Tuan ZC, Hidayah M, Chai L, Tunung R, Ghazali FM, Son R. The scenario of norovirus contamination in food and food handlers. J Microbiol Biotechnol. 2010;20(2):229–37.
123. Brassard J, Gagné M-J, Généreux M, Côté C. Detection of human food-borne and zoonotic viruses on irrigated, field-grown strawberries. Appl Environ Microbiol. 2012;78(10):3763–6.
124. El-Senousy WM, Costafreda MI, Pintó RM, Bosch A. Method validation for norovirus detection in naturally contaminated irrigation water and fresh produce. Int J Food Microbiol. 2013;167(1):74–9.
125. Baert L, Mattison K, Loisy-Hamon F, Harlow J, Martyres A, Lebeau B, et al. Norovirus prevalence in Belgian, Canadian and French fresh produce: a threat to human health? Int J Food Microbiol. 2011;151(3):261–9.
126. Stals A, Baert L, Jasson V, Van Coillie E, Uyttendaele M. Screening of fruit products for norovirus and the difficulty of interpreting positive PCR results. J Food Prot. 2011;74 (3):425–31.
127. De Keuckelaere A, Li D, Deliens B, Stals A, Uyttendaele M. Batch testing for noroviruses in frozen raspberries. Int J Food Microbiol. 2015;192:43–50.
128. Le Guyader FS, Bon F, DeMedici D, Parnaudeau S, Bertone A, Crudeli S, et al. Detection of multiple noroviruses associated with an international gastroenteritis outbreak linked to oyster consumption. J Clin Microbiol. 2006;44(11):3878–82.
129. Webby R, Carville KS, Kirk MD, Greening G, Ratcliff RM, Crerar S, et al. Internationally distributed frozen oyster meat causing multiple outbreaks of norovirus infection in Australia. Clin Infect Dis. 2007;44(8):1026–31.
130. Dame RF. Ecology of marine bivalves: an ecosystem approach. Boca Raton, FL: CRC Press; 2011.
131. Tian P, Engelbrektson AL, Jiang X, Zhong W, Mandrell RE. Norovirus recognizes histo-blood group antigens on gastrointestinal cells of clams, mussels, and oysters: a possible mechanism of bioaccumulation. J Food Prot. 2007;70(9):2140–7.
132. Maalouf H, Schaeffer J, Parnaudeau S, Le Pendu J, Atmar RL, Crawford SE, et al. Strain-dependent norovirus bioaccumulation in oysters. Appl Environ Microbiol. 2011;77 (10):3189–96.
133. Maalouf H, Zakhour M, Le Pendu J, Le Saux J-C, Atmar RL, Le Guyader FS. Norovirus genogroups I and II ligands in oysters: tissue distribution and seasonal variations. Appl Environ Microbiol. 2010;76(16):5621–30.
134. Le Guyader FS, Le Saux JC, Ambert-Balay K, Krol J, Serais O, Parnaudeau S, et al. Aichi virus, norovirus, astrovirus, enterovirus, and rotavirus involved in clinical cases from a French oyster-related gastroenteritis outbreak. J Clin Microbiol. 2008;46(12):4011–7.
135. Widdowson M-A, Sulka A, Bulens SN, Beard RS, Chaves SS, Hammond R, et al. Norovirus and foodborne disease, United States, 1991–2000. Emerg Infect Dis. 2005;11(1):95.
136. Green LR, Selman CA, Radke V, Ripley D, Mack JC, Reimann DW, et al. Food worker hand washing practices: an observation study. J Food Prot. 2006;69(10):2417–23.
137. Carpenter L, Green A, Norton D, Frick R, Tobin-D'Angelo M, Reimann D, et al. Food worker experiences with and beliefs about working while ill. J Food Prot. 2013;76(12):2146–54.
138. Matthews J, Dickey B, Miller R, Felzer J, Dawson B, Lee A, et al. The epidemiology of published norovirus outbreaks: a review of risk factors associated with attack rate and genogroup. Epidemiol Infect. 2012;140(7):1161–72.
139. Seitz SR, Leon JS, Schwab KJ, Lyon GM, Dowd M, McDaniels M, et al. Norovirus infectivity in humans and persistence in water. Appl Environ Microbiol. 2011;77(19):6884–8.
140. Brugha R, Vipond I, Evans MR, Sandifer Q, Roberts R, Salmon R, et al. A community outbreak of food-borne small round-structured virus gastroenteritis caused by a contaminated water supply. Epidemiol Infect. 1999;122(01):145–54.
141. Anderson AD, Heryford AG, Sarisky JP, Higgins C, Monroe SS, Beard RS, et al. A waterborne outbreak of Norwalk-like virus among snowmobilers—Wyoming, 2001. J Infect Dis. 2003;187(2):303–6.

142. Boccia D, Tozzi AE, Cotter B, Rizzo C, Russo T, Buttinelli G, et al. Waterborne outbreak of Norwalk-like virus gastroenteritis at a tourist resort, Italy. Emerg Infect Dis. 2002;8(6):563–8.
143. Carrique-Mas J, Andersson Y, Petersen B, Hedlund K-O, Sjögren N, Giesecke J. A Norwalk-like virus waterborne community outbreak in a Swedish village during peak holiday season. Epidemiol Infect. 2003;131(01):737–44.
144. Gallay A, De Valk H, Cournot M, Ladeuil B, Hemery C, Castor C, et al. A large multi-pathogen waterborne community outbreak linked to faecal contamination of a groundwater system, France, 2000. Clin Microbiol Infect. 2006;12(6):561–70.
145. Blackburn BG, Craun GF, Yoder JS, Hill V, Calderon RL, Chen N, et al. Surveillance for waterborne-disease outbreaks associated with drinking water—United States, 2001–2002. MMWR Surveill Summ. 2004;53(8):23–45.
146. Lawson H, Braun M, Glass R, Stine S, Monroe S, Atrash H, et al. Waterborne outbreak of Norwalk virus gastroenteritis at a southwest US resort: role of geological formations in contamination of well water. Lancet (London, England). 1991;337(8751):1200–4.
147. McAnulty JM, Rubin GL, Carvan CT, Huntley EJ, Grohmann G, Hunter R. An outbreak of Norwalk-like gastroenteritis associated with contaminated drinking water at a caravan park. Aust J Public Health. 1993;17(1):36–41.
148. Parshionikar SU, Willian-True S, Fout GS, Robbins DE, Seys SA, Cassady JD, et al. Waterborne outbreak of gastroenteritis associated with a norovirus. Appl Environ Microbiol. 2003;69(9):5263–8.
149. Schvoerer E, Bonnet F, Dubois V, Rogues A-M, Gachie J-P, Lafon M-E, et al. A hospital outbreak of gastroenteritis possibly related to the contamination of tap water by a small round structured virus. J Hosp Infect. 1999;43(2):149–54.
150. Papadopoulos VP, Vlachos O, Isidoridou E, Kasmeridis C, Pappa Z, Goutzouvelidis A, et al. A gastroenteritis outbreak due to Norovirus infection in Xanthi, northern Greece: management and public health consequences. J Gastrointestin Liver Dis. 2006;15(1):27.
151. Vantarakis A, Mellou K, Spala G, Kokkinos P, Alamanos Y. A gastroenteritis outbreak caused by noroviruses in Greece. Int J Environ Res Public Health. 2011;8(8):3468–78.
152. Katayama H, Okuma K, Furumai H, Ohgaki S. Series of surveys for enteric viruses and indicator organisms in Tokyo Bay after an event of combined sewer overflow. Water Sci Technol. 2004;50(1):259–62.
153. Beller M, Ellis A, Lee SH, Drebot MA, Jenkerson SA, Funk E, et al. Outbreak of viral gastroenteritis due to a contaminated well: international consequences. JAMA. 1997;278 (7):563–8.
154. Brieseman M, Hill S, Holmes J, Giles S, Ball A. A series of outbreaks of food poisoning? N Z Med J. 2000;113(1104):54–6.
155. Sinclair R, Jones E, Gerba CP. Viruses in recreational water-borne disease outbreaks: a review. J Appl Microbiol. 2009;107(6):1769–80.
156. Blevins L, Itani D, Burns A, Lohff C, Schoenfeld S, Knight W, et al. An outbreak of norovirus gastroenteritis at a swimming club-Vermont, 2004. Morb Mortal Wkly Rep. 2004;53 (34):793–7.
157. Maunula L, Kalso S, Von Bonsdorff C-H, Pönkä A. Wading pool water contaminated with both noroviruses and astroviruses as the source of a gastroenteritis outbreak. Epidemiol Infect. 2004;132(04):737–43.
158. Podewils L, Blevins LZ, Hagenbuch M, Itani D, Burns A, Otto C, et al. Outbreak of norovirus illness associated with a swimming pool. Epidemiol Infect. 2007;135(05):827–33.
159. Hoebe CJ, Vennema H, de Roda Husman AM, van Duynhoven YT. Norovirus outbreak among primary schoolchildren who had played in a recreational water fountain. J Infect Dis. 2004;189(4):699–705.
160. Kappus KD, Marks JS, Holman RC, Bryant JK, Baker C, Gary GW, et al. An outbreak of Norwalk gastroenteritis associated with swimming in a pool and secondary person-to-person transmission. Am J Epidemiol. 1982;116(5):834–9.

161. Baron RC, Murphy FD, Greenberg HB, Davis CE, Bregman DJ, Gary GW, et al. Norwalk gastrointestinal illness: an outbreak associated with swimming in a recreational lake and secondary person-to-person transmission. Am J Epidemiol. 1982;116(1):198.
162. Cheesbrough J, Barkess-Jones L, Brown D. Possible prolonged environmental survival of small round structured viruses. J Hosp Infect. 1997;35(4):325–6.
163. Doultree J, Druce J, Birch C, Bowden D, Marshall J. Inactivation of feline calicivirus, a Norwalk virus surrogate. J Hosp Infect. 1999;41(1):51–7.
164. Duizer E, Bijkerk P, Rockx B, De Groot A, Twisk F, Koopmans M. Inactivation of caliciviruses. Appl Environ Microbiol. 2004;70(8):4538–43.
165. Park GW, Boston DM, Kase JA, Sampson MN, Sobsey MD. Evaluation of liquid-and fog-based application of Sterilox hypochlorous acid solution for surface inactivation of human norovirus. Appl Environ Microbiol. 2007;73(14):4463–8.
166. Barker J, Vipond I, Bloomfield S. Effects of cleaning and disinfection in reducing the spread of Norovirus contamination via environmental surfaces. J Hosp Infect. 2004;58(1):42–9.
167. Isakbaeva ET, Widdowson M-A, Beard RS, Bulens SN, Mullins J, Monroe SS, et al. Norovirus transmission on cruise ship. Emerg Infect Dis. 2005;11(1):154–8.
168. Centers for Disease Control and Prevention. Updated norovirus outbreak management and disease prevention guidelines. MMWR Recomm Rep. 2011;60(RR-3):1–18.
169. Liu P, Hsiao HM, Jaykus LA, Moe C. Quantification of Norwalk virus inocula: comparison of endpoint titration and real-time reverse transcription-PCR methods. J Med Virol. 2010;82 (9):1612–6.
170. MacCannell T, Umscheid CA, Agarwal RK, Lee I, Kuntz G, Stevenson KB. Guideline for the prevention and control of norovirus gastroenteritis outbreaks in healthcare settings. Infect Control Hosp Epidemiol. 2011;32(10):939–69.
171. Macinga DR, Sattar SA, Jaykus L-A, Arbogast JW. Improved inactivation of nonenveloped enteric viruses and their surrogates by a novel alcohol-based hand sanitizer. Appl Environ Microbiol. 2008;74(16):5047–52.
172. Park GW, Barclay L, Macinga D, Charbonneau D, Pettigrew CA, Vinjé J. Comparative efficacy of seven hand sanitizers against murine norovirus, feline calicivirus, and GII. 4 norovirus. J Food Prot. 2010;73(12):2232–8.
173. Tuladhar E, Hazeleger WC, Koopmans M, Zwietering MH, Duizer E, Beumer RR. Reducing viral contamination from finger pads: handwashing is more effective than alcohol-based hand disinfectants. J Hosp Infect. 2015;90(3):226–34.
174. Park GW, Collins N, Barclay L, Hu L, Prasad BV, Lopman BA, et al. Strain-specific Virolysis patterns of human Noroviruses in response to alcohols. PLoS One. 2016;11(6):e0157787.
175. Hansen S, Stamm-Balderjahn S, Zuschneid I, Behnke M, Ruden H, Vonberg RP, et al. Closure of medical departments during nosocomial outbreaks: data from a systematic analysis of the literature. J Hosp Infect. 2007;65(4):348–53.
176. Girard M, Ngazoa S, Mattison K, Jean J. Attachment of noroviruses to stainless steel and their inactivation, using household disinfectants. J Food Prot. 2010;73(2):400–4.
177. Cromeans T, Park GW, Costantini V, Lee D, Wang Q, Farkas T, et al. Comprehensive comparison of cultivable norovirus surrogates in response to different inactivation and disinfection treatments. Appl Environ Microbiol. 2014;80(18):5743–51.
178. United States Food and Drug Administration. Food Code 2017. 2017. https://www.fda.gov/Food/GuidanceRegulation/RetailFoodProtection/FoodCode/ucm595139.htm
179. Rönnqvist M, Aho E, Mikkelä A, Ranta J, Tuominen P, Rättö M, et al. Norovirus transmission between hands, gloves, utensils, and fresh produce during simulated food handling. Appl Environ Microbiol. 2014;80(17):5403–10. https://doi.org/10.1128/AEM.01162-14.
180. Thornley C, Hewitt J, Perumal L, Van Gessel S, Wong J, David S, et al. Multiple outbreaks of a novel norovirus GII. 4 linked to an infected post-symptomatic food handler. Epidemiol Infect. 2013;141(8):1585–97.

181. Hlavsa MC, Cikesh BL, Roberts VA, Kahler AM, Vigar M, Hilborn ED, et al. Outbreaks associated with treated recreational water – United States, 2000-2014. MMWR Morb Mortal Wkly Rep. 2018;67(19):547–51.
182. Centers for Disease Control and Prevention (CDC). Model Aquatic Health Code. U.S. Department of Health and Human Services, Centers for Disease Control and Prevention. 2016. https://www.cdc.gov/mahc/pdf/2016-mahc-code-final.pdf.
183. Lemm D, Merettig N, Lucassen R, Bockmühl DP. Inactivation of human norovirus by common domestic laundry procedures. Tenside Surfactants Deterg. 2014;51(4):304–6.

Antinorovirus Drugs: Current and Future Perspectives 5

Armando Arias

5.1 Introduction

5.1.1 A Case for Antiviral Approaches Against Norovirus Infection

Human noroviruses (HuNoVs) account for about one fifth of the total number of cases of acute gastroenteritis worldwide. Despite their significant burden to global health, HuNoVs have historically received less attention than other pathogens causing gastrointestinal disorders [1]. Every year, there are aproximately 685 million episodes of diarrheal disease associated with norovirus infection that result in 900,000 hospitalizations and >200,000 deaths [2]. Noroviruses have a larger impact on low- and middle-income countries (LMICs) where gastroenteritis-associated diarrhea is responsible for large numbers of deaths, especially among children under the age of 5 and the elderly [1]. Norovirus is becoming the most common cause of pediatric gastroenteritis in high- and middle-income countries where efficient rotavirus vaccination programs have been recently implemented [2–4]. HuNoVs lead to lower number of severe cases and fatalities in developed countries, although the economic cost associated is highly significant. The economic burden associated with norovirus is related to a vast number of sick leaves caused by gastroenteritis episodes and the severe effect that nosocomial outbreaks have on the national health systems, leading to frequent ward closures and cancellation of surgery interventions [5, 6]. The annual global cost associated with HuNoV infections has been estimated to exceed $60 billion, with $4 billion directly linked to health care expenses, e.g. hospitalization and treatment [1].

A. Arias (✉)
Centro Regional de Investigaciones Biomédicas (CRIB), Universidad de Castilla La-Mancha (UCLM), Albacete, Spain

Technical University of Denmark (DTU), Kongens Lyngby, Denmark
e-mail: armando.arias@uclm.es; aaesteban2@gmail.com

N. M. Melhem (ed.), *Norovirus*,
https://doi.org/10.1007/978-3-030-27209-8_5

Noroviruses also cause other disorders of greater severity, especially in vulnerable individuals such as the elderly and immunocompromised people. In particular, hospital-acquired norovirus causing chronic diarrhea is a recurrent life-threatening condition affecting solid organ transplant (SOT) recipients [7]. Norovirus shedding in chronically-infected patients can last from few weeks to several years before resolving [7, 8]. Despite their significant impact on global health and wealth, there are yet neither drugs nor vaccines licenced to control norovirus infections [9].

5.1.2 Norovirus Genome Organization and Life Cycle

The noroviruses are positive-sense single-stranded RNA viruses that belong to the family *Caliciviridae*. Norovirus genome replication is catalyzed by the non-structural (NS) protein 7 (NS7) [10]. NS7 is an RNA-dependent RNA polymerase (RdRp) which copies noroviral RNA genomes into new viral genomes and also into subgenomic RNA. The synthesis of new viral RNA molecules is primed by the viral protein NS5, also known as VPg, which remains covalently attached to the 5′ end of both the genomic and subgenomic strands synthesized (Fig. 5.1). Norovirus subgenomic RNA is produced by the amplification of approximately the last third of

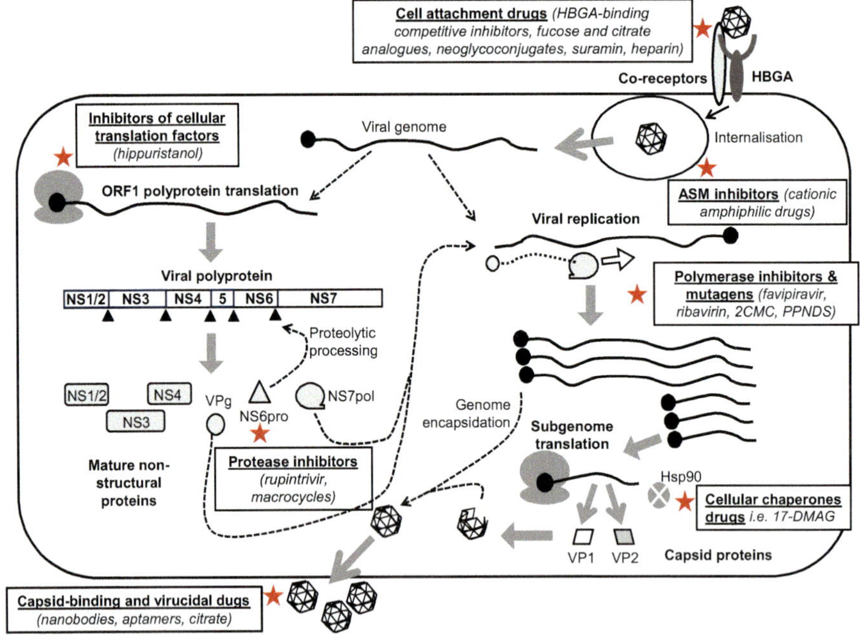

Fig. 5.1 General overview of the norovirus life cycle. The different steps of the life cycle are represented in different colors (white, grey and black). A red star symbol indicates a step that is the target of small molecule compounds. Some representative antiviral molecules identified for each target are indicated

the viral genome. The attachment of VPg to the viral RNA is critical for viral protein translation as it interacts with translation initiation factors that result in the recruitment of the ribosome to the 5′ end of viral RNA [11, 12]. HuNoV genome contains three different open reading frames (ORFs). Norovirus ORF1 translation leads to the synthesis of a polyprotein precursor (Fig. 5.1). A viral protease activity encoded by the NS6 region, also known as NS6pro, enables polyprotein self-processing, leading to the release of different non-structural proteins including NS6pro itself, the viral helicase (NS3), VPg (NS5), and the NS7 polymerase [13–16]. NS6 contains various pleiotropic activities as it seems to participate in the suppression of interferon-stimulated genes and the activation of caspases that contribute to cellular apoptosis [17]. The translation of the subgenomic RNA, containing ORF2 and ORF3, results in the synthesis of major (VP1) and minor capsid (VP2) proteins, respectively, which assemble around a single molecule of genomic viral RNA (Fig. 5.1). The mature virus particle is formed by the assembly of 180 copies of VP1 and a low number of VP2 copies [18, 19]. Two different domains can be identified in each monomer of VP1 (Fig. 5.2). The shell (S) domain, which is highly hydrophobic and occupies the internal surface of the capsid surrounding the viral genome, and the protruding (P) domain (external surface) which harbours the determinants for cell attachment and strain diversity [19, 20]. The P domain can be further subdivided into P1 and P2 subdomains. P2 is localized in the outer layer of the capsid where the epitopes for cellular attachment and mutations of resistance to neutralizing antibodies are generally found [19].

HuNoV viral particle interacts with histo-blood group antigens (HBGAs) expressed on the surface of target cells to get internalized in the cytoplasm and initiate viral replication in a process where other cellular co-receptors may also be implicated [21–23]. HuNoVs interact with different HGBAs including ABO and syalyl carbohydrates (e.g. Lewis X) that are typically found on cellular surfaces, epithelia, and body fluids. The different HBGAs are glycosylated by the fucosyl transferase 2 (FUT2) which adds fucose residues to these receptors. Individuals lacking a functional copy of *FUT2* gene, termed non-secretors, generally manifest lower susceptibility to infection [24], although they also can develop both asymptomatic and symptomatic infections. The factors underlying this variable susceptibility amongst non-secretor individuals remain unclear, although it has been suggested that gut microbiota diversity influences this phenotypic difference [24].

Until recently and owing to the absence of a cell culture model, we have had a limited understanding of the molecular steps following attachment and entry of HuNoV virions into the cell. Dynamin and cholesterol contribute to the internalization of murine norovirus (MNV) in a clathrin- and caveolae- independent mechanism. The fact that simvastatin (a cholesterol-lowering drug) enhances HuNoV replication in vitro and infectivity in vivo suggests that cholesterol may play a prominent role in viral entry, although further investigations are needed to elucidate the particle internalization process [25–28].

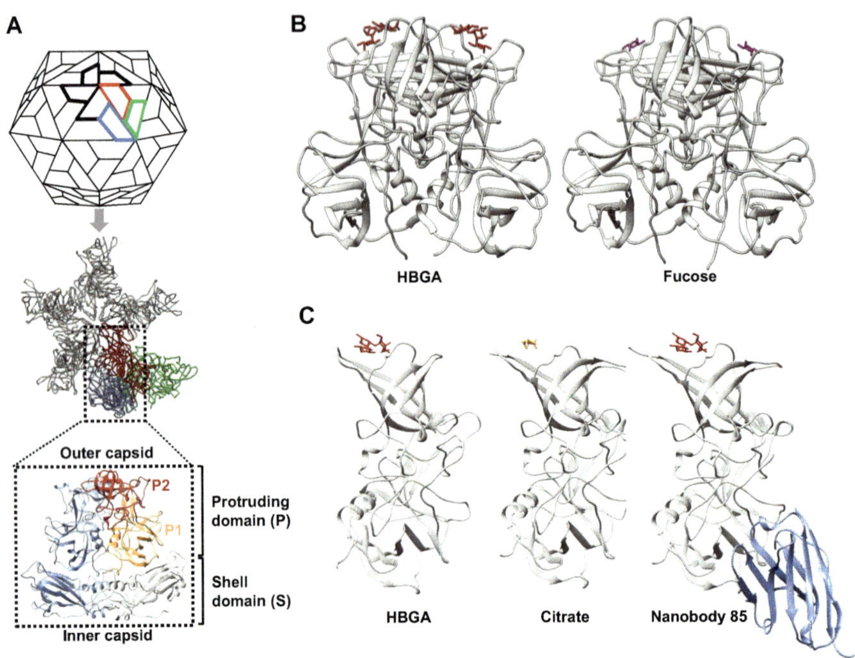

Fig. 5.2 Illustration of human norovirus capsid-complex with receptors or inhibitory molecules. (**a**) the top panel schematically represents the conformation of a norovirus capsid. In the middle panel a close-up view of the pentamer and trimer VP1 structural arrangements is shown. Bottom panel, a VP1 dimer is represented. The monomer in the left is represented entirely in blue while the molecule on the right corresponds to different structural domains: the inner shell surface (S) is depicted in grey; the protruding external domain (P) is shown in red (P2) and orange (P1). (**b**) Structure of a VP1 dimer bound to an HBGA receptor trisaccharide (left) and fucose molecules (right). The determinants of capsid binding to the HBGA receptor are located in the P2 subdomain. (**c**) Structure of VP1 monomers in complex with distinct molecules: HBGA trisaccharide (left), citrate (middle), and nanobody 85 (right)

5.1.3 Genetics and Evolution of Noroviruses

There are ten different norovirus genogroups (I–X) with HuNoV found in I, II, and IV, and MNV in V [29, 30, 31]. The recent pandemic spread of genogroup II noroviruses has displaced genogroup I as the most dominant genogroup globally [32, 33]. The epidemiological success of genogroup II noroviruses genotype 4 (GII.4) has been associated with larger intra-host genetic diversity and faster evolution rates. This larger genetic diversity in genogroup II HuNoV has been related to lower genome copying fidelity in their viral polymerases [32–36], implying that increased replication error rates are connected with faster intra-host genetic evolution and emergence of antigenic variants [37]. We have recently demonstrated that there is a direct relationship between norovirus intra-host genetic diversity and virus transmission in vivo using mouse model and reverse genetics methods available for MNV [38]. This finding supports the hypothesis suggesting that the recent epidemiological success of GII.4 is related to polymerase fidelity. In addition to viral polymerase fidelity, it has been also suggested that viral genetic diversity is influenced by host cellular factors driving genomic hypermutation in the rapid intrapatient evolution of HuNoV GII.4 [39].

5.2 Tools to Study Norovirus Biology and the Development of Antiviral Drugs

5.2.1 Establishment of Cell Culture Methods

The development of cell culture methods for the propagation of HuNoV constitutes a major advancement in the field of anti-norovirus drug research [21, 40, 41]. The successful infection and replication of HuNoV has been documented in two different cell culture systems, one of them involving immortalized B cell lines (BJAB) and another using ex vivo cultivations of human intestinal enteroids (HIEs). These novel tools may enable unprecedented studies on the life cycle and the molecular biology of noroviruses that can be applied in the identification of antivirals [21, 40, 42–44]. A possible advantage of ex vivo HIEs cultures is that these cells form three dimensional tissue which might recreate better the physiological context of infection in the intestine [21]. Organoids are typically derived from stem cells isolated from the organ of interest. In this case, HIEs are generally derived from intestine tissue [21, 41]. Unlike transformed cell lines, enteroid cultures show a complex organization of heterogeneous differentiated cells which resemble the cellular association observed in organs and tissues [41]. Additional advantages of enteroid systems with respect to traditional organotypic cultures is that they can be propagated [41]. The use of HIEs has been successfully employed in the study of infection by different enteric viruses other than HuNoV (e.g. rotavirus and echovirus 11) [41].This system has also been used to demonstrate that the interaction of HuNoV with HBGA receptors is critical for infection, thus promoting this cell culture model to evaluate inhibitors targeting entry [21, 45].

The development of cell culture systems also offers the possibility to investigate HuNoV biology by reverse genetics. Although the tools for the recovery of infectious virus from cDNA have not been yet established, viral particles containing encapsidated genomes have been obtained, suggesting that systems for the genetic manipulation of HuNoV could be soon available [46]. Reverse genetics systems have been already generated for murine norovirus (Sect. 5.2.3) which further encourage the establishment of the technique for human strains. Overall, these novel procedures are highly promising and open new possibilities to the field of anti-noroviral research. Further optimization of these methods would be required to facilitate their regular use in the identification of antivirals.

5.2.2 Animal Models

A mouse model of infection for HuNoV has also been recently established [47]. Mice deficient in the recombination activation gene (*RAG*) 1 or 2 and common gamma chain (γc) inoculated with clarified human stool filtrate from infected patients showed increased viral loads in feces and tissues (i.e. stomach, small and large intestines, mesenteric lymph nodes, spleen, liver, kidney and lung) [44, 47]. This model permits the evaluation of antiviral compounds in vivo with recent evidence demonstrating the efficacy of inhibitors against human norovirus [44]. In particular, in this study the authors have described the antiviral efficacy of a nucleoside analogue (2′-C-methylcytidine) against HuNoV replication in mice (further explained in Sect. 5.5.1.2). The advantages of mouse systems over models involving larger animals are their relative cost-effectiveness and the possibility to study disease in a genetically defined background. A major limitation of using mice, however, is that they are relatively far from representing an infection in the natural host; low virus titres are observed in tissues and feces possibly reflecting limited rounds of viral replication. It has been suggested that adaptation of HuNoV to mice, after serial passage of virus in vivo, could lead to a more robust system to study infection. Combining this prospective adapted virus-mouse model with cell culture systems abovementioned (Sect. 5.2.1.) can be exploited in the rapid screening of small molecule libraries [44].

Models of infection involving larger animals, such as gnotobiotic pigs, minipigs, chimpanzees and macaques, probably represent better surrogates to study infection in humans [25, 48–51]. In the pig model, the susceptibility to HuNoV infection can be further stimulated by simvastatin, a drug that lowers cholesterol levels in the organism [25]. Gnotobiotic pigs that are B-, T-, and natural killer cell-deficient (double knockout for *Rag2* and interleukin receptor 2 subunit gamma) have been recently generated. These animals showed higher viral levels in tissues and blood, and prolonged shedding in feces, supporting their use as a conceivable system to study antiviral therapies in the control of chronic infection in the immunocompromised [52]. Infection in the chimpanzee constitutes an additional approach that can be used in the development of antiviral treatments in vivo. HuNoV intravenously inoculated in these animals leads to the establishment of infection with viral loads detected in feces [53]. The use of chimpanzees as an infection model has also been suggested as a suitable model in the identification of efficient prophylactic compounds [54]. However, studies on chimpanzees have been restricted due to new ethical limits on the use of non-human primates [55]. A major recent advance has been recently published by Van Dyke and colleagues as a preprint manuscript where they show robust HuNoV replication in zebrafish larvae. This model can be of utmost relevance to advance in anti-noroviral research if this observation is finally confirmed [56].

These different animal models have advantages and limitations. Chimpanzees represent models of infection relatively similar to humans (orally susceptible), although a drawback of this system is that chimpanzees show no symptomatic infection. Other disadvantages are the elevated economic costs associated, and significant ethical concerns that can cause delays to experimental studies. The advantage of using gnotobiotic pigs or macaques is that, in addition to being susceptible to oral inoculation, both animals show clinical signs of infection. However, similarly to the chimpanzee model, these systems are also very costly. An additional disadvantage of using macaques is that they lack an IgA response to infection which limits their use in the development of vaccines [54]. In the case of the gnotobiotic pig model, several major limitations include the lack of microbiota (which can have an impact on HuNoV infection, Sect. 5.8.3.2) and their underdeveloped immunity which can affect the course of infection. To summarize, the main common disadvantage of large animal models is their elevated costs which hinders their use in testing and screening drug inhibitors [54]. Thus, to high-throughput analysis of antiviral drugs, the use of Rag/γ chain-deficient mice represents a more affordable and manageable alternative. It also permits the analysis of infection in a genetically defined background, and to investigate the role of a given host gene in knock-out models. Nonetheless, the main weakness associated with this mouse model is its limited resemblance with a natural infection in the host, e.g. only transient replication, no virus detected in feces, and lack of oral susceptibility. As an alternative surrogate approach, the use of mouse models for MNV has been suggested. The main advantage of MNV infection in mice is that it represents a cost-effective system for norovirus in its natural host (see below).

5.2.3 Murine Norovirus as a Surrogate System

The recently established methods to study HuNoV infection in cell culture and pathogenesis in mice (Sects. 5.2.1 and 5.2.2) represent a promising advancement in anti-noroviral research. These tools, however, may require further optimization to attain sufficiently robust viral replication that could be of use in antiviral studies. Owing to its close genetic relationship to HuNoV, many researchers employ MNV as surrogate system to examine the biology and genetics of noroviruses. There are several advantages associated with MNV such as the availability of highly efficient methods for the cultivation of the virus, and robust mouse models of infection in the natural host [57–60]. These models include both acute and persistent infection of MNV in mice that can be instrumental in the identification of antiviral compounds with specific efficacy in different scenarios [61–64]. Efficient reverse genetics tools are readily available for MNV permitting the identification of genetic determinants of infectivity and pathogenesis of noroviruses [38, 62, 65, 66]. The recent discovery of cellular protein family CD300 as the receptors for MNV has enabled investigations using this virus in a human context, by expressing the recombinant protein in human cell lines [17, 67–70]. These cellular systems can be relevant to the characterization of small compounds targeting human host factors required for norovirus replication.

In this manuscript, we provide a comprehensive review of progress made on the development of antivirals in the control of infection, transmission, and disease caused by these pathogens. Diverse molecules targeting different viral proteins and host cell factors, eliciting an inhibitory activity upon various steps of the life cycle (viral entry, internalization, RNA replication, translation, polyprotein self-processing, and other steps) are provided below (Tables 5.1 and 5.2).

5.3 Drugs Targeting the Viral Capsid

5.3.1 Molecules Inhibiting Capsid Binding to Cellular Receptors

5.3.1.1 Identification of Inhibitors Blocking the HBGA-Receptor Site in VP1

One of the most prolific areas in the identification of anti-norovirus drugs involves those compounds targeting the determinants of cellular receptor recognition in the viral particle (Fig. 5.2). Pioneering studies involving the screening of small molecule libraries led to the isolation of 14 compounds which inhibited binding to HBGA in a saliva carbohydrate-binding inhibition assay with median effective concentration (EC_{50}) values below 15 μM [71]. Some of these drugs displayed broad range activity against different HuNoV GI (Norwalk) and GII (MOH, VA387) strains that may be relevant to the design of effective anti-norovirus molecules.

Further studies involved the screening of molecules in silico to the identification of compounds predicted to inhibit the interaction between the HuNoV capsid and HBGA receptors [72, 73]. Docking simulations on the HuNoV capsid-HBGA receptor complex and virtual high throughput screening of >two million compounds led to the identification of >100 putative antiviral molecules. Several of these chemicals were further evaluated in assays measuring binding inhibition to HBGA trisaccharide molecules A and B in vitro [73, 74]. The top 20 best inhibitors showed a half maximal inhibitory concentration (IC_{50}) values below 40 μM. Many of these drugs identified above shared the presence of a cyclopenta [a] dimethyl phenanthrene scaffold. This particular subgroup of drugs exhibits good cell toxicity profiles encouraging further development of cyclopenta [a] dimethyl phenanthren derivatives as anti-norovirus drugs [73, 74]. In an unrelated in silico docking study of fragment-based virtual libraries, four molecules were identified with predicted high affinity for the HBGA-binding pocket in the viral capsid (compounds termed 23–26). These molecules can now be used as platforms in the development of optimized inhibitors targeting the interaction between the capsid and the receptor [75].

5.3.1.2 Fucose Mimetics

Structural and functional studies unravelled a critical role of fucose residues in the binding of HBGA to HuNoV capsid (Fig. 5.2b). Hence, the HuNoV capsid pocket implicated in the interaction with fucose constitutes an attractive target for antivirals. The residues in the viral capsid protein interacting with fucose have been identified (residue S380 in HuNoV GI; S343, S344, R345 and D374 in HuNoV GII.4) [76, 77]. For the identification of potential fucose-binding inhibitors, libraries of small molecule compounds were analyzed, using diverse ligand-based nuclear

Table 5.1 Antiviral compounds targeting norovirus proteins or viral RNA

Target	Type of compound	Compound name	Inhibitory activity observed	References
Viral capsid	HBGA-binding site inhibitors	Cyclopenta [a] dimethyl phenanthren-derivatives	Inhibit VLP binding to HBGAs in vitro	Ali et al. [73], Zhang et al. [74]
		Fucose mimetics	Inhibit VLP binding to substrate in vitro	Rademacher et al. [78]
		Natural and synthetic glycoconjugates	Inhibit VLP binding to HBGAs in vitro	Rocha-Pereira et al. [28], Han et al. [80], Shang et al. [81], Jiang et al. [82], Yazawa et al. [87]
		Citrate	Decreased infectivity and altered particle morphology	Koromyslova et al. [90]
	Heparan-sulphate analogues	Suramin	Reduced binding to intestinal cells	Tamura et al. [94]
		Heparin	Reduced binding to intestinal cells	Tamura et al. [94]
	Theaflavins	BFTC	Inhibits MNV infectivity in cell culture	Ohba et al. [108]
	Salicylic acid derivative	Bismuth subsalicylate	Reduces MNV infectivity	Pitz et al. [111]
	Antibodies	Nanobodies	Inhibition of particle binding to receptor in vitro	Koromyslova and Hansman [91], Garaicoechea et al. [123]
Viral protease NS6	Dipeptide inhibitors	Structure-based optimized peptide	Inhibition of HuNoV replicon in cell culture and MNV infection in the mouse model	Galasiti Kankamalage et al. [125]
	Tri-peptide inhibitors	Rupintrivir	Inhibition of HuNoV replication in cell culture	Rocha-Pereira et al. [138]
	Peptide macrocycles	Triazole- and oxidazole-based derivatives	Inhibition of HuNoV replication in cell culture	Damalanka et al. [130], Weerawarna et al. [132]

(continued)

Table 5.1 (continued)

Target	Type of compound	Compound name	Inhibitory activity observed	References
	Peptidomimetics	Oxazolidinone derivatives	Inhibition of HuNoV replication in cell culture	Damalanka et al. [142]
Viral polymerase NS7	Nucleoside analogues	Favipiravir	Antiviral activity in vivo against both HuNoV and MNV. Mechanism: Lethal mutagenesis. Inhibition of HuNoV NS7 polymerase in vitro	Jin et al. [149], Arias et al. [151], Ruis et al. [156]
		Ribavirin	Clearance of chronic norovirus infection in a subset of CVID patients	Woodward et al. [114], Woodward et al. [162]
		2′-C-methylcytidine	Inhibition of norovirus replication in mouse models of infection. Prophylactic protection in model of virus transmission.	Kolawole et al. [44], Rocha-Pereira et al. [158], Rocha-Pereira et al. [159]
		2′-F-2′-methylcytidine	Inhibit replication of MNV and HuNoV-replicon in cell culture	Costantini et al. [161]
		B-D-N (4) hydroxycytidine	Inhibit replication of MNV and HuNoV-replicon in cell culture	Costantini et al. [161]
		Neplanocin A derivatives	Inhibition of norovirus replication in cell culture	Ye and Schneller [174], Liu et al. [175]

	Non-nucleoside analogues	Suramin and other napthtalene derivatives	In vitro inhibition of HuNoV and MNV polymerase activity	Mastrangelo et al. [98], Mastrangelo et al. [180], Tarantino et al. [181], Croci et al. [183]
		Chromone derivatives	MNV inhibition in cell culture	Rocha-Pereira et al. [186]
		NIC02 (Phenylthiazole carboxamide)	Norovirus inhibition in cell culture and NS7 in vitro	Eltahla et al. [191]
		JTK-109	Inhibition of NS7 in vitro and MNV in cell culture	Netzler et al. [185]
		TMC-647055	Inhibition of NS7 in vitro	Netzler et al. [185]
		Beclabuvir	Inhibition of NS7 in vitro	Netzler et al. [185]
		NIC04 (Pyrazole acetamide)	Norovirus inhibition in cell culture and NS7 in vitro	Eltahla et al. [191]
		NIC10 (Triazole)	Norovirus inhibition in cell culture and NS7 in vitro	Eltahla et al. [191]
		NIC12 (Pyrazolidinedione)	Norovirus inhibition in cell culture and NS7 in vitro	Eltahla et al. [191]
Viral RNA	RNA binding molecules	Phosphorodiamidiate morpholino oligomers	Inhibition of norovirus replication in cell culture	Bok et al. [188]

Table 5.2 Host-directed drugs exhibiting antinoroviral activity

Target	Type of compound	Compound name	Inhibitory activity observed	References
Acyl-coenzyme A: cholesterol acyltransferase	Pyranobenzopyrones derivatives	Cl-976	Inhibits HuNoV-replicon	Chang [27]
		YIC-08-434	Inhibits HuNoV-replicon	Chang [27]
		Sandoz 58-035	Inhibits HuNoV-replicon	Chang [27]
		Quinolylmethyl derivatives	Inhibits HuNoV-replicon	Pokhrel et al. [206]
Acid sphingomyelinase	Cationic amphiphilic drugs	AY9944	Inhibition of MNV replication in cell culture	Shivanna et al. [211]
		Fluoxetine	Inhibition of MNV replication in cell culture	Shivanna et al. [211]
		Desipramine	Inhibition of MNV replication in cell culture	Shivanna et al. [211]
		Chlorpromazine	Inhibition of MNV replication in cell culture	Shivanna et al. [211]
Eukaryotic initiation factor-4A (eIF4A)	Natural compound obtained from coral	Hippuristanol	Inhibits MNV infection in cell culture (translation inhibition)	Chaudhry et al. [220]
Hsp90 chaperone	Geldanamycinm derivatives	17-DMAG	Inhibits MNV infection in mouse model	Vashist et al. [192]
Cellular deubiquitinases	Deubiquitinase inhibitors	WP1130	Inhibits MNV replication in the mouse model	Perry et al. [60]
		Several 2-cyano-3-acrylamide derivatives	Inhibits HuNoV-replicon and MNV replication in cell culture	Passalacqua et al. [200], Gonzalez-Hernandez et al. [202], Charbonneau et al. [205]

Modulating immunity	Interferons (IFNs)	IFN-α	Inhibits norovirus replication in cell culture	Chang and George [225], Changotra et al. [226]
		IFN-γ	Cures an MNV persistent infection in vivo	Nice et al. [58]
	Calcineurin targets	Cyclosporine	Inhibits HuNoV-replicon	Dang et al. [214]
		Tacrolimus	Inhibits HuNoV-replicon	Dang et al. [214]
	IMPDH inhibitors	Mycophenolic acid	Inhibits HuNoV-replicon	Dang et al. [214]
		Ribavirin	Inhibits HuNoV-replicon and MNV replication in cell culture	Arias et al. [151], Dang et al. [214]
Unknown target	Thiazolides	Nitazoxanide	Antiviral activity in vitro and in vivo	Rossignol [234], Dang et al. [238]

magnetic resonance (NMR) binding methods. These procedures led to the isolation of 26 competitive inhibitors of the fucose-binding site and four non-competitive ligands interacting with secondary sites in the proximity of the binding pocket [78]. This secondary site is <5 Å of distance to the main binding site although the precise location has not been yet identified. There are nine putative secondary binding sites. One of the sites most frequently identified in docking experiments of HuNoV GII.4 capsid with fucose is formed by residues A346, H347, K348, D374 and S441. It has been proposed that new antiviral molecules containing these two binding moieties in the same structure, both in contact with the fucose-binding site and the secondary site, can help generating inhibitors with larger affinity for HuNoV particles [73]. Additional improvements based on this study led to the design of multivalent polymer entry-inhibitors that simultaneously interact with several fucose-binding pockets in a single viral particle. These polymeric structures exhibited a 10^3–10^6-fold increase in their affinity for the viral capsid when compared to monomeric forms which supports the development of multivalent inhibitors as possible antivirals or prophylactic substances in the control of infection and spread [73, 78]. In subsequent NMR studies, the same authors characterized the binding between the fucose-binding pocket and active conformations of synthetic HBGA fragments which can now be exploited in the rational design of optimized inhibitors [78, 79].

5.3.1.3 Human Milk and Other Natural Resources for the Identification of Antiviral Drugs

The screening of natural compounds rich in carbohydrates, such as human milk, has been proposed as an alternative strategy in the discovery of norovirus receptor-binding inhibitors [28, 80–82]. Surface plasma resonance and mass spectrometry were used for the identification of several milk oligosaccharides with predicted binding to the capsid [28, 80, 81]. Further analysis found that milk glycans namely 2′- and 3-fucosyllactose can block the interaction between prevalent GII.4 noroviruses and HBGAs, by direct competition for the same binding site in the capsid [73, 83]. 2′-fucosyllactose can also inhibit different genogroup I and II noroviruses, supporting its use as a broad spectrum drug against different viral strains [84]. These observations stimulate additional studies to evaluate the therapeutic value of these molecules and the development of improved glycomimetic inhibitors [85].

Milk carbohydrates have also been instrumental in the design of glycoconjugates eliciting VLP-binding inhibition activities in vitro [81]. To the characterization of binding inhibition, a surface plasmon resonance imaging (SPRi) approach has been developed which allows the massive screening of glycan libraries in microarrays, and hence the rapid identification of high-affinity substrates [81]. To characterize receptor-binding inhibition in different HuNoV genogroups, there have been generated *P* particles of both Norwalk (GI) and VA387 (GII) strains [81]. *P* particles are soluble oligomeric complexes of the VP1-protruding (P) domain (each particle contains 12 P-domain dimers). These particles show similar binding activity to HBGA receptors as natural HuNoV and VLP capsids, with the advantages of being generally more stable than VLPs and that can be easily produced in *E. coli* [86]. Using this technique, neoglycoproteins and oligosaccharide-glycine derivatives with an antiviral potential have been identified [81]. While neoglycoproteins containing several saccharide moieties efficiently inhibited both HuNoV GI and GII binding to HBGA receptors, soluble oligosaccharides showed efficacy only against GI particles [81].

Porcine and squid blood group substances have also been investigated as alternative sources to isolate natural glycans with potential inhibitory activity. Different saccharides obtained from these animals abrogated VLP binding to human gastric mucosa and saliva. It has been suggested that these molecules can now be used as scaffolds for the development of improved synthetic drugs [87]. Glycerol is an additional natural compound with capsid-binding activity. The structural analysis of HuNoVs capsids (GII.13 and GII.21) has revealed that glycerol is a natural competitor of the HBGA-binding site, supporting the possible development of inexpensive derivatives. The crystal structure information available on this viral-capsid-glycerol interaction can be exploited in the design of specific inhibitors against these new lineages of viruses [88]. In summary, naturally-sourced carbohydrates with affinity for HuNoV capsid constitute a promising strategy to develop inhibitors that specifically compete with natural HBGA receptors. The identification of some of these molecules in different natural products, such as milk, can facilitate the isolation of inexpensive drugs by minor modifications of the carbohydrate isolated, e.g. fucosyllactose.

5.3.1.4 Citrate-Based Inhibitors

Citrate is a natural mimic of fucose that competitively inhibits the binding of HBGA trisaccharides to the HuNoV capsid with an inhibitor constant (*K*i) of 600 μM [89]. Structural studies confirmed the presence of citrate residues in the norovirus VP1 capsid pocket for HBGA receptor molecules (Fig. 5.2) [89, 90]. Additional crystallographic analysis also identified citrate molecules when VLPs were soaked in *PureGreen24*, an antimicrobial disinfectant approved against norovirus by the Environmental Protection Agency [90]. *PureGreen24* formulation is recommended for the disinfection of hard surfaces to several infectious agents, including noroviruses. Hence, it can be used in the disinfection of self-contained facilities following an outbreak (e.g. hospital, nurseries, restaurants, and others). *PureGreen24* formulation contains silver dihydrogen citrate as the active compound, further suggesting that citrate molecules in *PureGreen*24 are responsible for eliminating norovirus infectivity. The binding to citrate dramatically affects the morphology of norovirus VLPs, leading to significant increases in the diameter of particles. As a consequence, the protruding domain of VP1 becomes more exposed in the outer surface of the capsid suggesting that it may also facilitate its recognition by neutralizing antibodies [90]. These studies suggested that this citrate-binding site in the P2 subdomain is directly connected by a stretch of residues with a 'trigger region' located in the P1 subdomain with conformational alterations of the trigger region leading to capsid integrity loss [90, 91]. All these studies suggest that compounds mimicking citrate can be exploited as virucidal agents in the control of infection or as disinfectants.

5.3.1.5 Drugs Targeting Other Cellular Co-receptors

There is increasing evidence supporting the participation of diverse cellular glycans in the attachment and entry of the norovirus particle into susceptible cells in addition to HBGAs. Specifically, it has been documented that the interaction of HuNoV with heparan-sulphate and sialic acid (e.g. sialyl Lewis X glycoprotein, gangliosides) is required for viral attachment [92–96]. Small compounds such as heparin or ganglioside derivatives are competitive inhibitors of these cellular membrane glycans [93, 94]. Different heparan sulphate-inhibitory drugs such as suramin and heparin abrogate VLPs (GII) binding to intestinal cells (Caco cells), suggesting that heparan sulphate plays a role in viral attachment [94]. Treatment with chlorate also inhibited the binding of particles to cells which might indicate that sulphate residues in heparin-sulphate receptors are critical for binding. Suramin is a drug clinically approved for the treatment of other diseases (e.g. typically used in the treatment of sleeping sickness caused by a protozoan, i.e. *Trypanosoma*) which may encourage its potential therapeutic use against norovirus [97]. In addition to its inhibitory activity on the attachment of the viral particle to the cell, this molecule also inhibits the noroviral polymerase activity as explained below (Sect. 5.5.2) [98].

Several members of the CD300 family have been identified as cellular receptors for MNV [69]. The expression of murine CD300 molecules in human cells, that are non-permissive for MNV, results in a productive infection following viral inoculation [67, 68]. There is no evidence supporting that the homologous version of murine CD300 in humans can act as a receptor for HuNoV; however, should this molecule function as a co-receptor for HuNoV it could be exploited as a therapeutic target (e.g. structure-based drugs targeting the interacting surfaces). The availability of a crystal structure of the interaction between MNV protruding domain and the soluble CD300lf receptor could be instrumental in the design of drugs targeting highly conserved pockets in either the viral capsid or the receptor [69, 70].

5.3.2 Virucidal Compounds and Other Molecules Binding to the Viral Capsid

5.3.2.1 Phenolic Based Inhibitors

Some phenolic compounds extracted from plants have shown antiviral potential against feline calicivirus (FCV). Owing to the close genetic relationship between this calicivirus and noroviruses, it has been suggested that these drugs could also have an antiviral activity against HuNoV infection. The limited solubility of these molecules, however, have discouraged their possible clinical use [99]. An elegant methodology to circumvent solubility issues consists in the enzymatic conjugation of phenolic molecules with rutinose. The rutinosylation of phenolic molecules can be attained through the reverse hydrolysis reaction catalyzed by the rutinase. The rutinose derivative of sinapic acid attains significantly larger inhibition of FCV replication in cell culture than the same non-rutinoside molecule [100]. Improved antiviral activities were also observed when other rutinosylated derivatives were evaluated [100]. Several evidences support that the mechanism of action of these phenolic drugs involves a virucidal activity. The incubation of viral samples in the presence of phenolic molecules, preceding inoculation to cells, resulted in drastic losses in infectivity. On the contrary, pre-treatment of cell cultures with the same drugs had no effect on viral infection yields. However, the ultimate molecular interactions and processes underlying this inhibitory activity remain unclear [100].

5.3.2.2 Egg Lysozyme

Heat-denatured egg lysozyme has exhibited an unexpected antiviral behavior against MNV during infections in cell culture [101]. The antimicrobial activities of enzymatically active lysozymes on Gram-positive bacteria are well established, although recent studies suggest additional unexpected properties associated with the inactive forms of lysozyme, for instance the inhibition of Gram-negative bacteria. Lysozyme anti-noroviral activity seems to be related to viral particle alterations triggered by this protein, leading to reduced viral infectivity. Specifically, the incubation with heat-inactivated lysozyme led to significant increases in the diameter of the viral particle (from 35 nm in untreated particles to 51 nm in those treated with lysozyme). MNV and HuNoV capsids treated with heated lysozyme also showed increased permeability, as suggested by experiments using propidium monoazide (PMA). PMA is a photo-inducible agent that degrades RNA. RNA is protected from degradation in intact capsids as they are non-permeable to PMA. In MNV and HuNoV preparations treated with heated lysozyme, a total degradation of viral RNA was observed, suggesting loss of capsid integrity. Inhibition assays using several synthetic peptides spanning the entire amino acid sequence have permitted the identification of lysozyme epitopes responsible for its antiviral activity. The region exhibiting greater inhibition comprises the residues 23–57 of the lysozyme molecule [101]. Even though additional investigations are required, these findings stimulate further studies to the development of antivirals based on lysozyme inactivating peptides.

5.3.2.3 DNA Aptamers Directed Against the Viral Capsid

Aptamers are DNA or RNA ligands with high affinity for a target molecule that are generated after in vitro evolution, in a process known as SELEX [102]. SELEX evolution of nucleic acids entails consecutive cycles of PCR amplification and selection for molecules with increased affinity for the target product. Amplification errors introduced during PCR (and transcription for RNA aptamers) step (s) contribute to the expansion of the repertoire of ligands, and thus lead to the rapid evolution and selection of aptamers with increased affinity. Some DNA aptamers have demonstrated efficient antiviral binding to the capsids of both NoV GI and GII [103]. Although these studies pursue the development of improved aptamer-based detection methods, the specific binding of these molecules to noroviral capsids could also be exploited for antiviral purposes. Those DNA aptamer sequences exhibiting higher affinity for viral particles can be used as a guide in the design of cost-effective compounds using structural data available [103–105].

5.3.2.4 Synbodies

Synbodies are synthetic protein-binding ligands that are generally formed by the conjugation of two peptides with affinity for a target element. The presence of two peptides in a single molecule typically results in increased affinity for their target factor. Synbodies against HuNoV GII.4 VLP have been generated and proposed for the highly sensitive detection of noroviruses in a biological sample [106]. Four different synbodies were tested in vitro, and they showed high affinity and cross-reactivity [all dissociation constants (K_D) are in the nanomolar range] against different norovirus genogroups and strains [106]. The method to isolate synbodies resembles to some degree the selection process (SELEX) for aptamers mentioned above. In this study, a large number of random 20-mer peptides (~10,000) were screened, followed by optimization of those molecules exhibiting higher affinity to HuNoV VLPs. Synbodies of larger complexity and enhanced binding can be generated by coupling two high-affinity peptides to a scaffold molecule. Although these molecules were originally developed with diagnostic purposes, the elevated affinity of synbodies for HuNoV capsid protein is stimulating their use in the control of disease [107]. A simple platform of 100 variant peptides has been successfully used to identify molecules efficiently binding to different pathogens, including influenza virus and rotavirus. Those synbodies binding to influenza virus in vitro exhibited antiviral activity in cell culture experiments [107], which constitutes a proof-of-principle for their potential use in the treatment of infection.

5.3.2.5 Miscellaneous Compounds with Virucidal Activity

Several other molecules eliciting virucidal activity have been identified. Theaflavins exhibit a potent antiviral activity against MNV and several other caliciviruses [108]. These molecules abolish infectivity in a mechanism preceding viral entry into the cell. It has been suggested that their antiviral activity is provided by the hydroxyl residues located in the benzocycloheptenone group [108]. A synthetic heterocyclic carboxamide named 5-Bromo-N-(6-fluorobenzo[d]thiazol-2-yl)thiophene-2-carboxamide (BFTC) also shows anti-noroviral activity in cell culture evinced by reduced cytopathic effect during MNV infection. BFTC was identified during the screening of >2000 small molecules from an in-house library. Some new improved derivatives were subsequently obtained by structural optimization [109]. The molecular targets of BFTC and its derivatives remain unknown, although it appears that they act upon an intracellular step of the viral life cycle.

Bismuth subsalicylate (BSS) is a broad antimicrobial molecule generally used as a safe and effective therapeutic treatment of traveller's diarrhea caused by different enteric pathogens including bacteria, viruses and protozoans [110]. Recent studies have demonstrated that BSS has a virucidal activity on MNV, encouraging new investigations to assess BSS as a prophylactic to control spread during norovirus outbreaks in self-contained premises [111].

Lactoferrin is a multifunctional milk protein that also displays an inhibitory activity on several steps of MNV life cycle including cell attachment and subsequent viral replication. The inhibition of viral replication is possibly related to the induction of host cell antiviral molecules such as interferon (IFN) α and β [112]. The molecular mechanism leading to reduced MNV attachment to the cell is not clear yet. Moreover, it is not known whether this inhibition involves a direct interaction between the viral particle and lactoferrin. If the latter is the case, the molecular interaction between lactoferrin and the virion could be exploited in the design of capsid-targeting drugs.

5.3.3 Antiviral Strategies Based on Antibodies

5.3.3.1 Immunotherapeutic Approaches

Passive immunotherapy with monoclonal antibodies (MAbs) has been proposed for the control of infection in vulnerable cohorts of patients such as children under five, the elderly and immunocompromised people. There is clinical evidence supporting the therapeutic efficacy of antibodies in the treatment of chronic noroviral infection in solid organ transplantation (SOT) patients [113, 114]. Two patients who developed norovirus infection during small intestine transplantation resolved infection after oral treatment for 48 h with human serum immunoglobulin [113]. The method was also highly efficient on lung transplant recipients with 10 out of 11 patients recovering from infection [114]. Phage display libraries derived from infected chimpanzees permitted the identification of fragment antigen-binding (Fab) molecules with high affinity for HuNoV particles, in the range of 1 nM. All Fabs tested recognized different conformational epitopes in the P domain of VP1 but not in the S domain [115]. The binding of Fabs named B7, G4, and F11 to VP1 did not inhibit binding to any other Fab, suggesting that they recognize distinct epitopes. In contrast, D8 and E5 are possibly sharing the same target epitope which is probably different from those recognized by other Fabs mentioned above. Additional analysis of G4 binding to the capsid has identified Gly 365 in the P2 subdomain as a critical residue for this Fab-capsid interaction. This residue is not involved in the binding to any other Fab.

Preparations of MAbs based on high affinity Fabs also exhibited high affinity for HuNoV VLPs in carbohydrate-binding and hemagglutination assays. Pre-incubation of HuNoV with these MAbs prevented the establishment of an infection in the chimpanzee model, which is encouraging for the use of these molecules in the treatment of disease [115]. Recent findings suggest that a highly conserved blockade epitope in the viral capsid can be the target of broad-range anti-norovirus MAbs [116]. MAbs directed against this conserved motif efficiently bound to a broad range of GII.4 strain capsids, further supporting this possibility [116]. The authors proposed that HuNoV capsid can adopt two conformational states: one in which the blockade epitopes are exposed but there is reduced binding to ligands, and the opposite scenario where there is reduced access of blockade residues to MAbs but increased binding to ligands. This blockade epitope has not been identified but VP1 residues 310, 316, 484, and 493 regulate the accessibility of Mab to this region. An exciting possibility would be that locking the VLP in an epitope-accessible conformation may prevent receptor binding. If this is proven true, drugs triggering locked

HuNoV capsid conformation could be developed as broad antinoroviral treatments, as it has been previously described for unrelated rhinovirus [115].

The limited stability of antibodies during their transit through the gastrointestinal tract may constitute a major hindrance to the implementation of oral immunotherapeutic approaches. Different strategies to mitigate antibody loss during gastrointestinal transit are being investigated such as antibody encapsulation or the use of decoy proteins to buffer proteolytic degradation. MAbs can also be conjugated with protective agents such as polyethyleneglycol that can mask the cleavage sites recognized by the gastric proteases [117]. Further advances include strategies to reduce costs associated with antibody synthesis such as methods to scale up their production, such as methods based on the generation of chicken antibodies [118, 119]. Elevated antibody titres were detected for more than 3 months in the eggs laid by chickens previously immunized with HuNoV P particle. Recent studies have also documented the development of dual antibody preparations against both rotavirus and norovirus which can be potentially used as broad therapeutics against viral diarrhea in humans [120].

5.3.3.2 Nanobodies

Nanobodies are recombinant, single-chain antibodies with high binding affinities for specific targets. The use of nanobodies constitute a methodological advance in the development of immunotherapeutic approaches. Camelid antibodies are only formed by two heavy chains with a single variable antigen-binding domain in each, namely variable heavy chain fragments (VHH). In contrast to human MAbs, single-chained VHHs retain their binding capacity to the antigen, and therefore many bioengineering possibilities based on single high-affinity biomolecules have emerged [121]. Antibodies isolated from alpacas immunized with HuNoV VLPs were used to prepare nanobodies that showed efficient binding to viral particles in vitro, with dissociation constants in the range of 3–30 nM [91]. Two of these nanobodies exhibited specific interaction with an occluded region of the P domain (Fig. 5.2c). Subsequent studies identified a novel set of nanobodies that target different P domain regions in VP1 (bottom, top and side). One of these new molecules (Nano-26) and one of those previously identified (Nano-85) caused complete capsid disassembly in vitro, in addition to suppression of VP1 binding to HBGAs [91, 122]. Both Nano-85 and Nano-26 showed affinity for a broad number of HuNoV genotypes, supporting their possible use as novel therapeutic agents [122].

In addition to alpacas, llamas have also been employed for the generation of nanobodies against noroviruses. A total of 18 llama-derived VHHs were generated against both GI and GII HuNoVs with affinity constants in the range of 0.1–40 nM [123]. All the llama's VHHs isolated, except one, recognized conformational epitopes in the P domain of VP1 as determined in several binding and hemagglutination assays in vitro [123]. Two residues in VP1 (positions 340 and 376) were associated with conformational recognition by one of these VHH antibodies. These positions are in the P2 subdomain relatively close to the HBGA binding site. It was also identified a linear epitope (residues 517–528) recognized by the only VHH antibody in the assay which showed no binding to conformational epitopes. This region is located in the base of the P domain far from the receptor binding site and in a highly conserved region in genogroup II viruses.

Antiviral therapies based on nanobodies show several advantages over strategies involving conventional antibodies. Nanobodies are typically more resistant to high temperatures and pH alterations, and also to protein folding modifications by chemical compounds [121, 124]. VHHs can also be genetically improved in vitro to engineer molecules with increased resistance to gastrointestinal contents [121]. Further improvement includes the development of multimeric structures containing several different nanobody molecules that could potentially lead to enhanced antiviral efficacy [121]. These findings stimulate additional studies to better characterize nanobodies as possible therapeutic tools or to the development of structure-based inhibitors.

5.4 The Noroviral Protease NS6: a Major Target for Antiviral Compounds

The proteolytic processing of the viral polyprotein is a critical step for virus replication (Fig. 5.1). The viral protease NS6, also known as NS6pro, 3CL or 3C-like protein (from its similarity to picornavirus cysteine protease 3C), catalyses the proteolysis of ORF1 polyprotein, releasing all the mature non-structural proteins that participate in different intracellular steps of the viral cycle (Fig. 5.1). Hence, HuNoV NS6 has become a major and prolific antiviral target [125, 126]. HuNoV NS6 inhibitors encompass a broad variety of dipeptidyl and tripeptidyl-based compounds such as peptidyl α-ketoamides, α-ketoheterocycles, dipeptidyl α-hydroxyphosphonate transition state mimics and macrocyclic inhibitors, among others [127]. The availability of in vitro enzymatic assays and the resolution of the three-dimensional structure of HuNoV NS6pro allowed a detailed characterization of the critical interactions and molecular rearrangements regulating peptide-substrate recognition and catalytic cleavage. Cys139 drives the nucleophilic attack on the peptidyl substrate [128]. Residues surrounding the target polyprotein cleavage sites, the amino acids spanning positions P4 to P2′ (P4–P2′), control norovirus protease enzyme efficiency [129]. Residues P1 to P4 are accommodated respectively in binding pockets S1 to S4. The majority of the interactions take place at the S1 and S2 pockets. The S1 pocket is comprised by residues T134, H157 and A160 which interact with a conserved Glu residue at P1 position [128]. P1 is further stabilized by other hydrophobic interactions with I135, and P136, and A158. These interactions allow the correct positioning of P1 residue to facilitate the nucleophilic attack by C139. The S2 pocket, formed by amino acids I109, Q110, R112, and V114, surrounds a variable hydrophobic residue at position P2. The S3 pocket is not very well defined while the S4 pocket includes residues M107, R108, I109, T166, and V168 [128]. The availability of these structural data has contributed to the significant advances in the development of norovirus drugs with improved specificity, including some compounds tested in vivo [125, 130–132]. Further studies of NS6 protease combined with different peptidyl substrates have led to the identification of the residues regulating the cleavage order during polyprotein self-processing that can be leading to the rational design of better inhibitors [128, 129].

5.4.1 Di- and Tripeptidyl Inhibitors

A vast collection of dipeptidyl and tripeptidyl competitive inhibitors were generated based on natural protein substrates of NS6pro. These compounds can be subdivided into transition-state (TS) inhibitors, TS latent drugs and TS mimics. Several TS inhibitors have been proven efficient against a broad range of 3C-like viral proteases including those from noroviruses, picornaviruses, and coronaviruses [133]. Their mechanism of inhibition involves the irreversible covalent binding to the nucleophilic cysteine in the active site (Fig. 5.3a), which results in the complete inactivation of the proteolytic activity [134]. TS inhibitors are formed by a peptidyl molecule, based on sequences recognized by the viral protease, conjugated with a warhead residue e.g. aldehydes, ketones, esters, bisulphites [85, 126].

Despite their optimal activities in vitro, the therapeutic value of TS inhibitors is deterred by low pharmacokinetics and oral bioavailability. Modification of TS inhibitors to generate bisulphite adducts (e.g. peptidyl aldehydes and α-ketoamides) leads to latent derivatives which become active in the gastrointestinal tract and blood plasma [135]. To further optimise TS inhibitors, novel ester and carbamate prodrug forms of bisulphite adducts were synthesized [136]. The derivatives showed reduced toxicity and improved pharmacokinetics and physicochemical properties in different physiological environments, such as human serum [136]. Further supporting their potential value, these improved TS inhibitors elicited strong inhibition in vitro (0.1–10 μM) and in cell culture (0.01–1 μM).

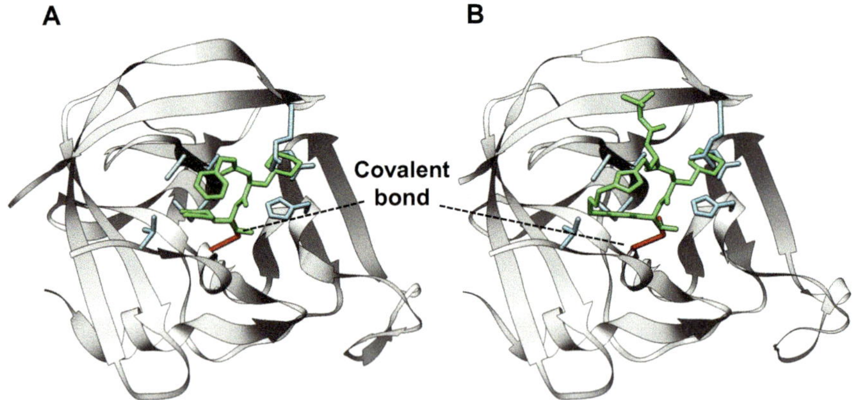

Fig. 5.3 Crystal structure of human norovirus protease NS6 in complex with two different peptidyl inhibitors. The peptidyl drug is represented in green and different residues in the catalytic site of NS6 are shown in light blue. Both compounds inhibit the protease by their irreversible covalent binding to catalytic cysteine 139 (represented in red). (**a**) dipeptide inhibitor; (**b**) cell-permeable macrocyclic inhibitor

TS mimics constitute another class of protease inhibitors where the peptidyl recognition element is coupled with a moiety resembling the transition state of the reaction. Their inhibitory activity is due to the irreversible binding of the moiety to the protease in a non-covalent manner [126]. The structural analysis of NS6 in complex with diverse dipeptidyl inhibitors was instrumental to the rational design of optimized compounds with improved activities. One of these new dipeptidyl molecules with high inhibitory activities in cell culture (IC_{50} of 0.02 μM) and in vitro (IC_{50} of 0.1 μM) also showed a significant reduction in MNV titres during infections in vivo, encouraging additional studies in this direction [125].

Tripeptide inhibitors were also successfully tested in the control of norovirus infection with some of them exhibiting potent inhibitory activity both in enzymatic assays in vitro and during infections in cells [137]. Though, a major limitation to the therapeutic use of tripeptidyl molecules is the large polar surfaces of these molecules which can compromise membrane permeability and therefore their efficacy at the clinical level [137]. Among these molecules, rupintrivir is a promising tripeptidyl compound which exhibits antinoroviral activity in cell culture tested clinically safe in humans [126, 138]. Rupintrivir showed efficient inhibition of norovirus replication in cell culture (EC_{50} is 0.3 μM), leading to the complete clearance of HuNoV RNA in replicon-bearing cells [138]. Rupintrivir was originally identified as an irreversible inhibitor of enterovirus 3C proteases. The compound was then tested in Phase I and Phase II trials to the treatment of human rhinovirus infection but no additional studies were performed due to unsatisfactory results [139, 140]. It remains however to be investigated its possible value in the treatment of noroviral disease in vivo. Its poor oral bioavailability suggests that this is not an optimal therapeutic candidate for the clinical treatment of infection, although the promising data obtained in cell culture encourage additional studies to the rational optimization of rupintrivir [138]. A recent study has revealed the rapid selection of resistance mutations in HuNoV NS6pro when replicon-bearing cells are cultivated in the presence of rupintrivir [141]. These findings cast reasonable doubt on the efficacy of therapeutic drugs based on protease inhibitors.

5.4.2 Peptide Macrocyclisation

The major disadvantages of peptidyl-based antivirals are their limited conformation stability, low cellular membrane permeability, and elevated sensitivity to proteolytic degradation. A recent improvement in the design of protease peptidyl drugs involves macrocyclization of the inhibitors (Fig. 5.3b). Peptide macrocycles typically present reduced flexibility leading to enhanced conformation stability and increased resistance to proteolytic degradation [126]. This increased molecular stability is also predicted to facilitate the correct positioning of the inhibitory moiety in the active site which generally leads to improved efficacy. A new series of HuNoV triazole- and oxadiazole-based macrocycles were recently identified [130, 132]. These molecules displayed increased cellular membrane permeability and conformation stability which might facilitate their practical use as antivirals [85, 130,

132]. Preliminary studies in cell culture showed satisfactory results with some of these compounds displaying inhibitory activities against the norovirus replicon, and the best five compounds showing EC_{50} values in the 2–10 μM range [130].

5.4.3 Peptidomimetic Inhibitors

In addition to peptide macrocyclization, an alternative approach to generate depeptidized inhibitors involves using peptidomimetic compounds. A series of peptodmimetics containing a functional heterocyclic ring (oxazolidinone) have been tested in vitro and in cell culture (replicon). These molecules showed efficient antiviral activity in these systems with inhibitory constants IC_{50} and EC_{50} in the micromolar range [142].

Altogether, a considerable number of molecules targeting norovirus NS6 have been developed and some of them show promising results in cell culture and in vivo models. Some of these molecules could possibly lead to the soon identification of therapeutic drugs. Owing to variations in peptide-substrate specificities amongst NS6 proteases from different HuNoV genogroups [143], further studies are needed to identify inhibitors with broader reactivity against different noroviruses.

5.5 Antiviral Drugs Targeting Norovirus Genome Replication

5.5.1 Viral RNA Replication as an Antiviral Target: Mutagens and Polymerase Chain-Terminators

Viral RNA synthesis, and specifically RdRps catalyzing it, constitutes a major target for the development of antiviral strategies. Several broad-spectrum antiviral nucleosides and nucleobases have been tested against noroviruses with promising results [57]. Nucleoside compounds are typically phosphorylated by cellular enzymes when entering into the cell (e.g. ribavirin). Nitrogen bases are generally converted into their nucleoside derivative that is then susceptible of being phosphorylated by the cellular machinery (e.g. 5-fluorouracil). The resulting nucleotide analogues can then compete with natural nucleotides for the active site of the virus polymerase. Even though the incorporation of these molecules into the viral RNA generally results in premature viral genome synthesis termination, there is a growing number of non-chain terminator nucleotides that exert their antiviral activity by increasing the mutation rates (Fig. 5.4). Antiviral strategies based on drugs that provoke increased virus mutation frequencies, also known as lethal mutagenesis of viruses, have been proposed as an alternative approach to conventional inhibitors [57, 144–146]. Lethal mutagenesis leads to a larger number of mutations randomly accumulated in the viral genome, which indirectly increases the possibility of inactivating viral enzymes and/or regulatory regions. The mechanism underlying mutagenesis typically involves the ambiguous base-pairing behaviour of these nucleotide analogues when they are incorporated into the nascent viral RNA or,

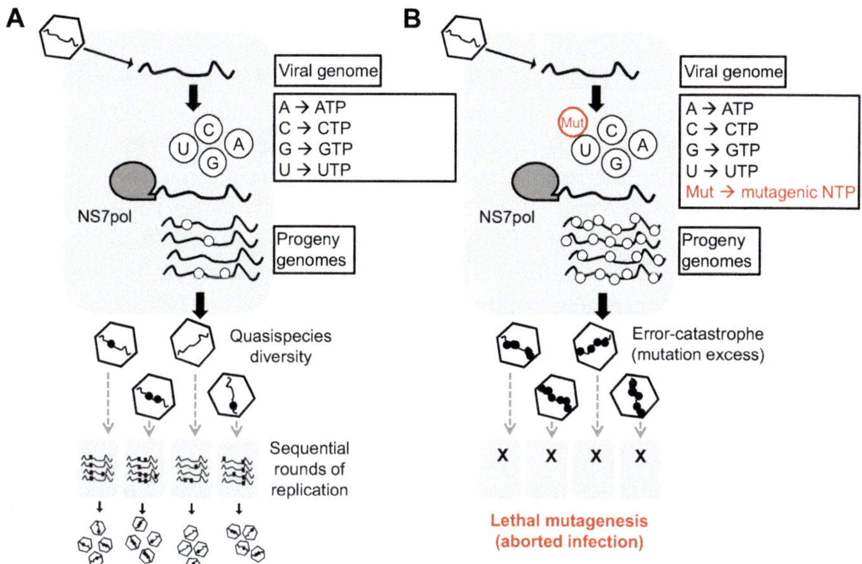

Fig. 5.4 Lethal mutagenesis of viruses. (**a**) Viral genome replication in the cell generally leads to a highly genetically diverse progeny that will be released to the extracellular environment where subsequent rounds of viral infection and replication will occur, leading to vast collections of genetically diverse populations, known as viral quasispecies. (**b**) Replication in the presence of a mutagenic nucleotide (Mut) leads to a non-tolerated number of mutations in newly synthesized genomes. The resulting populations show decreased or total lack of infectivity in new rounds of infection. The extinction of a virus population as a consequence of increased mutation rates elicited by mutagens is known as lethal mutagenesis of viruses

once incorporated in the viral RNA, when they are copied into new RNA strands [147–150]. Here we summarize several nucleoside analogues that can act as inhibitors or mutagens during norovirus replication.

5.5.1.1 Favipiravir

Despite extensive evidence from experiments in cell culture, there was insufficient data in vivo supporting lethal mutagenesis as a feasible therapeutic approach in the control of disease caused by RNA viruses. We demonstrated that favipiravir, a nucleoside analogue, (also known as T-705 or Avigan) can cure a persistent infection of MNV in vivo, and this activity was associated with increased mutation frequencies in virus populations isolated before complete viral clearance [151]. MNV isolated from treated mice also exhibited decreased viral fitness and specific infectivity relative to virus isolated from untreated animals. These are two general hallmarks of lethal mutagenesis of diverse viruses in cell culture [146, 151]. Our results supported that favipiravir-induced-mutagenesis was directly responsible for the loss of specific infectivity observed in MNV populations in vivo (Fig. 5.4). Favipiravir is a purine base analogue that is intracellularly converted into a nucleoside, and then metabolized into its triphosphate derivative. Several viral polymerases, including norovirus NS7, can recognize and incorporate FAV-triphosphate (FAV-TP) into newly synthesized RNA during activity assays performed in vitro [149, 150, 152]. FAV-TP is incorporated both opposite C and U in the template molecule, in agreement with increased transition rates observed during experiments in cell culture and in vivo with several viruses [149–151, 153–155]. Notwithstanding the incorporation of FAV-TP by NS7 strongly inhibits the incorporation of the next nucleotide, nascent RNA molecules can be successfully elongated to generate full-length products containing these mutagenic nucleotides. These results suggest that: (1) FAV nucleotides are incorporated during norovirus genome replication; (2) after incorporation FAV inhibits viral RNA elongation; (3) those viral RNA genomes that are elongated despite FAV-TP incorporation are prone to contain more mutations due to FAV ambiguous pairing behaviour [149, 151]. Overall, these data suggested that lethal mutagenesis by favipiravir could be applied in the control of HuNoV infection at the clinical level.

A recent study has examined the therapeutic value of favipiravir in the clinical treatment of infection [156]. A 48-year old patient with common variable immunodeficiency enteropathy (diarrhea, malabsorption, etc) and diagnosed with HuNoV infection responded positively to treatment with this drug (6000 mg on day 1 that were given in three doses, followed by 1200 mg twice a day). During treatment, the patient showed significant body weight increase and reduced norovirus levels in stools. The patient had previously not responded to high intravenous treatment with other antiviral agents generally used against norovirus infection (immunoglobulin, nitazoxanide, and ribavirin) further supporting that this compound is especially effective against noroviruses [156]. Unfortunately, due to a lack of drug availability, treatment had to be terminated and the patient died of a secondary lung infection. One major concern that has emerged during this case study is that treatment with favipiravir was associated with liver function affected during prolonged treatment periods. Overall, this constitutes a very promising result that encourages further investigation on this drug and potentially improved derivatives.

5.5.1.2 Valopicitabine (2′-C-methylcytidine)

Valopicitabine, a prodrug of the nucleoside 2′-C-methylcytidine (2CMC) is one of the most promising candidates for the treatment of norovirus infection [53]. 2CMC showed a strong activity against HuNoV replication both in the mouse model and B-cell culture infection systems recently developed with an EC_{50} value of 0.3 μM [44]. 2CMC was also effective in preventing the establishment of diarrhea symptoms and mortality in a mouse model of lethal infection for MNV [157]. Further analysis has proven this compound as an effective prophylactic agent, leading to complete protection in sentinel mice that were in contact with MNV-infected untreated animals [158]. These observations support using 2CMC as a prophylactic drug to control or to mitigate norovirus spread during nosocomial outbreaks in hospitals, a significant cause of concern in cohorts of vulnerable inpatients [53, 159]. The potential therapeutic value of 2CMC was demonstrated during clinical trials involving hepatitis C virus (HCV)-infected patients. Although, treatment with 2CMC led to significant decreases in HCV viraemia, its clinical use was put on hold by the *US*

Food & Drug Administration due to adverse side-effects reported in treated patients [160]. Several improved analogues of 2CMC have been developed [161]. Two of these derivatives [2′-F-2′-methylcytidine and B-D-N(4) hydroxycytidine] exhibit inhibitory activity on both MNV and HuNoV replicon (EC_{50} values in the range of 1–13 μM) which motivates additional investigations in vivo to assess their therapeutic potential [161].

5.5.1.3 Ribavirin

Ribavirin (RBV) has been used with relative success in Common Variable Immunodeficiency (CVID) patients chronically infected with norovirus. Prolonged treatment with RBV led to the elimination of viral shedding and the resolution of persistent diarrhea in a subset of treated patients. Specifically, in this study by Woodward and colleagues, eight individuals suffering from CVID (age average: 46 years) who tested positive for norovirus were enrolled on a 6-month trial. Of these, five patients were treated with RBV and the remaining three constituted the control group. At the end of the trial period, two out of these five patients completely cleared the infection. The treatment procedure consisted in an initial administration of 400 mg of RBV twice a day. This dosage was then increased as needed in each case until reaching 1500 ng/ml in serum. These results are encouraging to further investigate the possible therapeutic activity of RBV in other cohorts of immunocompromised individuals [114]. They also support further development of antivirals based on RBV, or drug cocktails including RBV and alternative inhibitors that could attain additive therapeutic efficacy in the treatment of chronic infection [114, 162].

RBV is a broad-range antiviral drug that has been successfully used in the control of several viral diseases. RBV displays several pleiotropic antiviral activities associated with alterations mediated upon different viral proteins (e.g. polymerase, capping enzymes, helicase-like protein), or cellular enzymes and metabolic routes (e.g. inhibition of inosine-5′-monophosphate dehydrogenase which regulates intracellular NTPs levels, and regulation of the immune response) [163–165]. In addition to these activities, ribavirin can cause lethal mutagenesis of different RNA viruses, including noroviruses, flaviviruses, picornaviruses, and other viruses [151, 166–168]. We and other investigators have recently demonstrated that ribavirin inhibits MNV replication in cell culture associated with increased mutagenesis of replicating virus genomes [151, 169]. It has not been yet determined whether the antiviral activity observed in CVID-treated patients is associated with increased mutagenesis or any other pleiotropic antiviral activities associated with this drug.

5.5.1.4 Uracil Analogues

Some uracil derivatives have been studied as potential inhibitors of norovirus NS7 polymerase, including 5-fluorouracil (5FU) and 2-thiouridine (2TU) [127]. The crystal structures of MNV polymerase bound to 5FU and 2TU revealed the presence of these two inhibitors in the catalytic site of NS7 [170, 171]. 5FU interacts with highly conserved residues in the active site of NS7 which supports that this molecule is recognized as a substrate during viral RNA synthesis [170]. Experiments in cell culture have demonstrated that both 5FU and 2TU inhibit norovirus replication

[38, 171]. MNV infection is inhibited by 2TU with larger efficacy than ribavirin, suggesting that this molecule can be considered for further studies in vivo [171]. We have recently described that an MNV variant containing a high fidelity polymerase shows lower sensitivity to 5FU [38]. This fact may be suggestive of an antiviral activity connected to mutagenesis as documented for other viruses [172]. 5FU is a drug typically used in cancer chemotherapy which may encourage its possible use in the treatment of severe chronic norovirus disease in immunocompromised patients [172, 173].

5.5.1.5 Other Nucleoside Analogues

Neplanocin, a carbocyclic nucleoside originally isolated from an actinomycete, and some enantiomeric derivatives exert broad-spectrum activity against different viruses (e.g., Ebola virus, human immunodeficiency virus, dengue, measles), including noroviruses. This supports their rational optimization in the development of anti-norovirus molecules [174–177]. The antiviral mechanism of action of neplanocins is not well understood, although it is probably related to its inhibitory activity on S-adenosylhomocysteine hydrolase which affects methylation reactions, and thus the maturation of viral RNA at the 5′-end [176]. Structural studies are also predicting that 5-nitrocytidine (5NC) may act as a competitive inhibitor of HuNoV [14, 178]. The analysis of NS7 in complex with 5NC-triphosphate revealed that this drug is occupying the catalytic site preventing the incorporation of natural nucleotides [14]. A related compound to 5NC is 2′-amino-2′-deoxycytidine. Its triphosphate derivative also interacts with the active site of the viral polymerase, stimulating additional investigations to clarify its potential antiviral activity [179].

5.5.2 Non-nucleoside Inhibitors of Viral Replication

5.5.2.1 Suramin and Naphthalene Derivatives

Suramin constitutes an attractive molecule to the development of antivirals given its inhibitory activities on different steps of norovirus life cycle. Apart from inhibiting binding to cell receptors (described earlier), suramin abrogates replication by specifically targeting the viral NS7 polymerase [98]. This molecule was originally identified as a putative inhibitor of NS7 during an in silico-docking screening using a virtual library of commercially available drugs. Despite that suramin is clinically used in the treatment of sleeping sickness [85], its use against norovirus seems unlikely due to its poor pharmacokinetics, low membrane permeability and elevated toxicity in vivo. To overcome membrane permeability issues, the efficacy of suramin encapsulated in liposomes was evaluated in cell culture. This approach led to significant increases in the antiviral efficacy of suramin, as liposomes facilitated the release of the drug in the cytoplasmic environment [180]. Several suramin derivatives with lower toxicity and improved permeability profiles were tested as anti-norovirus drugs [10, 98, 181, 182]. NF023, which was also identified during the in silico screening mentioned above, exerted an efficient inhibitory activity on norovirus RdRp. However, similarly to suramin, this compound has membrane permeability impediments that compromise its possible therapeutic value [98].

Pyridoxal-50-phosphate-6-(20-naphthylazo-60-nitro-40,80-disulfonate), namely PPNDS, is another related compound showing a strong inhibitory activity in vitro (IC_{50} < 1 μM) on both HuNoV and MNV polymerases [181, 183]. PPNDS is chemically related to both suramin and NF023, with all these drugs containing a naphthalene di-sulfonate (NAF2) head in the molecule structure. Despite their structural homology, PPNDS binds to a different site in the polymerase, named site B (Fig. 5.5), than suramin and NF023 which bind to a different pocket termed A [181, 182]. Structural studies suggest that two PPNDS molecules are binding to the thumb subdomain in NS7 polymerase in an antiparallel stacking conformation which resembles the positioning of RNA bases in the replication complex [181, 182]. This conformational arrangement is predicted to obstruct the viral RNA exit channel in the polymerase, and hence to block viral replication [183]. Since all these three inhibitors contain a NAF2 head it has been investigated whether the inhibitory activity is directly associated with this molecule [98, 181]. NAF2 alone exhibits an inhibitory activity with an IC_{50} of 14 μM. NAF2 contacts the same binding site in the viral polymerase as PPNDS [181]. Altogether these studies support the development

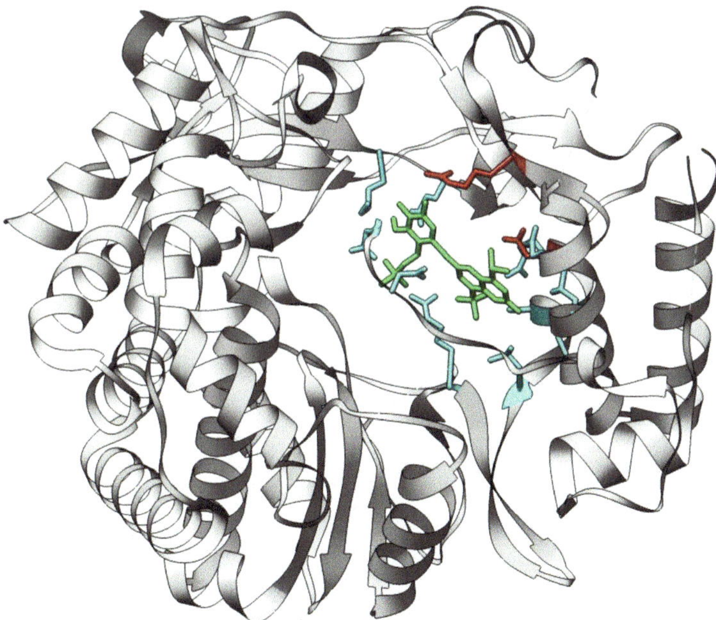

Fig. 5.5 PPNDS occupies Site-B in HuNoV polymerase NS7. PPNDS inhibitor is represented in green. Those residues in the viral polymerase interacting with the inhibitor are represented in light blue. Those highly conserved residues in Site-B which are particularly relevant to the interaction are represented in red (Q414 and R419)

of a new class of antiviral compounds targeting the norovirus polymerase based on the chemical structure of PPNDS and other NAF2 derivatives.

A recent study has led to the identification of a strong non-nucleoside inhibitor (IC_{50} of 5.6 μM) directed against site B of HuNoV polymerase [184]. The former was employed in in silico virtual screening methods to analyse drug-like compounds available (around 300,000 molecules) that were targeted against suramin- and PPNDS-binding sites in NS7. The analysis resulted in the identification of 62 molecules that were then characterized in vitro, leading to the identification of two strong antiviral compounds. These molecules were then employed in the design of derivatives, leading to the identification of this strong non-nucleoside compound abovementioned [184].

5.5.2.2 Broad-Spectrum Inhibitors Isolated for Other Viral Polymerases

An alternative tactic to the identification of norovirus antivirals involves the screening of inhibitors available for other RNA viruses [185]. In silico docking studies predicted that JTK-109, an HCV RdRp inhibitor, can bind norovirus polymerase sites A and B mentioned above [181, 182, 185]. In the same study, the authors demonstrated the ability of JTK-109 to inhibit norovirus replication in vitro and in cell culture (IC_{50} of 4–10 μM) and that this inhibition is antagonistic of PPNDS. This evidence supports that JTK-109 also targets the site B in norovirus polymerases (Fig. 5.5). In addition to JTK-109, two other non-nucleoside inhibitors acting on HCV also manifested inhibitory activity against HuNoV: TMC-647055 and Beclabuvir, with IC_{50} values in vitro ranging between 15–30 μM and 20–100 μM, respectively. These molecules are currently evaluated in clinical trials against HCV and, if successful, they could be considered for additional tests to the treatment of norovirus infection [185].

5.5.2.3 Styrylchromone Derivatives

A variety of 2-styrylchromone derivatives were shown to inhibit MNV in cell culture. Six molecules from a collection of 12 different 2-styrylchromone compounds inhibited replication with IC_{50} values <40 μM [186]. To this aim, a plaque reduction assay in RAW264.7 macrophage cells was used, involving the infection of cell monolayers followed by plaque development under semisolid media for 48 h [186]. Although the mechanism of action is not yet known, chromone derivatives are broad-spectrum inhibitors of nucleic acid replication which suggests that 2-styrylchromones may be acting against viral genome synthesis [186, 187].

5.5.2.4 Molecules Targeting the Viral RNA

The use of compounds directly targeting the viral genomic RNA such as small interfering RNAs (siRNAs) or phosphorodiamidate morpholino oligomers (PMOs) has also been investigated as alternatives to classic drugs against viral proteins. PMOs are synthetic DNA molecules that inhibit replication and translation by specific complementary annealing to a target sequence. PMOs directed against highly conserved regions in the norovirus genome abolished replication in cell culture [188]. The main difference between PMOs and natural DNA molecules is the presence of a morpholine ring in place of the deoxyribose sugar which improves molecule stability and increases the therapeutic activity. The limited solubility of PMOs, however, constitutes a major hindrance to their potential use for clinical purposes. PMOs solubility can be partially improved by conjugation with arginine-rich peptides, enhancing permeability and cellular uptake [188]. Several peptide-conjugated PMOs (PPMOs) tested in cell culture against MNV showed antiviral efficacy which further stimulated the development of optimized derivatives [64, 127, 188]. Additional studies in vivo are needed to determine the therapeutic potential of PPMOs in the control of norovirus disease.

An additional technique entails the intracellular delivery of a single-chain variable antibody fragment, termed 3D8 scFv, which has affinity for nucleic acids and the ability to hydrolyze them without any sequence specificity requirement [189]. It has been found that 3D8 scFv can directly penetrate into mammalian cells via a caveolae-lipid raft pathway, and once inside the cell target and degrade nucleic acids. This activity has stimulated further investigations to examine possible antiviral approaches based on its oral delivery to control infection in the gut. A proof-of-principle for this approach was obtained with MNV-infected mice; mice received orally *Lactobacillus* bacteria secreting soluble 3D8 scFv. resulting in significant reductions in the amounts of viral RNA detected in the feces of treated mice [190]. It was confirmed that 3D8 scFv, expressed by *Lactobacillus*, is internalized in intestinal epithelial cells, suggesting that viral replication inhibition is caused by direct degradation of MNV RNA in vivo [190].

5.5.2.5 High Throughput Screening for the Identification of Novel Inhibitors of Norovirus Polymerase

High-throughput screening approaches have been used in the identification of new small molecule compounds directed against the HuNoV genogroup II polymerase. The inhibitory activities of ~20,000 lead-like compounds were evaluated using fluorescent-based RNA synthesis methods. Four compounds showing larger inhibitory activities ($IC_{50} < 10\ \mu M$) were further analyzed in enzymatic assays and experiments in cell culture, using the HuNoV replicon (genogroup I) and MNV. To test the inhibitory activities in vitro, the authors used different noroviral polymerases recombinantly expressed, and evaluated how drugs affected their enzymatic activities (incorporation of GTP). One of these compounds, NIC02, exhibit a broad range of antiviral activity against different genogroup polymerases (GI, GII and GV) [191]. These inhibitors also led to reduced number of viral RNA copies in cell culture in assays involving Huh-7 cells containing the HuNoV replicon (GI) and also RAW264.7 cells infected with MNV. They also demonstrated that infected cells treated with these compounds develop reduced MNV plaque sizes in titration assays. These novel molecules may constitute potential scaffolds to the design of antinoroviral drugs with activity against the NS7 polymerase.

5.6 Antiviral Molecules Directed Against Host Factors Regulating Norovirus Replication

Noroviruses are dependent on a complex network of interactions with the host cell to complete infection. In particular, virus replication entails complex events in which multiple viral and host factors are engaged. An extensive number of cellular proteins interacting with the norovirus life cycle have been identified and characterized, with some of them being critical for infectivity [192]. Targeting host cellular proteins with small molecule inhibitors has been thus suggested as an alternative strategy to virus-directed drugs. A major advantage of compounds inhibiting cellular proteins is that they are less likely to select drug-resistant viruses as their target is not on the virus side. Additional advantages include potential broader activity against diverse genotypes and serotypes, and the existence of a larger number of protein targets than those found in the viral genome [193]. There are many examples of host-targeting drugs successfully used in the control of infection, with recent data supporting their use against the noroviruses [60, 192].

5.6.1 Cellular Chaperones as an Antinoroviral Target

Host cell heat shock protein 90 (Hsp90) assists to the correct folding of norovirus proteins after translation, and is specifically critical to the stability of VP1 [192]. This dependence on cellular chaperones to viral protein synthesis is not exclusive to noroviruses [194, 195]. Alvespimycin, also known as 17-dimethylaminoethylamino-17-demethoxygeldanamycin (17-DMAG), is a water soluble selective inhibitor of Hsp90. This compound strongly inhibits MNV replication in cell culture (IC_{50} of 60 nM), with reduced viral titres also observed during mouse infection in vivo. Mice intraperitoneally inoculated 1 h before infection with 17-DMAG showed reduced virus titres in the intestinal ileum [192].

Treatment with 17-DMAG leads to reduced synthesis of several viral proteins in MNV-infected cells including major capsid protein VP1, suggestive of a critical role for Hsp90 during the assembly of viral particles [192]. The transient expression of HuNoV VP1 in cell culture is also inhibited by 17-DMAG, evincing that in the absence of enzymatically active Hsp90 viral capsid proteins are incorrectly folded and rapidly degraded. Hsp90 was originally identified as a norovirus-associated factor during the screening of cellular proteins interacting with the 5′-viral genome extremity [196]. This would be in agreement with specific sequences in the viral genome recruiting Hsp90 to translating viral RNA to assist in the correct folding of viral proteins.

Hsp90 assists in the correct folding and stabilisation of many cellular proteins. The inhibition of Hsp90 affects various oncogenic and inflammatory pathways, leading to the degradation of their client proteins. Alvespimycin has been or is currently being tested in trials against a wide variety of cancers (breast cancer, cervical cancer, melanoma, leukaemia, lymphoma and others) with some results showing therapeutic promise [197]. Novel Hsp90 inhibitors exhibiting improved toxicity profiles have been recently identified which could also be considered for the treatment of HuNoV infection [198].

5.6.2 Deubiquitinases and Unfolded Protein Response

Cellular deubiquitinases (DUBs) are important actors during the course of norovirus infection. DUBs regulate cellular protein degradation pathways that are dependent on the ubiquitine-proteasome system. During infection, protein degradation is activated as a consequence of different stress responses, e.g. unfolded protein response. DUBs specifically reverse ubiquitination and the proteasomal fate of different target proteins. DUB functions have been described in regulating innate immunity. Viruses have evolved mechanisms to manipulate DUBs for their own benefit, and hence they constitute attractive targets for therapies [199].

WP1130, a DUB inhibitor also known as Degrasyn, inhibits norovirus replication both during infections in cell culture and in vivo [60, 200]. Treatment with WP1130 led to reduced titres in RAW264.7 cells and bone marrow derived macrophages that were infected with MNV. Treatment of HuNoV replicon-bearing Huh-7 cells rendered lower viral RNA levels as compared to untreated cultures, suggesting broad anti-norovirus activity associated with this compound. WP1130 also manifests its antiviral activity in vivo with significantly lower titres detected in the intestine (duodenum/jejunum) of MNV-infected mice treated with the drug. To identify the molecular target of WP1130, pull-down experiments of cellular factors interacting with biotinylated-WP1130 led to the isolation of a DUB named USP14. Specifically, uninfected cells were incubated with either biotinylated-WP1130 (WP) or an inactive biotinylated-analogue. The interacting proteins were isolated by affinity to streptavidin beads which permitted to identify a single distinct band associated with active WP1130, but not with the inactive analogue, which corresponded to USP14. This protein contributes to regulate proteasome degradation and modulate the unfolded protein response, a complex cellular mechanism triggered by different types of stress which include intracellular infection by pathogens [60, 201]. In addition to these studies, USP14 was previously isolated as a norovirus-associated factor during cellular protein pull downs using viral RNA probes, further demonstrating the connection of this protein with the viral life cycle [196].

Owing to their function as modulators of the host response to different types of stress, including infection by pathogens, DUBs have been proposed as possible broad-spectrum targets for antimicrobial drugs. Other RNA viruses and bacteria are also affected by treatment with WP1130 (i.e. viruses: La Crosse, vesicular stomatitis virus, encephalomyocarditis virus, Sindbis virus, Tulane viruses; bacteria: *Listeria monocytogenes*) [60, 202]. The intracellular proliferation of pathogens generally triggers autophagy in the host cell modulated by DUBs. Alterations in the control of ubiquitination have also been related to cancer and other human diseases, which has encouraged targeting DUBs with WP1130 and the development of new compounds with improved properties in cancer treatment [203, 204]. WP1130 shows limited solubility and bioavailability in vivo [60] which has stimulated investigations on the optimization of this compound [202, 205]. Initial screenings using a small library of WP1130-based compounds resulted in the identification of molecules eliciting improved antiviral activity against both HuNoV-replicon and MNV [202, 205]. Further modifications led to the discovery of 2-cyano-3-acrylamide-based compounds with lower toxicity in cell culture and improved stability in mice [200]. Their possible antiviral efficacy in vivo against norovirus remains to be tested.

5.6.3 Drugs Modulating Host Cell Lipid Metabolism to Control Norovirus Replication

Pioneering studies demonstrated that the expression of genes participating in the metabolism of cholesterol is significantly affected in cells carrying the HuNoV replicon [27]. Those genes required for the synthesis of cholesterol (e.g. hydroxymethylglutaryl-coenzyme A synthase, squalene epoxidase) were typically downregulated while those involved in cholesterol catabolism such as acyl-coenzyme A:cholesterol acyltransferase (ACAT) were upregulated [27]. These observations suggested that HuNoV replication is linked to reduced cellular levels of cholesterol. Further insight indicated that norovirus replication is dependent on the expression of low density lipoprotein receptor (LDLR); the expression of the latter is repressed by cholesterol levels [27]. The authors suggested that LDLR might be acting as a proviral factor in the viral replication complex. In agreement with this critical role for cholesterol in the norovirus life cycle, treatment with cholesterol-lowering drug simvastatin resulted in increased replication during infection in cell culture and in vivo, using the gnotobiotic pig model [25, 27]. Therefore, cholesterol-catabolic enzymes such as ACAT represent tantalising targets for potential anti-norovirus drug therapies. Treatment with several ACAT inhibitors (Cl-976, YIC-C8-434, and Sandoz 58-035) led to reduced HuNoV replication in replicon-containing cells (Huh-7) that were connected to reduced levels of LDLR [27]. Additional studies with a series of pyranobenzopyrone derivatives with specific anti-ACAT activity identified two promising drugs eliciting antiviral activities in cell culture with EC_{50} values ranging from 2 to 4 μM [127, 206].

In addition to its role in virus replication, cholesterol was suggested to participate in norovirus particle entry into the cell. Until recently and owing to the absence of cell culture systems, the process of HuNoV particle internalization remains poorly investigated. MNV enters the cell in a mechanism mediated by dynamin and cholesterol, and independent of clathrin and caveolae vesicles [207–209]. These among other evidences support that the MNV particle may be interacting with lipid rafts in the cellular membrane to reach the intracellular environment [209]. Lipid rafts are thought to be formed by dynamic ternary complexes of sphingomyelin, cholesterol, and ceramide with these molecules displacing each other [210, 211]. These complexes have been implicated in the regulation of different cellular processes such as the fluidity of the cellular membrane, traffic of proteins and the response to different types of stress [210, 211]. The enzymatic route for the synthesis of ceramide is critical to the formation of small rafts and therefore constitutes an attractive target to inhibit norovirus entry [211]. The specific downregulation of host cell acid sphingomyelinase (ASM), which is a ceramide-synthesis enzyme, results in reduced norovirus replication and the retention of MNV particles in the endosome, suggestive of ceramide participating in the internalization of noroviruses [211]. Different ASM drugs include cationic amphiphilic compounds such as fluoxetine, AY9944, desipramine, and chlorpromazine. Some of these molecules are clinically used to for the treatment of depression and therefore can be considered as conceivable compounds to be clinically used against norovirus. Further evidence supporting that ASM plays a major role in the process of viral entry comes from studies with genetically related porcine enteric calicivirus (PEC). The endosomal escape of PEC is facilitated when cells are treated with molecules that stimulate the synthesis of ceramide [212]. These results invite additional investigations to identify new drugs targeting cellular enzymes implicated in the viral internalization process. Although not related to the metabolism of lipids, other host cell factors contributing to endosomal function can be the target of anti-norovirus drugs. In particular, cysteine protease cathepsin L is a cellular protease that is needed for norovirus replication and thus could be the target of inhibitors [213].

5.6.4 Immunosuppressant Drugs Targeting Calcineurin and Inosine-5′-Monophosphate Dehydrogenase (IMPDH)

Since Solid Organ Transplant (SOT) patients need to be treated with immunosuppressive drugs to avoid organ rejection, a treatment strategy based on molecules exhibiting both immunosuppressive and anti-norovirus activities has been proposed [214]. Dang and colleagues tested the antiviral activity of several immunosuppressive molecules against norovirus in vitro, including both calcineurin inhibitors, such as cyclosporine and tacrolimus, and inosine-5′-monophosphate dehydrogenase (IMPDH) drugs, such as ribavirin and mycophenolic acid [214]. Cyclosporine is a natural immunosuppressant used in the treatment of autoimmune disease and in organ transplants to prevent rejection. Cyclosporine binds to the cellular protein cyclophilin, and this complex inhibits calcineurin, a factor which regulates interleukin-2 expression, leading to reduced T-cell function. Tacrolimus (known as FK506) binds to FKBP12 and the complex, similarly to cyclosporine-cyclophilin, inhibits calcineurin and T-cell activation. Treatment with different calcineurin inhibitors such as cyclosporine, voclosporin or tacrolimus led to reduced viral RNA levels in HuNoV-replicon containing cells, supporting their therapeutic use in SOT patients chronically infected with HuNoV.

Mycophenolic acid (MPA) is a potent inhibitor of IMPDH, a cellular target for immunosuppressive drugs in organ transplant patients. The inhibition of IMPDH by MPA causes drastic decreases in intracellular levels of GTP and dGTP nucleotides that indirectly lead to immunosuppression [214, 215]. MPA also exhibits antiviral activities against a broad range of viruses (norovirus, Lassa virus, HCV, dengue virus, coronavirus) which are possibly related to alterations in the intracellular nucleotide levels associated with IMPDH inhibition. In addition to other antiviral effects explained above (i.e. lethal mutagenesis), ribavirin is also a potent inhibitor of IMPDH [216, 217]. The combination of ribavirin with MPA and other calcineurin inhibitors (e.g. tacrolimus and cyclosporine) resulted in an additive inhibitory activity against norovirus replication [214]. This observation encourages the potential use of combination therapies based on multiple immunosuppressants with antiviral activity in the treatment of SOT infected patients. It has been suggested that the ribavirin broad-range antiviral activity is possibly related to the combined effects of imbalanced GTP levels and mutagenesis. Notably, several studies suggest that imbalances in the intracellular NTP levels observed in ribavirin-treated cells are boosting the mutagenic activity associated with this drug [215, 216].

5.6.5 Inhibitors of Viral Protein Translation

Viral protein translation is mediated by the VPg molecule covalently attached to the 5′ end of viral RNA genome and subgenome molecules [11, 12]. VPg redirects cellular ribosome translation machinery to viral RNA molecules by directly interacting with different eukaryotic initiation factors of translation or eIFs (Fig. 5.1). HuNoV VPg interacts with various members of the eIF4F complex and with eIF3 which are involved in the recruitment of the small subunit of the ribosome [218, 219]. MNV translation initiation entails the contact of VPg with eIF4G, an important scaffold protein of the eIF4F complex that also binds other factors associated with translation such as eIF3 [11, 12].

Since norovirus translation requires a different array of interactions than host cellular RNAs, this step was proposed as a conceivable target for specific antiviral strategies. A natural inhibitor of norovirus translation is hippuristanol. This molecule, which is extracted from coral *Isis hippuris*, specifically targets cellular RNA helicase eIF4A which forms part of the eIF4F complex. Treatment with hippuristanol in cell culture resulted in decreased MNV and FCV progeny production providing a proof-of-principle for the development of antivirals targeting norovirus translation [220]. Hippuristanol has been recently used to control tumor growth in mice, thus supporting its possible clinical use for other purposes. However, its poor solubility and limited potency advised against this possibility [220–222]. Several derivatives together with novel eIF4A inhibitors, such as silvestrol and pateamine A derivatives, constitute plausible compounds deserving further investigation [222, 223].

5.6.6 Interferon Treatment

Interferons (IFNs) are a class of peptides generated by the host organism in response to infection. IFNs are generally used in the clinical treatment of disease caused by different infectious agents [224]. Pegylated IFN-α, alone or in combination with ribavirin, has been extensively used in the clinical treatment of other viruses such as

HCV which underpins their use against norovirus [224]. Type I (α and β), type II (γ) and type III (λ) IFNs all contribute to the control of norovirus during infection in the host [42, 57, 225–227]. Type I IFNs appear to control the systemic spread of MNV in mice, while IFN-λ is relevant to the control and elimination of persistently replicating virus in intestinal epithelial cells which are thought to be acting as a reservoir in vivo [58, 228, 229]. Additional evidence suggests that the establishment of persistence may be associated with alterations in IFN-λ–dependent immunity triggered by the commensal microbiota [230]. The treatment of both immunocompetent mice and mice deficient in IFN α receptor (i.e., *Ifnar −/−*) with IFN-λ led to the complete elimination of a persistent MNV infection which stimulates the implementation of therapies based on IFNs [58]. Additional data have demonstrated that IFN-λ specifically protect uninfected mice against transmission from MNV-infected mice. Sentinel AG129 mice previously inoculated with an IFN-λ expression plasmid exhibited reduced susceptibility to infection transmitted by littermates [231].

IFN-containing nanogels based on cross-linked polyethyleneimine-polyethylenglycol (PEG), which can be used in the oral treatment of patients, have elicited antiviral activity against MNV and HuNoV replicon in cell culture [232]. Recent clinical studies have shown however limited efficacy of IFN-α against chronic norovirus infection. CVID patients that did not respond to ribavirin monotherapy and were subsequently treated with ribavirin and IFN-α remained unresponsive [162, 233]. Additional studies involving treatment-naïve patients would be desirable to clarify whether IFN-α displays any activity in vivo.

5.7 Norovirus Inhibitors with Unknown Targets

5.7.1 Nitazoxanide

Nitazoxanide has arisen as a highly promising compound in the clinical treatment of norovirus. Nitazoxanide is a broad-range antimicrobial compound that was originally identified in the control of protozoal infection [234]. It has been suggested that its therapeutic activities could be partly associated with the stimulation of interferon production [234]. Treatment with this molecule led to significant antiviral activities in norovirus-infected patients, including decreased severity and duration of disease [234]. Specifically, its therapeutic activity was demonstrated in a phase 2 randomized, double-blind, placebo-controlled study involving 50 individuals of at least 12 years of age with gastroenteritis caused by rotavirus or norovirus. The patients were treated twice a day with 500 mg of nitazoxanide, which resulted in a significant reduction of the clinical symptoms for both types of viral infections. Further evidence supporting the therapeutic use of this drug came from a clinical case report where chronic infection was cured in a patient suffering from leukaemia [235]. Nonetheless, a different case study has cast doubt on the antiviral efficacy of nitazoxanide, as it failed to cure chronic norovirus infection in an immunodeficient patient suffering from X-linked agammaglobulinemia [236, 237]. The administration of nitazoxanide (500 mg twice a day) during 12 months showed no antiviral efficacy, and thus treatment was discontinued. These apparently contradictory observations in two different clinical cases may convey that the efficacy of this drug is both virus and host specific, and possibly modulated by the immunological status of each patient [236].

Recent studies have demonstrated direct antiviral activity elicited by nitozaxonide against HuNoV replicon in cell culture [238]. MNV seems however unaffected by treatment with nitazoxanide suggesting that murine and human strains have different requirement for the factors targeted by this drug. While the mechanism of action is not fully understood, it is suggested, similarly to other thiazolide compounds, that nitazoxanide targets different host-regulated processes. A recent investigation suggests that nitazoxanide inhibits HuNoV replication by stimulating a subset of interferon-stimulated genes, including interferon regulatory factor 1, namely IRF-1 [239]. Further evidence was obtained from overexpression of IRF-1, which inhibited replication, and knockdown of IRF-1, which reduced the sensitivity of HuNoV replicon to nitazoxanide [239]. Nitazoxanide limited solubility and short elimination half-life have prompted investigations on the identification of other related thiazolides as possible alternatives to the treatment of viral infections [234, 240].

5.7.2 Other Compounds

Other antinoroviral drugs with unknown mechanisms of action include cyclic and acyclic sulfamides, and piperazine derivatives. The viral or cell host targets for these compounds also remain to be elucidated [126, 241–245]. The antiviral properties of cyclic sulfamides were identified as part of a program to identify anti-norovirus drugs using an in-house library (~2000 compounds). To this aim, an array of peptidomimetics were evaluated using the HuNoV-replicon system which resulted in the identification of several hits for molecules having in common a cyclosulfamide scaffold [242]. The initial hit compound, with an ED_{50} of 4 μM, has been further derived to generate novel drugs. Further studies demonstrated that acyclic sulfamide drugs were also effective against noroviruses [242].

A scaffold hopping strategy was used to identify novel cyclosulfamide derivatives with anti-norovirus activity [241]. These investigations led to the identification of a cyclosulfamide-based piperazine with a potent activity. The piperazine scaffold is a flexible structure that can bind to multiple receptors with high affinity. This is possibly related to the fact that this structural motif is repeatedly found in many biologically active molecules. Based on this result, a series of piperazine derivatives were designed that showed inhibition of HuNoV-replicon in the micromolar range [244].

An oxysterol compound named 25-hydroxycholesterol has also been found to elicit antiviral activity against MNV during infection in RAW264.7 cells [246]. This activity seems to be specific since other related molecules, such as cholesterol or another oxysterol compound named 22-S-hydroxycholesterol, did not affect replication. Although the mechanism of action is not yet known, 25-hydroxycholesterol led to enhanced apoptosis mediated by caspase 3/7 stimulation.

5.8 Final Remarks and Future Perspectives on the Development of Antinoroviral Strategies

5.8.1 Recent Developments in the Treatment of Disease and Prospective Therapies

Despite the absence of licenced drugs for the treatment of norovirus infection, a major burden to global public health, recent advances in the development of antivirals against these agents are promising [2]. Current strategies for the development of antivirals are not only aiming at mitigating the duration and severity of gastroenteritis, but also procuring prophylaxis to vulnerable patients and health care staff during outbreaks in hospitals and other semi-closed settings. In addition to antivirals to control acute infection, there is also a growing need to design strategies to control chronic infection in the immunocompromised.

Recent progress in the field of norovirus antivirals includes the identification of compounds exhibiting efficacy in vivo, as evinced in several clinical trials and in studies involving animal models of infection. Several nucleoside analogues such as favipiravir (T-705), ribavirin, and 2CMC, have shown relative success in vivo inspiring further investigations on their possible clinical value or potential derivatives with improved properties [151, 156, 158, 162]. The antiviral activity displayed by favipiravir seems to be associated with lethal mutagenesis, a process where a virus population is directed to extinction by increased mutational loads induced with mutagenic compounds [146, 151]. Favipiravir and ribavirin are considered safe mutagens as they are not utilised by the cellular DNA polymerases and therefore their activity is limited to the viral genome [152, 247, 248]. The poor solubility and oral bioavailability of favipiravir constitutes a caveat for its therapeutic use, although several clinical trials in the treatment of influenza virus and Ebola virus disease, and a clinical case study on a norovirus chronically-infected patient encourage their possible use in therapies [156, 249, 250]. New derivatives of this compound with improved properties could lead to the identification of highly efficient anti-noroviral drugs [151, 251]. 2CMC is another nucleoside which exhibits a potent inhibitory activity in vivo although some toxicity associated during trials on HCV-infected patients prevented from further clinical studies against viral disease. Derivatives of 2CMC with reduced toxicity are currently being investigated which may lead to further investigation on the noroviruses [160, 161, 252, 253].

In addition to nucleoside analogues, several protease inhibitors have also been synthesized and constitute new prospective therapeutic candidates. Rupintrivir is an ample spectrum inhibitor tested clinically safe that shows anti-noroviral activity in cell culture. Other norovirus protease inhibitors recently developed had shown positive activity in mice, supporting the use of molecules against the viral protease as a plausible antiviral strategy.

5.8.2 Combination Therapies in the Control of Noroviral Infection

Multi-drug therapies have been typically proven to be more effective in the control of viral infections as compared to monotherapy regimes [254–257]. Prospective anti-norovirus drug cocktails could benefit from including mutagenic nucleosides such as ribavirin and favipiravir, with evidence suggesting that combinations including mutagens are more effective than multi-drug therapies based only on non-mutagenic drugs [166, 254]. Woodward and colleagues evaluated a combination of ribavirin and pegylated-interferon in patients that were not responding to monotherapy with ribavirin [114, 162]. The efficacy of this cocktail, however, remains to be tested on treatment-naïve patients, and therefore it cannot be excluded that this combination could lead to enhanced antiviral activities. Additional data have demonstrated that combinations of rupintrivir with favipiravir or 2CMC exhibit an additive activity against Norwalk virus replicon which stimulates further investigations in this direction [138]. Ribavirin combined with MPA or calcineurin inhibitors also displayed an additive inhibitory activity, encouraging the potential use of immunosuppressive molecule cocktails in the treatment of SOT-infected patients [214].

Owing to the lower risk for viral resistance, host factors are regarded as safer targets for multi-drug therapies involving combinations with mutagens. Noroviruses as other obligate intracellular parasites require the interaction with different cellular proteins to replicate their genomes and complete the life cycle. Some promising host factors that can be considered therapeutic targets in the control of norovirus infection include cellular DUBs and chaperones, with evidence demonstrating the strong antiviral efficacy of inhibitors against these proteins during infection in cell culture and in vivo [60, 192]. The combination of these molecules together with viral RNA synthesis inhibitors, such as 2CMC, or mutagens, such as favipiravir and ribavirin, may be pondered as a potential strategy to control disease. The use of interferons also constitutes a conceivable alternative in the development of multidrug therapies such as those aforementioned in the treatment of HCV. IFN-λ led to the complete clearance of a persistent infection of norovirus in mice which should be considered in future clinical studies. Some other compounds such as nitazoxanide also show promise in the treatment of chronic disease, although additional investigations are needed to clarify its clinical efficacy [234, 236, 237].

5.8.3 New Prospective Host Targets for the Design of Antiviral Compounds

5.8.3.1 Targeting Host Cell Factors Contributing to Viral Genome Replication and Translation

Several cellular factors associated with the metabolism of RNA such as poly(rC)-binding protein 2 (PCBP2), heterogeneous nuclear ribonucleoprotein A1 (hnRNP A1), and ATP-dependent RNA helicase DDX3 are essential for norovirus replication [196, 258]. Some of these proteins have been isolated in co-immunoprecipitation experiments as factors interacting with genomic and subgenomic noroviral RNA and may be necessary for genome circularization and replication [196, 258]. Mutations in the viral genome affecting these RNA-protein interactions or the specific downregulation of the cellular factors led to reduced viral replication in cell culture [258]. Although not tested yet as antivirals, several hnRNPA1 drugs such as camptothecine and derivatives are readily available. These compounds are used in the treatment of cancer which supports the identification of derivatives with therapeutic potential [259]. DDX3 is a multifunctional human RNA helicase contributing to the replication of multiple viruses including noroviruses. The specific knockdown of DDX3 leads to significant inhibition of MNV replication in cell culture suggesting that it is a suitable target for antinoroviral strategies. Alterations in the expression and activation of DDX3 are also connected with different types of cancer, and an increasing number of inhibitors have been recently developed as potential chemotherapeutic molecules that could be evaluated against norovirus [260–262].

VAMP (Vesicle Associated Membrane Protein)-associated protein A (VAPA) is also known to assist norovirus replication, with MNV showing delayed replication in cells lacking VAPA gene [263]. VAMP is a SNARE (Soluble N-ethylmaleimide Sensitive Factor Attachment Protein Receptor) protein involved in vesicle fusion. VAPA is an endoplasmic reticulum (ER) protein which interacts with VAMP; this interaction has been implicated in several intracellular membrane trafficking processes such as lipid transport, ER homeostasis, and vesicle fusion [264]. It has been demonstrated that VAPA plays a proviral role for many viruses which supports studies to the identification of broad-range antivirals targeting this host protein [263]. Although the mechanism of action in noroviruses is not yet understood, it appears that NS1/2 interaction with VAPA provokes Golgi and vesicular trafficking disruption [265]. Viral NS1/2 plays an important role in vivo as it modulates the establishment of a persistent infection in mice, although the possible relevance of VAPA to this phenotype has not been investigated [266].

5.8.3.2 Targeting the Commensal Microbiota in the Control of HuNoV Infection

HuNoV interacts with commensal bacteria in the intestine; this relationship is important to the establishment of an infection. Thus, a tantalising possibility may involve the use of therapeutic drugs targeting this interaction [40, 42, 43]. Experimental infection of B-cell cultures was stimulated by the exogenous addition of *Enterobacter cloacae*, suggesting that enteric bacteria promote viral particle internalization. Recent findings suggest that HuNoV can interact with multiple commensal bacteria types in the gut by binding to HBGA-like molecules expressed on bacteria [267, 268]. These evidences support that commensal bacteria may enable HuNoV particle transport to the gut mucosa, facilitating particle penetration into the cell [40, 43, 269]. The elimination of commensal bacteria with antibiotics prevents MNV infection in mice, hinting at the interaction of HuNoV with gut microbiota as a possible target for antivirals. The mechanism by which bacteria stimulate norovirus infection is not well known as it has been associated with direct participation of bacteria in facilitating epithelia infection, and also by triggering alterations in the host immunity which makes mice susceptible to infection, e.g. limiting the efficacy of IFN-λ mediated immunity [230]. Treatment of mice with commensal *Lactobacillus* leads to reduced susceptibility to infection which endorses the use of lactic-derived products (e.g. drinking yogurt) in the control or mitigation of infection [270]. Thus, these reports stimulate new studies to investigate the development of antivirals targeting the interaction between the viral particle and commensal bacteria, or probiotics leading to reduced susceptibility to infection that could be used in the treatment of norovirus infection.

References

1. Bartsch SM, Lopman BA, Ozawa S, Hall AJ, Lee BY. Global economic burden of norovirus gastroenteritis. PLoS One. 2016;11:e0151219.
2. Lopman BA, Grassly NC. Editorial commentary: pediatric norovirus in developing countries: a picture slowly comes into focus. Clin Infect Dis. 2016;62:1218–20.
3. Ballard S-B, Saito M, Mirelman AJ, Bern C, Gilman RH. Tropical and travel-associated norovirus: current concepts. Curr Opin Infect Dis. 2015;28:408–16.
4. Lopman BA, Steele D, Kirkwood CD, Parashar UD. The vast and varied global burden of norovirus: prospects for prevention and control. PLoS Med. 2016;13:e1001999.
5. Brown JR, Shah D, Breuer J. Viral gastrointestinal infections and norovirus genotypes in a paediatric UK hospital, 2014–2015. J Clin Virol. 2016;84:1–6.
6. Sadique Z, Lopman B, Cooper BS, Edmunds WJ. Cost-effectiveness of ward closure to control outbreaks of norovirus infection in United Kingdom National Health Service Hospitals. J Infect Dis. 2016;213:S19–26.
7. Angarone MP, Sheahan A, Kamboj M. Norovirus in transplantation. Curr Infect Dis Rep. 2016;18:17.
8. Costantini VP, Cooper EM, Hardaker HL, Lee LE, Bierhoff M, Biggs C, Cieslak PR, Hall AJ, Vinjé J. Epidemiologic, virologic, and host genetic factors of norovirus outbreaks in long-term care facilities. Clin Infect Dis. 2016;62:1–10.
9. De Clercq E, Li G. Approved antiviral drugs over the past 50 years. Clin Microbiol Rev. 2016;29:695–747.
10. Deval J, Jin Z, Chuang YC, Kao CC. Structure(s), function(s), and inhibition of the RNA-dependent RNA polymerase of noroviruses. Virus Res. 2017;234:21–33.
11. Chung L, Bailey D, Leen EN, Emmott EP, Chaudhry Y, Roberts LO, Curry S, Locker N, Goodfellow IG. Norovirus translation requires an interaction between the C terminus of the genome-linked viral protein VPg and eukaryotic translation initiation factor 4G. J Biol Chem. 2014;289:21738–50.
12. Leen EN, Sorgeloos F, Correia S, et al. A conserved interaction between a C-terminal motif in norovirus VPg and the HEAT-1 domain of eIF4G is essential for translation initiation. PLoS Pathog. 2016;12:e1005379.
13. Zeitler CE, Estes MK, Venkataram Prasad BV. X-ray crystallographic structure of the Norwalk virus protease at 1.5-A resolution. J Virol. 2006;80:5050–8.
14. Zamyatkin DF, Parra F, Alonso JMM, Harki DA, Peterson BR, Grochulski P, Ng KK-S. Structural insights into mechanisms of catalysis and inhibition in Norwalk virus polymerase. J Biol Chem. 2008;283:7705–12.
15. Medvedev A, Viswanathan P, May J, Korba B. Regulation of human norovirus VPg nucleotidylylation by ProPol and nucleoside triphosphate binding by its amino terminal sequence in vitro. Virology. 2017;503:37–45.

16. Li T-F, Hosmillo M, Schwanke H, Shu T, Wang Z, Yin L, Curry S, Goodfellow IG, Zhou X. Human norovirus NS3 has RNA helicase and chaperoning activities. J Virol. 2017;92: e01606–17.
17. Emmott E, Sorgeloos F, Caddy SL, Vashist S, Sosnovtsev S, Lloyd R, Heesom K, Locker N, Goodfellow I. Norovirus-mediated modification of the translational landscape via virus and host-induced cleavage of translation initiation factors. Mol Cell Proteomics. 2017;16: S215–29.
18. Donaldson EF, Lindesmith LC, Lobue AD, Baric RS. Viral shape-shifting: norovirus evasion of the human immune system. Nat Rev Microbiol. 2010;8:231–41.
19. Prasad BV, Hardy ME, Dokland T, Bella J, Rossmann MG, Estes MK. X-ray crystallographic structure of the Norwalk virus capsid. Science. 1999;286:287–90.
20. Hansman GS, Biertümpfel C, Georgiev I, McLellan JS, Chen L, Zhou T, Katayama K, Kwong PD. Crystal structures of GII.10 and GII.12 norovirus protruding domains in complex with histo-blood group antigens reveal details for a potential site of vulnerability. J Virol. 2011;85:6687–701.
21. Ettayebi K, Crawford SE, Murakami K, et al. Replication of human noroviruses in stem cell-derived human enteroids. Science (80-). 2016;353:1387–93.
22. Marionneau S, Ruvoën N, Le Moullac-Vaidye B, Clement M, Cailleau-Thomas A, Ruiz-Palacois G, Huang P, Jiang X, Le Pendu J. Norwalk virus binds to histo-blood group antigens present on gastroduodenal epithelial cells of secretor individuals. Gastroenterology. 2002;122:1967–77.
23. Guix S, Asanaka M, Katayama K, Crawford SE, Neill FH, Atmar RL, Estes MK. Norwalk virus RNA is infectious in mammalian cells. J Virol. 2007;81:12238–48.
24. Nordgren J, Sharma S, Kambhampati A, Lopman B, Svensson L. Innate resistance and susceptibility to norovirus infection. PLoS Pathog. 2016;12:e1005385.
25. Bui T, Kocher J, Li Y, et al. Median infectious dose of human norovirus GII.4 in gnotobiotic pigs is decreased by simvastatin treatment and increased by age. J Gen Virol. 2013;94:2005–16.
26. Jung K, Wang Q, Kim Y, Scheuer K, Zhang Z, Shen Q, Chang K-O, Saif LJ. The effects of simvastatin or interferon-α on infectivity of human norovirus using a gnotobiotic pig model for the study of antivirals. PLoS One. 2012;7:e41619.
27. Chang K-O. Role of cholesterol pathways in norovirus replication. J Virol. 2009;83:8587–95.
28. Rocha-Pereira J, Neyts J, Jochmans D. Norovirus: targets and tools in antiviral drug discovery. Biochem Pharmacol. 2014;91:1–11.
29. Karst SM, Zhu S, Goodfellow IG. The molecular pathology of noroviruses. J Pathol. 2015;235:206–16.
30. White PA. Evolution of norovirus. Clin Microbiol Infect. 2014;20:741–5.
31. Chhabra P, de Graaf M, Parra GI, Chan MC, Green K, Martella V, Wang Q, White PA, Katayama K, Vennema H, Koopmans MPG, Vinjé J. Updated classification of norovirus genogroups and genotypes. J Gen Virol. 2019;100(10):1393–406.
32. de Graaf M, Villabruna N, Koopmans MP. Capturing norovirus transmission. Curr Opin Virol. 2017;22:64–70.
33. Bull RA, Eden J-S, Rawlinson WD, White PA. Rapid evolution of pandemic noroviruses of the GII.4 lineage. PLoS Pathog. 2010;6:e1000831.
34. De Graaf M, Van Beek J, Vennema H, et al. Emergence of a novel GII.17 norovirus – end of the GII.4 era? Eur Secur. 2015;20:1–8.
35. Pan L, Xue C, Fu H, Liu D, Zhu L, Cui C, Zhu W, Fu Y, Qiao S. The novel norovirus genotype GII.17 is the predominant strain in diarrheal patients in Shanghai, China. Gut Pathog. 2016;8:49.
36. Lu J, Fang L, Zheng H, et al. The evolution and transmission of epidemic GII.17 Noroviruses. J Infect Dis. 2016;214:556–64.

37. Debbink K, Lindesmith LC, Ferris MT, Swanstrom J, Beltramello M, Corti D, Lanzavecchia A, Baric RS. Within-host evolution results in antigenically distinct GII.4 noroviruses. J Virol. 2014;88:7244–55.
38. Arias A, Thorne L, Ghurburrun E, Bailey D, Goodfellow I. Norovirus polymerase fidelity contributes to viral transmission in vivo. mSphere. 2016;1:e00279–16.
39. Cuevas JM, Combe M, Torres-Puente M, Garijo R, Guix S, Buesa J, Rodríguez-Díaz J, Sanjuán R. Human norovirus hyper-mutation revealed by ultra-deep sequencing. Infect Genet Evol. 2016;41:233–9.
40. Jones MK, Grau KR, Costantini V, et al. Human norovirus culture in B cells. Nat Protoc. 2015;10:1939–47.
41. Ramani S, Crawford SE, Blutt SE, Estes MK. Human organoid cultures: transformative new tools for human virus studies. Curr Opin Virol. 2018;29:79–86.
42. Baldridge MT, Turula H, Wobus CE. Norovirus regulation by host and microbe. Trends Mol Med. 2016;22:1047–59.
43. Jones MK, Watanabe M, Zhu S, et al. Enteric bacteria promote human and mouse norovirus infection of B cells. Science (80-). 2014;346:755–9.
44. Kolawole AO, Rocha-Pereira J, Elftman MD, Neyts J, Wobus CE. Inhibition of human norovirus by a viral polymerase inhibitor in the B cell culture system and in the mouse model. Antivir Res. 2016;132:46–9.
45. Tan M, Jiang X. Norovirus-host interaction: multi-selections by human histo-blood group antigens. Trends Microbiol. 2011;19:382–8.
46. Katayama K, Murakami K, Sharp TM, Guix S, Oka T, Takai-Todaka R, Nakanishi A, Crawford SE, Atmar RL, Estes MK. Plasmid-based human norovirus reverse genetics system produces reporter-tagged progeny virus containing infectious genomic RNA. Proc Natl Acad Sci. 2014;111:E4043–52.
47. Taube S, Kolawole AO, Höhne M, Wilkinson JE, Handley SA, Perry JW, Thackray LB, Akkina R, Wobus CE. A mouse model for human norovirus. MBio. 2013;4:e00450–13.
48. Farkas T. Natural norovirus infections in rhesus macaques. Emerg Infect Dis. 2016;22:1272–4.
49. Bok K, Parra GI, Mitra T, et al. Chimpanzees as an animal model for human norovirus infection and vaccine development. Proc Natl Acad Sci U S A. 2011;108:325–30.
50. Seo DJ, Jung D, Jung S, Ha S-K, Ha S-D, Choi I-S, Myoung J, Choi C. Experimental miniature piglet model for the infection of human norovirus GII. J Med Virol. 2018;90:655–62.
51. Todd K, Tripp R. Human norovirus: experimental models of infection. Viruses. 2019;11:151.
52. Lei S, Ryu J, Wen K, et al. Increased and prolonged human norovirus infection in RAG2/IL2RG deficient gnotobiotic pigs with severe combined immunodeficiency. Sci Rep. 2016;6:25222.
53. Rocha-Pereira J, Van Dycke J, Neyts J. Norovirus genetic diversity and evolution: implications for antiviral therapy. Curr Opin Virol. 2016;20:92–8.
54. Wobus CE, Cunha JB, Elftman MD, Kolawole AO. Animal models of norovirus infection. In: Viral gastroenteritis. Amsterdam: Elsevier; 2016. p. 397–422.
55. Cohen J. Committee Recommends NIH Continue Chimpanzee Research, With New Limits. Science (80-). 2011;334:1484–5.
56. Van Dycke J, Ny A, Conceição-Neto N, et al. A robust human norovirus replication model in zebrafish larvae clinical and epidemiological virology. bioRxiv. 2019; https://doi.org/10.1101/528364.
57. Thorne L, Arias A, Goodfellow I. Advances toward a norovirus antiviral: from classical inhibitors to lethal mutagenesis. J Infect Dis. 2016;213(Suppl):S27–31.
58. Nice TJ, Baldridge MT, McCune BT, Norman JM, Lazear HM, Artyomov M, Diamond MS, Virgin HW. Interferon-λ cures persistent murine norovirus infection in the absence of adaptive immunity. Science. 2015;347:269–73.

59. Arias A, Ureña L, Thorne L, Yunus MA, Goodfellow I. Reverse genetics mediated recovery of infectious murine norovirus. J Vis Exp. 2012;64:1–8.
60. Perry JW, Ahmed M, Chang K-O, Donato NJ, Showalter HD, Wobus CE. Antiviral activity of a small molecule deubiquitinase inhibitor occurs via induction of the unfolded protein response. PLoS Pathog. 2012;8:e1002783.
61. Thackray LB, Wobus CE, Chachu KA, Liu B, Alegre ER, Henderson KS, Kelley ST, Virgin HW. Murine noroviruses comprising a single genogroup exhibit biological diversity despite limited sequence divergence. J Virol. 2007;81:10460–73.
62. Arias A, Bailey D, Chaudhry Y, Goodfellow I. Development of a reverse-genetics system for murine norovirus 3: long-term persistence occurs in the caecum and colon. J Gen Virol. 2012;93:1432–41.
63. Karst SM, Wobus CE, Lay M, Davidson J, Virgin HW. STAT1-dependent innate immunity to a Norwalk-like virus. Science. 2003;299:1575–8.
64. Arias A, Emmott E, Vashist S, Goodfellow I. Progress towards the prevention and treatment of norovirus infections. Future Microbiol. 2013;8:1475–87.
65. McFadden N, Arias A, Dry I, Bailey D, Witteveldt J, Evans DJ, Goodfellow I, Simmonds P. Influence of genome-scale RNA structure disruption on the replication of murine norovirus—similar replication kinetics in cell culture but attenuation of viral fitness in vivo. Nucleic Acids Res. 2013;41:6316–31.
66. Thorne L, Bailey D, Goodfellow I. High-resolution functional profiling of the norovirus genome. J Virol. 2012;86:11441–56.
67. Haga K, Fujimoto A, Takai-Todaka R, Miki M, Doan YH, Murakami K, Yokoyama M, Murata K, Nakanishi A, Katayama K. Functional receptor molecules CD300lf and CD300ld within the CD300 family enable murine noroviruses to infect cells. Proc Natl Acad Sci U S A. 2016;113:E6248–55.
68. Orchard RC, Wilen CB, Doench JG, et al. Discovery of a proteinaceous cellular receptor for a norovirus. Science. 2016;353:933–6.
69. Kilic T, Koromyslova A, Malak V, Hansman GS. Atomic structure of the murine norovirus protruding domain and soluble CD300lf receptor complex. J Virol. 2018;92:e00413–8.
70. Nelson CA, Wilen CB, Dai Y-N, Orchard RC, Kim AS, Stegeman RA, Hsieh LL, Smith TJ, Virgin HW, Fremont DH. Structural basis for murine norovirus engagement of bile acids and the CD300lf receptor. Proc Natl Acad Sci. 2018;115:E9201–10.
71. Feng X, Jiang X. Library screen for inhibitors targeting norovirus binding to histo-blood group antigen receptors. Antimicrob Agents Chemother. 2007;51:324–31.
72. Tan M, Jiang X. Histo-blood group antigens: a common niche for norovirus and rotavirus. Expert Rev Mol Med. 2014;16:e5.
73. Ali ES, Rajapaksha H, Carr JM, Petrovsky N. Norovirus drug candidates that inhibit viral capsid attachment to human histo-blood group antigens. Antivir Res. 2016;133:14–22.
74. Zhang X-F, Tan M, Chhabra M, Dai Y-C, Meller J, Jiang X. Inhibition of histo-blood group antigen binding as a novel strategy to block norovirus infections. PLoS One. 2013;8:e69379.
75. Lundborg M, Ali E, Widmalm G. An in silico virtual screening study for the design of norovirus inhibitors: fragment-based molecular docking and binding free energy calculations. Carbohydr Res. 2013;378:133–8.
76. Bu W, Mamedova A, Tan M, Xia M, Jiang X, Hegde RS. Structural basis for the receptor binding specificity of Norwalk virus. J Virol. 2008;82:5340–7.
77. Koromyslova AD, Leuthold MM, Bowler MW, Hansman GS. The sweet quartet: binding of fucose to the norovirus capsid. Virology. 2015;483:203–8.
78. Rademacher C, Guiard J, Kitov PI, Fiege B, Dalton KP, Parra F, Bundle DR, Peters T. Targeting norovirus infection-multivalent entry inhibitor design based on NMR experiments. Chemistry. 2011;17:7442–53.
79. Fiege B, Rademacher C, Cartmell J, Kitov PI, Parra F, Peters T. Molecular details of the recognition of blood group antigens by a human norovirus as determined by STD NMR spectroscopy. Angew Chem Int Ed Engl. 2012;51:928–32.

80. Han L, Kitova EN, Tan M, Jiang X, Klassen JS. Identifying carbohydrate ligands of a norovirus P particle using a catch and release electrospray ionization mass spectrometry assay. J Am Soc Mass Spectrom. 2014;25:111–9.
81. Shang J, Piskarev VE, Xia M, Huang P, Jiang X, Likhosherstov LM, Novikova OS, Newburg DS, Ratner DM. Identifying human milk glycans that inhibit norovirus binding using surface plasmon resonance. Glycobiology. 2013;23:1491–8.
82. Jiang X, Huang P, Zhong W, Tan M, Farkas T, Morrow AL, Newburg DS, Ruiz-Palacios GM, Pickering LK. Human milk contains elements that block binding of noroviruses to human histo-blood group antigens in saliva. J Infect Dis. 2004;190:1850–9.
83. Weichert S, Koromyslova A, Singh BK, Hansman S, Jennewein S, Schroten H, Hansman GS. Structural basis for norovirus inhibition by human milk oligosaccharides. J Virol. 2016;90:4843–8.
84. Koromyslova A, Tripathi S, Morozov V, Schroten H, Hansman GS. Human norovirus inhibition by a human milk oligosaccharide. Virology. 2017;508:81–9.
85. Prasad BV, Shanker S, Muhaxhiri Z, Deng L, Choi J-M, Estes MK, Song Y, Palzkill T, Atmar RL. Antiviral targets of human noroviruses. Curr Opin Virol. 2016;18:117–25.
86. Tan M, Fang P, Chachiyo T, Xia M, Huang P, Fang Z, Jiang W, Jiang X. Noroviral P particle: structure, function and applications in virus-host interaction. Virology. 2008;382:115–23.
87. Yazawa S, Yokobori T, Ueta G, et al. Blood group substances as potential therapeutic agents for the prevention and treatment of infection with noroviruses proving novel binding patterns in human tissues. PLoS One. 2014;9:e89071.
88. Liu W, Chen Y, Jiang X, Xia M, Yang Y, Tan M, Li X, Rao Z. A unique human norovirus lineage with a distinct HBGA binding interface. PLoS Pathog. 2015;11:e1005025.
89. Hansman GS, Shahzad-Ul-Hussan S, McLellan JS, Chuang GY, Georgiev I, Shimoike T, Katayama K, Bewley CA, Kwong PD. Structural basis for norovirus inhibition and fucose mimicry by citrate. J Virol. 2012;86:284–92.
90. Koromyslova AD, White PA, Hansman GS. Treatment of norovirus particles with citrate. Virology. 2015;485:199–204.
91. Koromyslova AD, Hansman GS. Nanobody binding to a conserved epitope promotes norovirus particle disassembly. J Virol. 2015;89:2718–30.
92. Rydell GE, Nilsson J, Rodriguez-Diaz J, Ruvoën-Clouet N, Svensson L, Le Pendu J, Larson G. Human noroviruses recognize sialyl Lewis x neoglycoprotein. Glycobiology. 2009;19:309–20.
93. Han L, Tan M, Xia M, Kitova EN, Jiang X, Klassen JS. Gangliosides are ligands for human noroviruses. J Am Chem Soc. 2014;136:12631–7.
94. Tamura M, Natori K, Kobayashi M, Miyamura T, Takeda N. Genogroup II noroviruses efficiently bind to heparan sulfate proteoglycan associated with the cellular membrane. J Virol. 2004;78:3817–26.
95. Taube S, Perry JW, Yetming K, et al. Ganglioside-linked terminal sialic acid moieties on murine macrophages function as attachment receptors for murine noroviruses. J Virol. 2009;83:4092–101.
96. Tan M, Wei C, Huang P, et al. Tulane virus recognizes sialic acids as cellular receptors. Sci Rep. 2015;5:11784.
97. McGeary RP, Bennett AJ, Tran QB, Cosgrove KL, Ross BP. Suramin: clinical uses and structure-activity relationships. Mini Rev Med Chem. 2008;8:1384–94.
98. Mastrangelo E, Pezzullo M, Tarantino D, Petazzi R, Germani F, Kramer D, Robel I, Rohayem J, Bolognesi M, Milani M. Structure-based inhibition of norovirus RNA-dependent RNA-polymerases. J Mol Biol. 2012;419:198–210.
99. Chattopadhyay D, Naik TN. Antivirals of ethnomedicinal origin: structure-activity relationship and scope. Mini Rev Med Chem. 2007;7:275–301.
100. Katayama S, Ohno F, Yamauchi Y, Kato M, Makabe H, Nakamura S. Enzymatic synthesis of novel phenol acid rutinosides using rutinase and their antiviral activity in vitro. J Agric Food Chem. 2013;61:9617–22.

101. Takahashi H, Nakazawa M, Ohshima C, Sato M, Tsuchiya T, Takeuchi A, Kunou M, Kuda T, Kimura B. Heat-denatured lysozyme inactivates murine norovirus as a surrogate human norovirus. Sci Rep. 2015;5:11819.
102. Sharma TK, Bruno JG, Dhiman A. ABCs of DNA aptamer and related assay development. Biotechnol Adv. 2017;35:275–301.
103. Moore MD, Escudero-Abarca BI, Suh SH, Jaykus L-A. Generation and characterization of nucleic acid aptamers targeting the capsid P domain of a human norovirus GII.4 strain. J Biotechnol. 2015;209:41–9.
104. Beier R, Pahlke C, Quenzel P, Henseleit A, Boschke E, Cuniberti G, Labudde D. Selection of a DNA aptamer against norovirus capsid protein VP1. FEMS Microbiol Lett. 2014;351:162–9.
105. Moore MD, Bobay BG, Mertens B, Jaykus L-A. Human norovirus aptamer exhibits high degree of target conformation-dependent binding similar to that of receptors and discriminates particle functionality. mSphere. 2016;1:e00298–16.
106. Gupta N, Lainson JC, Belcher PE, Shen L, Mason HS, Johnston SA, Diehnelt CW. Cross-reactive synbody affinity ligands for capturing diverse noroviruses. Anal Chem. 2017;89:7174–81.
107. Johnston SA, Domenyuk V, Gupta N, et al. A simple platform for the rapid development of antimicrobials. Sci Rep. 2017;7:17610.
108. Ohba M, Oka T, Ando T, et al. Antiviral effect of theaflavins against caliciviruses. J Antibiot (Tokyo). 2017;70:443–7.
109. Ohba M, Oka T, Ando T, et al. Discovery and synthesis of heterocyclic carboxamide derivatives as potent anti-norovirus agents. Chem Pharm Bull (Tokyo). 2016;64:465–75.
110. DuPont HL, Sullivan P, Evans DG, Pickering LK, Evans DJ, Vollet JJ, Ericsson CD, Ackerman PB, Tjoa WS. Prevention of traveler's diarrhea (emporiatric enteritis). Prophylactic administration of subsalicylate bismuth. JAMA. 1980;243:237–41.
111. Pitz AM, Park GW, Lee D, Boissy YL, Vinjé J. Antimicrobial activity of bismuth subsalicylate on Clostridium difficile, Escherichia coli O157:H7, norovirus, and other common enteric pathogens. Gut Microbes. 2015;6:93–100.
112. Ishikawa H, Awano N, Fukui T, Sasaki H, Kyuwa S. The protective effects of lactoferrin against murine norovirus infection through inhibition of both viral attachment and replication. Biochem Biophys Res Commun. 2013;434:791–6.
113. Florescu DF, Hill LA, McCartan MA, Grant W. Two cases of Norwalk virus enteritis following small bowel transplantation treated with oral human serum immunoglobulin. Pediatr Transplant. 2008;12:372–5.
114. Woodward J, Gkrania-Klotsas E, Kumararatne D. Chronic norovirus infection and common variable immunodeficiency. Clin Exp Immunol. 2017;188:363–70.
115. Chen Z, Sosnovtsev SV, Bok K, Parra GI, Makiya M, Agulto L, Green KY, Purcell RH. Development of Norwalk virus-specific monoclonal antibodies with therapeutic potential for the treatment of Norwalk virus gastroenteritis. J Virol. 2013;87:9547–57.
116. Lindesmith LC, Donaldson EF, Beltramello M, Pintus S, Corti D, Swanstrom J, Debbink K, Jones TA, Lanzavecchia A, Baric RS. Particle conformation regulates antibody access to a conserved GII.4 norovirus blockade epitope. J Virol. 2014;88:8826–42.
117. Jones RGA, Martino A. Targeted localized use of therapeutic antibodies: a review of non-systemic, topical and oral applications. Crit Rev Biotechnol. 2016;36:506–20.
118. Dai Y-C, Wang Y-Y, Zhang X-F, Tan M, Xia M, Wu X-B, Jiang X, Nie J. Evaluation of anti-norovirus IgY from egg yolk of chickens immunized with norovirus P particles. J Virol Methods. 2012;186:126–31.
119. Dai Y-C, Zhang X-F, Tan M, Huang P, Lei W, Fang H, Zhong W, Jiang X. A dual chicken IgY against rotavirus and norovirus. Antivir Res. 2013;97:293–300.
120. Ghosh S, Malik YS, Kobayashi N. Therapeutics and immunoprophylaxis against noroviruses and rotaviruses: the past, present, and future. Curr Drug Metab. 2018;19:170–91.
121. Vanlandschoot P, Stortelers C, Beirnaert E, Ibañez LI, Schepens B, Depla E, Saelens X. Nanobodies®: new ammunition to battle viruses. Antivir Res. 2011;92:389–407.

122. Koromyslova AD, Hansman GS. Nanobodies targeting norovirus capsid reveal functional epitopes and potential mechanisms of neutralization. PLoS Pathog. 2017;13:e1006636.
123. Garaicoechea L, Aguilar A, Parra GI, Bok M, Sosnovtsev SV, Canziani G, Green KY, Bok K, Parreño V. Llama nanoantibodies with therapeutic potential against human norovirus diarrhea. PLoS One. 2015;10:e0133665.
124. Dumoulin M, Conrath K, Van Meirhaeghe A, Meersman F, Heremans K, Frenken LGJ, Muyldermans S, Wyns L, Matagne A. Single-domain antibody fragments with high conformational stability. Protein Sci. 2002;11:500–15.
125. Galasiti Kankanamalage AC, Kim Y, Weerawarna PM, et al. Structure-guided design and optimization of dipeptidyl inhibitors of norovirus 3CL protease. Structure-activity relationships and biochemical, X-ray crystallographic, cell-based, and in vivo studies. J Med Chem. 2015;58:3144–55.
126. Kim Y, Galasiti Kankanamalage AC, Chang K-O, Groutas WC. Recent advances in the discovery of norovirus therapeutics. J Med Chem. 2015;58:9438–50.
127. Weerasekara S, Prior AM, Hua DH. Current tools for norovirus drug discovery. Expert Opin Drug Discov. 2016;11:529–41.
128. Muhaxhiri Z, Deng L, Shanker S, Sankaran B, Estes MK, Palzkill T, Song Y, Prasad BVV. Structural basis of substrate specificity and protease inhibition in Norwalk virus. J Virol. 2013;87:4281–92.
129. May J, Viswanathan P, Ng KK-S, Medvedev A, Korba B. The p4–p2' amino acids surrounding human norovirus polyprotein cleavage sites define the core sequence regulating self-processing order. J Virol. 2014;88:10738–47.
130. Damalanka VC, Kim Y, Alliston KR, et al. Oxadiazole-based cell permeable macrocyclic transition state inhibitors of norovirus 3CL protease. J Med Chem. 2016;59:1899–913.
131. Galasiti Kankanamalage AC, Kim Y, Rathnayake AD, et al. Structure-based exploration and exploitation of the S4 subsite of norovirus 3CL protease in the design of potent and permeable inhibitors. Eur J Med Chem. 2016;126:502–16.
132. Weerawarna PM, Kim Y, Galasiti Kankanamalage AC, et al. Structure-based design and synthesis of triazole-based macrocyclic inhibitors of norovirus protease: structural, biochemical, spectroscopic, and antiviral studies. Eur J Med Chem. 2016;119:300–18.
133. Kaufman SS, Green KY, Korba BE. Treatment of norovirus infections: moving antivirals from the bench to the bedside. Antivir Res. 2014;105:80–91.
134. Kim Y, Lovell S, Tiew K-C, Mandadapu SR, Alliston KR, Battaile KP, Groutas WC, Chang K-O. Broad-spectrum antivirals against 3C or 3C-like proteases of picornaviruses, noroviruses, and coronaviruses. J Virol. 2012;86:11754–62.
135. Mandadapu SR, Gunnam MR, Tiew K-C, Uy RAZ, Prior AM, Alliston KR, Hua DH, Kim Y, Chang K-O, Groutas WC. Inhibition of norovirus 3CL protease by bisulfite adducts of transition state inhibitors. Bioorg Med Chem Lett. 2013;23:62–5.
136. Galasiti Kankanamalage AC, Kim Y, Rathnayake AD, Alliston KR, Butler MM, Cardinale SC, Bowlin TL, Groutas WC, Chang K-O. Design, synthesis, and evaluation of novel prodrugs of transition state inhibitors of norovirus 3CL protease. J Med Chem. 2017;60:6239–48.
137. Prior AM, Kim Y, Weerasekara S, Moroze M, Alliston KR, Uy RAZ, Groutas WC, Chang K-O, Hua DH. Design, synthesis, and bioevaluation of viral 3C and 3C-like protease inhibitors. Bioorg Med Chem Lett. 2013;23:6317–20.
138. Rocha-Pereira J, Nascimento MSJ, Ma Q, Hilgenfeld R, Neyts J, Jochmans D. The enterovirus protease inhibitor rupintrivir exerts cross-genotypic anti-norovirus activity and clears cells from the norovirus replicon. Antimicrob Agents Chemother. 2014;58:4675–81.
139. Hayden FG, Turner RB, Gwaltney JM, Chi-Burris K, Gersten M, Hsyu P, Patick AK, Smith GJ, Zalman LS. Phase II, randomized, double-blind, placebo-controlled studies of ruprintrivir nasal spray 2-percent suspension for prevention and treatment of experimentally induced rhinovirus colds in healthy volunteers. Antimicrob Agents Chemother. 2003;47:3907–16.

140. Patick AK, Brothers MA, Maldonado F, et al. In vitro antiviral activity and single-dose pharmacokinetics in humans of a novel, orally bioavailable inhibitor of human rhinovirus 3C protease. Antimicrob Agents Chemother. 2005;49:2267–75.
141. Kitano M, Hosmillo M, Emmott E, Lu J, Goodfellowa I. Selection and characterization of rupintrivir-resistant norwalk virus replicon cells in vitro. Antimicrob Agents Chemother. 2018;62:e00201–18.
142. Damalanka VC, Kim Y, Galasiti Kankanamalage AC, et al. Structure-guided design, synthesis and evaluation of oxazolidinone-based inhibitors of norovirus 3CL protease. Eur J Med Chem. 2018;143:881–90.
143. Emmott E, Sweeney TR, Goodfellow I. A cell-based FRET sensor reveals inter- and intragenogroup variation in norovirus protease activity and polyprotein cleavage∗. J Biol Chem. 2015;290:27841–53.
144. Perales C, Martín V, Domingo E. Lethal mutagenesis of viruses. Curr Opin Virol. 2011;1:419–22.
145. Andino R, Domingo E. Viral quasispecies. Virology. 2015;479–480:46–51.
146. Perales C, Domingo E. Antiviral strategies based on lethal mutagenesis and error threshold. Curr Top Microbiol Immunol. 2016;392:323–39.
147. Agudo R, Arias A, Pariente N, Perales C, Escarmís C, Jorge A, Marina A, Domingo E. Molecular characterization of a dual inhibitory and mutagenic activity of 5-fluorouridine triphosphate on viral RNA synthesis. Implications for lethal mutagenesis. J Mol Biol. 2008;382:652–66.
148. Crotty S, Maag D, Arnold JJ, Zhong W, Lau JY, Hong Z, Andino R, Cameron CE. The broad-spectrum antiviral ribonucleoside ribavirin is an RNA virus mutagen. Nat Med. 2000;6:1375–9.
149. Jin Z, Tucker K, Lin X, et al. Biochemical evaluation of the inhibition properties of favipiravir and 2'-C-methyl-cytidine triphosphates against human and mouse norovirus RNA polymerases. Antimicrob Agents Chemother. 2015;59:7504–16.
150. Jin Z, Smith LK, Rajwanshi VK, Kim B, Deval J. The ambiguous base-pairing and high substrate efficiency of T-705 (Favipiravir) ribofuranosyl 5′-triphosphate towards influenza A virus polymerase. PLoS One. 2013;8:e68347.
151. Arias A, Thorne L, Goodfellow I. Favipiravir elicits antiviral mutagenesis during virus replication in vivo. elife. 2014;3:e03679.
152. Furuta Y, Gowen BB, Takahashi K, Shiraki K, Smee DF, Barnard DL. Favipiravir (T-705), a novel viral RNA polymerase inhibitor. Antivir Res. 2013;100:446–54.
153. Baranovich T, Wong S-S, Armstrong J, Marjuki H, Webby RJ, Webster RG, Govorkova EA. T-705 (Favipiravir) induces lethal mutagenesis in influenza A H1N1 viruses in vitro. J Virol. 2013;87:3741–51.
154. de Ávila AI, Gallego I, Soria ME, Gregori J, Quer J, Esteban JI, Rice CM, Domingo E, Perales C. Lethal mutagenesis of hepatitis C virus induced by favipiravir. PLoS One. 2016;11: e0164691.
155. de Avila AI, Moreno E, Perales C, Domingo E. Favipiravir can evoke lethal mutagenesis and extinction of foot-and-mouth disease virus. Virus Res. 2017;233:105–12.
156. Ruis C, Brown L-AK, Roy S, et al. Mutagenesis in norovirus in response to favipiravir treatment. N Engl J Med. 2018;379:2173–6.
157. Rocha-Pereira J, Jochmans D, Debing Y, Verbeken E, Nascimento MSJ, Neyts J. The viral polymerase inhibitor 2'-C-methylcytidine inhibits Norwalk virus replication and protects against norovirus-induced diarrhea and mortality in a mouse model. J Virol. 2013;87:11798–805.
158. Rocha-Pereira J, Jochmans D, Neyts J. Prophylactic treatment with the nucleoside analogue 2'-C-methylcytidine completely prevents transmission of norovirus. J Antimicrob Chemother. 2015;70:190–7.

159. Rocha-Pereira J, Kolawole AO, Verbeken E, Wobus CE, Neyts J. Post-exposure antiviral treatment of norovirus infections effectively protects against diarrhea and reduces virus shedding in the stool in a mortality mouse model. Antivir Res. 2016;132:76–84.
160. Carroll SS, Koeplinger K, Vavrek M, et al. Antiviral efficacy upon administration of a HepDirect prodrug of 2'-C-methylcytidine to hepatitis C virus-infected chimpanzees. Antimicrob Agents Chemother. 2011;55:3854–60.
161. Costantini VP, Whitaker T, Barclay L, Lee D, McBrayer TR, Schinazi RF, Vinjé J. Antiviral activity of nucleoside analogues against norovirus. Antivir Ther. 2012;17:981–91.
162. Woodward JM, Gkrania-Klotsas E, Cordero-Ng AY, et al. The role of chronic norovirus infection in the enteropathy associated with common variable immunodeficiency. Am J Gastroenterol. 2015;110:320–7.
163. Graci JD, Cameron CE. Mechanisms of action of ribavirin against distinct viruses. Rev Med Virol. 2006;16:37–48.
164. Agudo R, de la Higuera I, Arias A, Grande-Pérez A, Domingo E. Involvement of a joker mutation in a polymerase-independent lethal mutagenesis escape mechanism. Virology. 2016;494:257–66.
165. Hong Z, Cameron CE. Pleiotropic mechanisms of ribavirin antiviral activities. Prog Drug Res. 2002;59:41–69.
166. Pariente N, Airaksinen A, Domingo E. Mutagenesis versus inhibition in the efficiency of extinction of foot-and-mouth disease virus. J Virol. 2003;77:7131–8.
167. Bassi MR, Sempere RN, Meyn P, Polacek C, Arias A. Extinction of Zika virus and Usutu virus by lethal mutagenesis reveals different patterns of sensitivity to three mutagenic drugs. Antimicrob Agents Chemother. 2018;62:e00380–18.
168. Domingo E, Perales C, Agudo R, Arias A, Escarmis C, Ferrer-Orta C, Verdaguer N. Mutation, quasispecies and lethal mutagenesis. In: Ehrenfeld E, Domingo E, Roos RP, editors. Picornaviruses. 1st ed. Washington DC: ASM Press; 2010. p. 197–211.
169. Julian TR, Baugher JD, Rippinger CM, Pinekenstein R, Kolawole AO, Mehoke TS, Wobus CE, Feldman AB, Pineda FJ, Schwab KJ. Murine norovirus (MNV-1) exposure in vitro to the purine nucleoside analog Ribavirin increases quasispecies diversity. Virus Res. 2016;211:165–73.
170. Lee J-H, Alam I, Han KR, Cho S, Shin S, Kang S, Yang JM, Kim KH. Crystal structures of murine norovirus-1 RNA-dependent RNA polymerase. J Gen Virol. 2011;92:1607–16.
171. Alam I, Lee J-H, Cho KJ, Han KR, Yang JM, Chung MS, Kim KH. Crystal structures of murine norovirus-1 RNA-dependent RNA polymerase in complex with 2-thiouridine or ribavirin. Virology. 2012;426:143–51.
172. Agudo R, Arias A, Domingo E. 5-fluorouracil in lethal mutagenesis of foot-and-mouth disease virus. Future Med Chem. 2009;1:529–39.
173. Lee JJ, Beumer JH, Chu E. Therapeutic drug monitoring of 5-fluorouracil. Cancer Chemother Pharmacol. 2016;78:447–64.
174. Ye W, Schneller SW. The enantiomers of the 1′,6′-isomer of neplanocin A: synthesis and antiviral properties. Bioorg Med Chem. 2014;22:5315–9.
175. Liu C, Chen Q, Schneller SW. Enantiomeric 3-deaza-1′,6′-isoneplanocin and its 3-bromo analogue: synthesis by the Ullmann reaction and their antiviral properties. Bioorg Med Chem Lett. 2016;26:928–30.
176. De Clercq E. S-adenosylhomocysteine hydrolase inhibitors as broad-spectrum antiviral agents. Biochem Pharmacol. 1987;36:2567–75.
177. Chen Q, Davidson A. Synthesis, conformational study and antiviral activity of l -like neplanocin derivatives. Bioorg Med Chem Lett. 2017;27:4436–9.
178. Harki DA, Graci JD, Edathil JP, Castro C, Cameron CE, Peterson BR. Synthesis of a universal 5-nitroindole ribonucleotide and incorporation into RNA by a viral RNA-dependent RNA polymerase. Chembiochem. 2007;8:1359–62.

179. Zamyatkin DF, Parra F, Machín A, Grochulski P, Ng KK-S. Binding of 2′-amino-2-′-deoxycytidine-5′-triphosphate to norovirus polymerase induces rearrangement of the active site. J Mol Biol. 2009;390:10–6.
180. Mastrangelo E, Mazzitelli S, Fabbri J, Rohayem J, Ruokolainen J, Nykänen A, Milani M, Pezzullo M, Nastruzzi C, Bolognesi M. Delivery of suramin as an antiviral agent through liposomal systems. ChemMedChem. 2014;9:933–9.
181. Tarantino D, Pezzullo M, Mastrangelo E, Croci R, Rohayem J, Robel I, Bolognesi M, Milani M. Naphthalene-sulfonate inhibitors of human norovirus RNA-dependent RNA-polymerase. Antivir Res. 2014;102:23–8.
182. Croci R, Pezzullo M, Tarantino D, et al. Structural bases of norovirus RNA dependent RNA polymerase inhibition by novel suramin-related compounds. PLoS One. 2014;9:e91765.
183. Croci R, Tarantino D, Milani M, Pezzullo M, Rohayem J, Bolognesi M, Mastrangelo E. PPNDS inhibits murine norovirus RNA-dependent RNA-polymerase mimicking two RNA stacking bases. FEBS Lett. 2014;588:1720–5.
184. Ferla S, Netzler NE, Ferla S, Veronese S, Tuipulotu DE, Guccione S, Brancale A, White PA, Bassetto M. In silico screening for human norovirus antivirals reveals a novel non-nucleoside inhibitor of the viral polymerase. Sci Rep. 2018;8:4129.
185. Netzler NE, Enosi Tuipulotu D, Eltahla AA, et al. Broad-spectrum non-nucleoside inhibitors for caliciviruses. Antivir Res. 2017;146:65–75.
186. Rocha-Pereira J, Cunha R, Pinto DCGA, Silva AMS, Nascimento MSJ. (E)-2-styrylchromones as potential anti-norovirus agents. Bioorg Med Chem. 2010;18:4195–201.
187. Gomes A, Freitas M, Fernandes E, Lima JLFC. Biological activities of 2-styrylchromones. Mini Rev Med Chem. 2010;10:1–7.
188. Bok K, Cavanaugh VJ, Matson DO, González-Molleda L, Chang K-O, Zintz C, Smith AW, Iversen P, Green KY, Campbell AE. Inhibition of norovirus replication by morpholino oligomers targeting the 5′-end of the genome. Virology. 2008;380:328–37.
189. Kim Y-R, Kim J-S, Lee S-H, Lee W-R, Sohn J-N, Chung Y-C, Shim H-K, Lee S-C, Kwon M-H, Kim Y-S. Heavy and light chain variable single domains of an anti-DNA binding antibody hydrolyze both double- and single-stranded DNAs without sequence specificity. J Biol Chem. 2006;281:15287–95.
190. Hoang PM, Cho S, Kim KE, Byun SJ, Lee T-K, Lee S. Development of *Lactobacillus paracasei* harboring nucleic acid-hydrolyzing 3D8 scFv as a preventive probiotic against murine norovirus infection. Appl Microbiol Biotechnol. 2015;99:2793–803.
191. Eltahla AA, Lim KL, Eden J-S, Kelly AG, Mackenzie JM, White PA. Nonnucleoside inhibitors of norovirus RNA polymerase: scaffolds for rational drug design. Antimicrob Agents Chemother. 2014;58:3115–23.
192. Vashist S, Urena L, Gonzalez-Hernandez MB, Choi J, de Rougemont A, Rocha-Pereira J, Neyts J, Hwang S, Wobus CE, Goodfellow I. Molecular chaperone Hsp90 is a therapeutic target for noroviruses. J Virol. 2015;89:6352–63.
193. Lin K, Gallay P. Curing a viral infection by targeting the host: the example of cyclophilin inhibitors. Antivir Res. 2013;99:68.
194. Geller R, Vignuzzi M, Andino R, Frydman J. Evolutionary constraints on chaperone-mediated folding provide an antiviral approach refractory to development of drug resistance. Genes Dev. 2007;21:195–205.
195. Taguwa S, Maringer K, Li X, Bernal-Rubio D, Rauch JN, Gestwicki JE, Andino R, Fernandez-Sesma A, Frydman J. Defining Hsp70 subnetworks in dengue virus replication reveals key vulnerability in flavivirus infection. Cell. 2015;163:1108–23.
196. Vashist S, Urena L, Chaudhry Y, Goodfellow I. Identification of RNA-protein interaction networks involved in the norovirus life cycle. J Virol. 2012;82:4449–60.
197. Mellatyar H, Talaei S, Pilehvar-Soltanahmadi Y, Barzegar A, Akbarzadeh A, Shahabi A, Barekati-Mowahed M, Zarghami N. Targeted cancer therapy through 17-DMAG as an Hsp90 inhibitor: overview and current state of the art. Biomed Pharmacother. 2018;102:608–17.

198. Shrestha L, Bolaender A, Patel HJ, Taldone T. Heat shock protein (HSP) drug discovery and development: targeting heat shock proteins in disease. Curr Top Med Chem. 2016;16:2753–64.
199. Gu Z, Shi W. Manipulation of viral infection by deubiquitinating enzymes: new players in host–virus interactions. Future Microbiol. 2016;11:1435–46.
200. Passalacqua KD, Charbonneau M-E, Donato NJ, Showalter HD, Sun D, Wen B, He M, Sun H, O'Riordan MXD, Wobus CE. Anti-infective activity of 2-cyano-3-acrylamide inhibitors with improved drug-like properties against two intracellular pathogens. Antimicrob Agents Chemother. 2016;60:4183–96.
201. Janssens S, Pulendran B, Lambrecht BN. Emerging functions of the unfolded protein response in immunity. Nat Immunol. 2014;15:910–9.
202. Gonzalez-Hernandez MJ, Pal A, Gyan KE, Charbonneau M-E, Showalter HD, Donato NJ, O'Riordan M, Wobus CE. Chemical derivatives of a small molecule deubiquitinase inhibitor have antiviral activity against several RNA viruses. PLoS One. 2014;9:e94491.
203. D'Arcy P, Wang X, Linder S. Deubiquitinase inhibition as a cancer therapeutic strategy. Pharmacol Ther. 2015;147:32–54.
204. Wang S, Kollipara RK, Srivastava N, et al. Ablation of the oncogenic transcription factor ERG by deubiquitinase inhibition in prostate cancer. Proc Natl Acad Sci U S A. 2014;111:4251–6.
205. Charbonneau M-E, Gonzalez-Hernandez MJ, Showalter HD, Donato NJ, Wobus CE, O'Riordan MXD. Small molecule deubiquitinase inhibitors promote macrophage anti-infective capacity. PLoS One. 2014;9:e104096.
206. Pokhrel L, Kim Y, Nguyen TDT, Prior AM, Lu J, Chang K-O, Hua DH. Synthesis and anti-norovirus activity of pyranobenzopyrone compounds. Bioorg Med Chem Lett. 2012;22:3480–4.
207. Perry JW, Wobus CE. Endocytosis of murine norovirus 1 into murine macrophages is dependent on dynamin II and cholesterol. J Virol. 2010;84:6163–76.
208. Gerondopoulos A, Jackson T, Monaghan P, Doyle N, Roberts LO. Murine norovirus-1 cell entry is mediated through a non-clathrin-, non-caveolae-, dynamin- and cholesterol-dependent pathway. J Gen Virol. 2010;91:1428–38.
209. Aybeke EN, Belliot G, Lemaire-Ewing S, Estienney M, Lacroute Y, Pothier P, Bourillot E, Lesniewska E. HS-AFM and SERS analysis of murine norovirus infection: Involvement of the lipid rafts. Small. 2017;13:1600918.
210. García-Arribas AB, Alonso A, Goñi FM. Cholesterol interactions with ceramide and sphingomyelin. Chem Phys Lipids. 2016;199:26–34.
211. Shivanna V, Kim Y, Chang K-O. Ceramide formation mediated by acid sphingomyelinase facilitates endosomal escape of caliciviruses. Virology. 2015;483:218–28.
212. Shivanna V, Kim Y, Chang K-O. The crucial role of bile acids in the entry of porcine enteric calicivirus. Virology. 2014;456–457:268–78.
213. Shivanna V, Kim Y, Chang K-O. Endosomal acidification and cathepsin L activity is required for calicivirus replication. Virology. 2014;464–465:287–95.
214. Dang W, Yin Y, Wang Y, et al. Inhibition of calcineurin or IMP dehydrogenase exerts moderate to potent antiviral activity against norovirus replication. Antimicrob Agents Chemother. 2017;61:e01095–17.
215. Ölschläger S, Neyts J, Günther S. Depletion of GTP pool is not the predominant mechanism by which ribavirin exerts its antiviral effect on Lassa virus. Antivir Res. 2011;91:89–93.
216. Airaksinen A, Pariente N, Menéndez-Arias L, Domingo E. Curing of foot-and-mouth disease virus from persistently infected cells by ribavirin involves enhanced mutagenesis. Virology. 2003;311:339–49.
217. Wray SK, Gilbert BE, Noall MW, Knight V. Mode of action of ribavirin: effect of nucleotide pool alterations on influenza virus ribonucleoprotein synthesis. Antivir Res. 1985;5:29–37.
218. Daughenbaugh KF, Fraser CS, Hershey JWB, Hardy ME. The genome-linked protein VPg of the Norwalk virus binds eIF3, suggesting its role in translation initiation complex recruitment. Eur Mol Biol Organ J. 2003;22:2852–9.

219. Royall E, Locker N. Translational control during calicivirus infection. Viruses. 2016;8:104.
220. Chaudhry Y, Nayak A, Bordeleau M-E, Tanaka J, Pelletier J, Belsham GJ, Roberts LO, Goodfellow IG. Caliciviruses differ in their functional requirements for eIF4F components. J Biol Chem. 2006;281:25315–25.
221. Tsumuraya T, Ishikawa C, Machijima Y, Nakachi S, Senba M, Tanaka J, Mori N. Effects of hippuristanol, an inhibitor of eIF4A, on adult T-cell leukemia. Biochem Pharmacol. 2011;81:713–22.
222. Malina A, Mills JR, Pelletier J. Emerging therapeutics targeting mRNA translation. Cold Spring Harb Perspect Biol. 2012;4:a012377.
223. Low W-K, Li J, Zhu M, Kommaraju SS, Shah-Mittal J, Hull K, Liu JO, Romo D. Second-generation derivatives of the eukaryotic translation initiation inhibitor pateamine A targeting eIF4A as potential anticancer agents. Bioorg Med Chem. 2014;22:116–25.
224. Friedman RM. Clinical uses of interferons. Br J Clin Pharmacol. 2008;65:158–62.
225. Chang K-O, George DW. Interferons and ribavirin effectively inhibit Norwalk virus replication in replicon-bearing cells. J Virol. 2007;81:12111–8.
226. Changotra H, Jia Y, Moore TN, Liu G, Kahan SM, Sosnovtsev SV, Karst SM. Type I and type II interferons inhibit the translation of murine norovirus proteins. J Virol. 2009;83:5683–92.
227. Hwang S, Maloney NS, Bruinsma MW, et al. Nondegradative role of Atg5-Atg12/ Atg16L1 autophagy protein complex in antiviral activity of interferon gamma. Cell Host Microbe. 2012;11:397–409.
228. Lee S, Wilen CB, Orvedahl A, McCune BT, Kim K-W, Orchard RC, Peterson ST, Nice TJ, Baldridge MT, Virgin HW. Norovirus cell tropism is determined by combinatorial action of a viral non-structural protein and host cytokine. Cell Host Microbe. 2017;22:449–459.e4.
229. Nice TJ, Robinson BA, Van Winkle JA. The role of interferon in persistent viral infection: insights from murine norovirus. Trends Microbiol. 2018;26:510–24.
230. Baldridge MT, Nice TJ, McCune BT, Yokoyama CC, Kambal A, Wheadon M, Diamond MS, Ivanova Y, Artyomov M, Virgin HW. Commensal microbes and interferon-λ determine persistence of enteric murine norovirus infection. Science. 2015;347:266–9.
231. Rocha-Pereira J, Jacobs S, Noppen S, Verbeken E, Michiels T, Neyts J. Interferon lambda (IFN-λ) efficiently blocks norovirus transmission in a mouse model. Antivir Res. 2018;149:7–15.
232. Kim Y, Thapa M, Hua DH, Chang K-O. Biodegradable nanogels for oral delivery of interferon for norovirus infection. Antivir Res. 2011;89:165–73.
233. Brown LAK, Clark I, Brown JR, Breuer J, Lowe DM. Norovirus infection in primary immune deficiency. Rev Med Virol. 2017;27:e1926.
234. Rossignol J-F. Nitazoxanide: a first-in-class broad-spectrum antiviral agent. Antivir Res. 2014;110:94–103.
235. Siddiq DM, Koo HL, Adachi JA, Viola GM. Norovirus gastroenteritis successfully treated with nitazoxanide. J Infect. 2011;63:394–7.
236. Thorne L, Goodfellow I. Reply to Kempf et al. J Infect Dis. 2017;215:487–8.
237. Kempf B, Edgar JD, Mc Caughey C, Devlin LA. Nitazoxanide is an ineffective treatment of chronic norovirus in patients with X-linked agammaglobulinemia and may yield false-negative polymerase chain reaction findings in stool specimens. J Infect Dis. 2017;215:486–7.
238. Dang W, Yin Y, Peppelenbosch MP, Pan Q. Opposing effects of nitazoxanide on murine and human norovirus. J Infect Dis. 2017;216:780–2.
239. Pan Q, Chen S, Ma B, Peppelenbosch MP, Xu L, Dang W, Yin Y, Chang K-O. Nitazoxanide inhibits human norovirus replication and synergizes with ribavirin by activation of cellular antiviral response. Antimicrob Agents Chemother. 2018;62:e00707–18.
240. Rossignol J-F. Thiazolides: a new class of antiviral drugs. Expert Opin Drug Metab Toxicol. 2009;5:667–74.
241. Dou D, Mandadapu SR, Alliston KR, Kim Y, Chang K-O, Groutas WC. Design and synthesis of inhibitors of noroviruses by scaffold hopping. Bioorg Med Chem. 2011;19:5749–55.

242. Dou D, Tiew K-C, He G, Mandadapu SR, Aravapalli S, Alliston KR, Kim Y, Chang K-O, Groutas WC. Potent inhibition of Norwalk virus by cyclic sulfamide derivatives. Bioorg Med Chem. 2011;19:5975–83.
243. Dou D, Mandadapu SR, Alliston KR, Kim Y, Chang K-O, Groutas WC. Cyclosulfamide-based derivatives as inhibitors of noroviruses. Eur J Med Chem. 2012;47:59–64.
244. Dou D, He G, Mandadapu SR, Aravapalli S, Kim Y, Chang K-O, Groutas WC. Inhibition of noroviruses by piperazine derivatives. Bioorg Med Chem Lett. 2012;22:377–9.
245. Dou D, Tiew K-C, Mandadapu SR, Gunnam MR, Alliston KR, Kim Y, Chang K-O, Groutas WC. Potent norovirus inhibitors based on the acyclic sulfamide scaffold. Bioorg Med Chem. 2012;20:2111–8.
246. Shawli G, Adeyemi O, Stonehouse N, Herod M. The oxysterol 25-hydroxycholesterol inhibits replication of murine norovirus. Viruses. 2019;11:97.
247. Eriksson B, Helgstrand E, Johansson NG, et al. Inhibition of influenza virus ribonucleic acid polymerase by ribavirin triphosphate. Antimicrob Agents Chemother. 1977;11:946–51.
248. Furuta Y, Takahashi K, Kuno-Maekawa M, Sangawa H, Uehara S, Kozaki K, Nomura N, Egawa H, Shiraki K. Mechanism of action of T-705 against influenza virus. Antimicrob Agents Chemother. 2005;49:981–6.
249. Sissoko D, Laouenan C, Folkesson E, et al. Experimental treatment with favipiravir for ebola virus disease (the JIKI trial): a historically controlled, single-arm proof-of-concept trial in Guinea. PLoS Med. 2016;13:e1001967.
250. Nagata T, Lefor AK, Hasegawa M, Ishii M. Favipiravir: a new medication for the Ebola virus disease pandemic. Disaster Med Public Health Prep. 2015;9:79–81.
251. Furuta Y, Takahashi K, Shiraki K, Sakamoto K, Smee DF, Barnard DL, Gowen BB, Julander JG, Morrey JD. T-705 (favipiravir) and related compounds: novel broad-spectrum inhibitors of RNA viral infections. Antivir Res. 2009;82:95–102.
252. Rocha-Pereira J, Jochmans D, Dallmeier K, Leyssen P, Cunha R, Costa I, Nascimento MSJ, Neyts J. Inhibition of norovirus replication by the nucleoside analogue 2'-C-methylcytidine. Biochem Biophys Res Commun. 2012;427:796–800.
253. Afdhal N, O'Brien C, Godofsky E, et al. Valopicitabine (NM 283), alone or with peg-interferon, compared to peg-interferon/ribavirin (PEGIFN/RBV) retreatment in patients with HCV-1 infection and prior non-response to PEGIFN/RBV: one year results. J Hepatol. 2007;46:S5.
254. Perales C, Agudo R, Tejero H, Manrubia SC, Domingo E. Potential benefits of sequential inhibitor-mutagen treatments of RNA virus infections. PLoS Pathog. 2009;5:e1000658.
255. Sharma PL, Nurpeisov V, Hernandez-Santiago B, Beltran T, Schinazi RF. Nucleoside inhibitors of human immunodeficiency virus type 1 reverse transcriptase. Curr Top Med Chem. 2004;4:895–919.
256. Gelman MA, Glenn JS. Mixing the right hepatitis C inhibitor cocktail. Trends Mol Med. 2011;17:34–46.
257. Pariente N, Sierra S, Lowenstein PR, Domingo E. Efficient virus extinction by combinations of a mutagen and antiviral inhibitors. J Virol. 2001;75:9723–30.
258. López-Manríquez E, Vashist S, Ureña L, Goodfellow I, Chavez P, Mora-Heredia JE, Cancio-Lonches C, Garrido E, Gutiérrez-Escolano AL. Norovirus genome circularization and efficient replication are facilitated by binding of PCBP2 and hnRNP A1. J Virol. 2013;87:11371–87.
259. Manita D, Toba Y, Takakusagi Y, et al. Camptothecin (CPT) directly binds to human heterogeneous nuclear ribonucleoprotein A1 (hnRNP A1) and inhibits the hnRNP A1/topoisomerase I interaction. Bioorg Med Chem. 2011;19:7690–7.
260. Maga G, Falchi F, Radi M, et al. Toward the discovery of novel anti-HIV drugs. Second-generation inhibitors of the cellular ATPase DDX3 with improved anti-HIV activity: synthesis, structure-activity relationship analysis, cytotoxicity studies, and target validation. ChemMedChem. 2011;6:1371–89.

261. Radi M, Falchi F, Garbelli A, et al. Discovery of the first small molecule inhibitor of human DDX3 specifically designed to target the RNA binding site: towards the next generation HIV-1 inhibitors. Bioorg Med Chem Lett. 2012;22:2094–8.
262. Brai A, Fazi R, Tintori C, et al. Human DDX3 protein is a valuable target to develop broad spectrum antiviral agents. Proc Natl Acad Sci U S A. 2016;113:5388–93.
263. McCune BT, Tang W, Lu J, et al. Noroviruses co-opt the function of host proteins VAPA and VAPB for replication via a phenylalanine–phenylalanine- acidic-tract-motif mimic in nonstructural viral protein NS1/2. MBio. 2017;8:e00668–17.
264. Sengupta S, Miller KK, Homma K, Edge R, Cheatham MA, Dallos P, Zheng J. Interaction between the motor protein prestin and the transporter protein VAPA. Biochim Biophys Acta, Mol Cell Res. 2010;1803:796–804.
265. Ettayebi K, Hardy ME. Norwalk virus nonstructural protein p48 forms a complex with the SNARE regulator VAP-A and prevents cell surface expression of vesicular stomatitis virus G protein. J Virol. 2003;77:11790–7.
266. Nice TJ, Strong DW, McCune BT, Pohl CS, Virgin HW. A single-amino-acid change in murine norovirus NS1/2 is sufficient for colonic tropism and persistence. J Virol. 2013;87:327–34.
267. Almand EA, Moore MD, Outlaw J, Jaykus L-A. Human norovirus binding to select bacteria representative of the human gut microbiota. PLoS One. 2017;12:e0173124.
268. Karst SM, Wobus CE. A working model of how noroviruses infect the intestine. PLoS Pathog. 2015;11:e1004626.
269. Miura T, Sano D, Suenaga A, Yoshimura T, Fuzawa M, Nakagomi T, Nakagomi O, Okabe S. Histo-blood group antigen-like substances of human enteric bacteria as specific adsorbents for human noroviruses. J Virol. 2013;87:9441–51.
270. Lee H, Ko G. Antiviral effect of vitamin A on norovirus infection via modulation of the gut microbiome. Sci Rep. 2016;6:25835.

Norovirus Correlates of Protection

6

Nada M. Melhem and Farouk F. Abou Hassan

6.1 Introduction

Human Norovirus (HuNoV), one of the leading causes of acute gastroenteritis (AGE), affects all age groups with higher infection rates observed in children less than 5 years old [1, 2]. Nevertheless, severe NoV-associated illness is also common among adults aged 65 years and above [3]. NoV infection results in symptomatic or asymptomatic infection with an estimated incubation period of 24–72 h in symptomatic infections. NoV infection is usually self-limited; however, NoV-associated gastroenteritis lasts for weeks or months among immunocompromised individuals. The clinical presentation of NoV-gastroenteritis typically includes non-bloody diarrhea and vomiting; nausea, abdominal pain, and fever could also be observed. Little is known about the immune responses following NoV infection and their ability to prevent and protect against secondary infections. Preclinical, clinical, and challenge studies have been essential models promoting our current understanding of correlates of protection from NoV-associated illness. The identification of the former is critical for the evaluation and the development of NoV vaccines [4]. This chapter advances the most recent findings on innate and adaptive immune responses, correlates of protection as well as the future directions in our understanding of immune mechanisms against noroviruses especially with the recent development of reproducible culture systems [5–7].

N. M. Melhem (✉) · F. F. Abou Hassan
Medical Laboratory Sciences Program, Division of Health Professions, Faculty of Health Sciences, American University of Beirut, Beirut, Lebanon
e-mail: melhemn@aub.edu.lb

N. M. Melhem (ed.), *Norovirus*,
https://doi.org/10.1007/978-3-030-27209-8_6

6.2 Innate Immune Responses

Few studies characterized the innate immune responses against noroviruses in humans. Recently, the acute cytokine responses were described in human volunteers challenged with Norwalk virus GI.1 NV [8]. While a T helper type 1 (Th1)-biased cytokine response was predominant, Th2 cytokines as well as interleukin-8 (IL-8) and monocyte chemoattractant protein (MCP)-1 increased in infected subjects as compared to uninfected individuals. The concentration of most cytokines increased post-challenge in both symptomatic and asymptomatic individuals. In general, most cytokines peaked 2 days post-challenge in infected volunteers. While the signs and symptoms caused by noroviruses have been described to be mediated by intestinal damage [9], symptomatic individuals participating in this study had a significant increase in serum interferon (IFN)-γ, tumor necrosis factor (TNF)-β, IL1b, interleukin-1 receptor antagonist (IL-1Ra), IL-2, IL-4, IL-8, and IL-10 four days after challenge, suggesting immune-mediated symptoms [8]. This study also reported significantly elevated levels of IL-6 and decreased levels of IL-12p40 associated with increased same day viral shedding. The level of viral shedding was similar between symptomatic and asymptomatic individuals 5 days post-challenge suggesting the lack of association between symptoms and viral shedding. This finding is supported by previous reports showing that fecal shedding is similar among symptomatic and asymptomatic individuals with norovirus detected in stool up to 8 weeks following the onset of symptoms [10–13].

Similar results were previously reported whereby a surge in serum Th1 cytokines (IFN-γ, IL-2, and TNF-α) was also detected in an earlier human volunteer study 2 days following challenge with GII.2 Snow Mountain virus (SMV); with contradictory results pertinent to IL-6 levels [14]. While these results are similar to those observed in gnotobiotic (Gn) pigs [15, 16], they differ in the timing and the level of cytokine responses generated following challenge. These studies suggest an increase in inflammatory cytokines following infection; however, further investigations are needed to identify the specific role of cytokine responses in gastrointestinal pathology and symptomatic illness.

Elevated sera levels of Th1-related cytokines such as IL-18, chemokine (C-X-C motif) ligand 9 (CXCL9) and CXCL10, soluble IL-2 receptor alpha (sIL-2Rα) and macrophage immigration factor (MIF) were also detected among older Swedish patients (age range: 51–86 years) admitted to hospitals with suspected AGE [17]. Moreover, lower levels of chemokine (C-C-motif) ligand 5 (CCL5) were associated with an extended viral shedding and slow viral clearance. These results should be cautiously interpreted especially due to the small sample size ($N = 93$), and the age range of study participants excluding children as the most common group affected by noroviruses.

While the IFN response is known to control murine norovirus (MNV) replication [18–22], it is unclear whether similar results are observed following infection with HuNoV [23, 24]. In the era of absence of a NoV cell culture infection system, the 293FT cells, derived from human embryonic kidney HEK293 cells, were transfected with NoV RNA isolated from stool of a volunteer following challenge with Norwalk virus GI.1 [25]. Replication was confirmed by detection of VP1. The replication of GI.1 and GII.3 in 293FT mammalian epithelial cells did not induce an IFN response. Moreover, the neutralization or knockdown of essential factors involved in the stimulation of IFN pathways did not enhance viral replication. These results suggest that epithelial IFN responses do not play a role in HuNoV replication, and thus contradict data reported in the mouse model [20, 22, 24] (described below). Importantly, the validation of these results is needed using other genogroups and genotypes. Further studies are needed to investigate the role of cytokines in pathogenesis and/or protection following NoV infection.

More studies were conducted in animal models to investigate the role of innate immunity in the control of NoV infection [26]. These models include Gn pigs, Gn calves and mice. Cytokine responses as well as intestinal and mucosal immunity were assessed following infection of Gn pigs with human GII.4 NoV (HS66 strain) [16]. The results show an increasse in the innate IFN-α in all infected animals; the former remained detectable until 28 days post-inoculation with HuNoV-HS66. Th1 IFN-γ and the pro-inflammatory IL-6 remained detectable until day 10 post-inoculation. Serum Th1 IL-12 levels also significantly increased following infection as compared to control pigs and followed the same trend of IFN-α. On the other hand, IL-4 and IL-10 (Th2 cytokines) were detected but at lower levels as compared to Th1 cytokines. When compared to control, a significant increase in IFN-α in the intestinal contents of infected pigs was detected on day 2 post-inoculation (PID 2), the day of the onset of diarrhea. Moreover, IL-12 was also elevated. In a later study, HuNoV-HS66 was used to infect Gn calves [15]. Data from this study showed increased serum levels of IFN-γ (Th1), IL-12 (Th1), TNF-α (pro-inflammatory) as well as the Th2/regulatory T cell (Treg) cytokines, IL-4 and IL-10. These findings along with those reported in earlier studies in humans [14] confirm the early release of IFN-γ associated with early innate immune responses.

Since MNV and HuNoV share a similar mechanism of RNA replication, the mouse model has been used to study the early responses generated following NoV infection. The inoculation of mice lacking IFN-α/β and IFN-γ receptors with MNV-1 strain resulted in lethal infection [26]. Moreover, signal transducer and activator of transcription (STAT)-1 deficient mice showed higher levels of MNV-1 RNA in the intestine, liver, and spleen with a considerable pathology detected on days 3 and 7 following infection and resulted in death, thus suggesting the role of STAT1 in innate immunity following NoV infection. The same study reported the efficient and persistent replication of MNV-1 in mice lacking the recombination-activating gene (RAG). RAG plays a crucial role in variable (V), diversity (D) and joining (J) genes (V(D)J) recombination of T-cell receptor (TCR) and B-cell receptor (BCR) [27]. This observation led the authors to further suggest that the role of innate immune responses is sufficient to control MNV-1 infection. These results were recorded in immunosuppressed mice and highlighted the potential role of human noroviruses in persons with immunodeficiencies.

In addition to the ability of MNV-1 to infect the intestinal tract of mice and the significant role of STAT1 and IFN receptors in enhancing resistance to infection in vitro [26], STAT1 along with IFN-αβ receptor inhibited growth of MNV-1 in dendritic cells (DCs) and macrophages [24]. Moreover, STAT1-dependent responses inhibited MNV-1 replication in the intestines of mice, limited MNV-1 dissemination to peripheral tissues such as lungs and liver but not spleen, and prevented clinical disease while leading to mild inflammation of the lamina propria in immunocompetent mice [28]. These results suggest that STAT1-dependent interferon responses may play a role in reducing the severity of the disease and direct interest in defining the mechanisms associated with IFN-based NoV treatment.

In an attempt to understand the mechanisms defining the IFN-α/β responses during MNV-1 infection, IFN-α/β-deficient (IFN-α/β$^{-/-}$) mice were inoculated with MNV [22]. Significant levels of replication were detected in the liver, spleen, mesenteric lymph nodes (MLNs), and lungs. These results suggest the role of IFN-α/β responses in the control of infection in vivo. Since MNV was previously shown to replicate in macrophages and DCs [24], the role of IFN-α/β responses was also tested in these cells in vivo and results mirrored those detected in vitro. The role of interferon regulatory factor-3 (IRF-3) and IRF-7, transcription factors regulating IFN-α/β responses following viral infections [29], was also tested and results showed that viral titers were significantly higher in tissues of IRF-3$^{-/-}$ and IRF-7$^{-/-}$ mice. Moreover, the presence of IRF-3 and IRF-7 resulted in inhibition of viral replication in primary macrophages and DCs. Importantly, data from this study also showed that IRF-3, IRF-7, and IFN-α/βR induce higher antiviral responses in tissues compared to the intestinal tract. Authors suggested other factors such as IFN-γ to play a role in controlling viral replication in the latter.

Following infection with MNV-1.CW3, melanoma differentiation-associated protein-5 (MDA5)-deficient mice and toll-like receptor-3 deficient (TLR-3$^{-/-}$) mice showed higher MNV-1 viral titers as compared to wild-type (WT) mice [20]. MDA-5 is a RIG-I-like receptor dsRNA helicase enzyme. The higher viral load was mainly detected in MLN, spleen, or proximal intestine of infected mice. Moreover, the treatment of bone marrow-derived DC of MDA5$^{-/-}$ mice infected with MNV-1 with type I IFN in vitro, prevented the increase in viral titers and reconstituted the WT phenotype. MDA5 was accordingly reported to function as a sensor of MNV-1 infection stimulating antigen-presenting cells to produce type I interferon, IL-6, MCP-1, and TNF-α. These cytokines are important in initiating antiviral pathways; consequently, MDA5 deficiency led to increased viral replication in mice and deficient cells as well as suppression of cytokines production. Further studies are needed to investigate the role of these sensors as well as others in mediating innate antiviral responses.

As described above, type I IFN (IFN-α and IFN-β) and type II IFN (IFN-γ) play a role in the control of HuNoV and MNV replication. Recently, the role of type III IFN-λ was explored [21]. Type I IFNs signal through interferon alpha and beta receptor subunit 1 (IFNAR1) and IFNAR2; type II IFN (IFN-γ) signals through interferon gamma receptor subunit 1 (IFNGR1) and IFNGR2; and type III through a receptor composed of interferon lambda receptor 1 (IFNLR1) and IL10R2 (IFNLR) [30, 31]. Mice deficient in *Ifnar1*, *Ifngr1*, *Ifnlr1*, or *STAT1* were inoculated with MNV CR6 strain. The results of this study highlighted the role of *STAT1* in controlling viral replication in colon. In addition, *Ifnar1* and *STAT1* played a role in limiting replication in tissues specifically spleen [22] and MLNs as previously reported [20, 24, 26]. The stimulation of IFN-λ and IFN-β in MLN, colon, and Peyer's patches (PP) were compared in mice inoculated with CR6 (a norovirus persistent strain) or the nonpersistent strain CW3. The latter induced IFN-λ and IFN-β in MLNs and PPs whereas CR6 did not. One day following oral inoculation with MNV CR6 strain, mice were treated with IFN-λ through the intraperitoneal route. This treatment inhibited the establishment of persistent MNV infection suggesting the role of IFN-λ in controlling systemic murine NoV infection. In a later study, authors investigated the role of type I IFN responses in control of CW3 systemic persistent infection [32]. DCs of mice deficient in type 1 IFN (DC11c-*Ifnar1*$^{-/-}$) allowed the persistence of CW3 MNV, a nonpersistent strain. This study linked persistence to increased viral titers in the intestines, spleen, liver, and MLNs. Moreover, an increased expression of DC activation markers including major histocompatibility complex (MHC) class I and class II, as well as co-stimulatory molecules (CD40, CD80, and CD86) was reported in addition to activation of CD8$^+$ T cell responses and antibody production. The stimulation of the adaptive immune responses in an environment of innate immune deficiency did not clear the infection thus suggesting the impact of the innate immune responses, specifically type I IFN on viral persistence. This model has been previously described in persistent hepatitis C virus infection in humans [33].

The role of IFN-λ was also tested in RAG1$^{-/-}$ mice inoculated with CR6 and treated with IFN-λ [18]. Intestinal persistence of CR6 was treated in these mice suggesting that the role of IFN-λ is independent of B and T cells. The concept of sterilizing innate immunity at mucosal surfaces was entertained by the authors despite the role played by adaptive immune responses in inducing long-lasting immunity. Mouse model depleted of *Ifnlr1* showed that the expression of *Ifnlr1* on intestinal epithelial cells (IECs) of mice was essential for clearing persistent MNV infection. In addition to *Ifnlr1*, the nucleotide-binding oligomerization domain-like receptor (Nlrp6) has been also shown recently to limit MNV-1 infection in mice gut [34, 35]. Briefly, Nlrp6 is highly expressed in the intestines and plays a role in maintaining intestinal homeostasis and responding to viral infection through inflammasome-dependent manner by serving as viral RNA sensor. Wang et al. [35] showed that *Nlrp6*$^{-/-}$ infected with MNV-1 had higher viral loads in the intestines, spleen, and MLNs as compared to wild-type mice. Moreover, WT mice were able to clear MNV-1 infection more rapidly than *Nlrp6*$^{-/-}$ mice. Nlrp6 was suggested to interact with DEAH (Asp-Glu-Ala-His) box helicase 15 (Dhx15) as a viral RNA sensor inducing type I and type II IFNs as well as IFN-stimulated genes (ISG) and consequently stimulating antiviral responses against enteric viruses namely MNV1 and encephalomyocarditis virus (EMCV).

McFadden et al. described and characterized a new open reading frame (ORF4) encoded by the MNV subgenomic RNA [36]. The protein encoded by ORF4 gene, referred to as virulence factor 1 (VF1) and localized in the mitochondria, was able to antagonize the innate immune response. Murine leukemia macrophage cell line (RAW264.7) infected with MNV-1 lacking VF1 produced higher IFN-β, CXCL10, and ISG54 mRNA in vitro compared to cells infected with WT MNV-1.

STAT1$^{-/-}$ mice were infected with MNV CW1 strain lacking the ability to express ORF4; these mice did not mount clinical symptoms. Viral RNA was detected in MLNs of WT mice but not in tissues of the strain lacking VF1 confirming further the significant role played by the latter. Therefore, authors proposed that infection with the virus lacking VF1 leads to the induction of ISGs; the mechanism behind this was suggested to be either via STAT2 or another unknown mechanism. Further studies are needed to understand the immune regulatory function of VF1 in mouse model and whether a counterpart gene exists in humans facilitating NoV escape from the immune system.

Future studies are needed to explain the role of innate antiviral responses in control of NoV infection. Understanding these responses will help elucidate their role in predicting infection and promote our understanding of acute immune responses. The development of an effective vaccine depends on understanding the interplay between pathogenesis, immune responses, and clinical outcomes.

6.3 Adaptive Immune Responses

The success of potential NoV vaccines relies on our understanding of the magnitude and the breadth of immune responses they generate. In an attempt to assess the best correlates of protection, studies in animal and human models as well as volunteer challenge studies advanced results on the role of NoV-specific protective antibodies (sera-specific antibodies, salivary IgA, and fecal IgA), the increased significance of T cell responses in NoV-associated gastroenteritis, and importantly the emergence of genetic variants and its impact on these responses. However, more is known about serum histo-blood group antigen (HBGA)-blocking antibodies as correlates of protection. The rest of this chapter summarizes the current knowledge on the development of norovirus vaccine in relation to surrogate measures of protection including HBGA and adaptive immune responses (Table 6.1).

GI and GII noroviruses have been recognized to be responsible for most human infections [56]. Specifically, GII.4 NoVs are known to widely circulate and cause sporadic acute viral gastroenteritis worldwide [57]. GII.4 strains are also responsible for 80% of NoV infections globally [58]. The reason behind the persistence and the wide spread of GII.4 globally is not well understood yet linked to viral fitness as a result of mutations in antigenic epitopes, the host's susceptibility, and the duration of GII.4-specific antibodies and NoV cross-reactive antibodies [57]. In addition, recent structural analysis reports investigating the binding of norovirus with the putative receptor HBGA using crystallography revealed differences in the binding interface between GI and GII noroviruses [59, 60]. Other studies demonstrated that GII.4 P domains can bind to multiple HBGA types [61], and antigenic variation of neutralizing epitopes of GII.4 significantly influence the evolution and fitness of GII.4 variants [58]. This comes in accordance with reports suggesting that inter- and intra-genotype recombination in addition to herd immunity serve as driving forces for the evolution of GII.4 lineage resulting in the emergence of new GII.4 strains [62]. Moreover, it has been recently demonstrated that the viral particle dynamics

Table 6.1 Norovirus correlates of protection

Study design	Immune responses (antibodies, T cell responses)	HBGA[a]/HAI[b]	References
Preclinical	Homotypic and heterotypic responses were mounted in mice immunized with Norwalk virus, NV; Snow Mountain virus, SM; Hawaii virus, HV; or Lordsdale virus, LV. It was noted that while homotypic responses were stronger, heterotypic ones were more robust within rather than between genogroups.	Sera from mice receiving the trivalent VRP[c] vaccine (NV, SM, HV) or the tetravalent VRP vaccine (NV, SM, HV, LV) blocked H1 binding to NV-VLP[d] and H3 binding to LV-VLP. H3-LV-VLP blocking ability was significantly higher in mice receiving the tetravalent vaccine.	LoBue et al. [37]
Preclinical	Monovalent and multivalent vaccine (with NV and/or LV VLPs) induced IgG2a titers in mice. Cross-reactive IgG responses were detected following immunization of BALB/c mice with multivalent vaccines.	Mice immunized with multivalent vaccines mounted cross-reactive and receptor blocking IgG responses against heterologous strains.	LoBue et al. [38]
Clinical	A dose-dependent response with increased sera titers was observed with increased vaccine (GI.1 VLPs) dosage. NoV-specific IgA and IgG ASCs[e] were detected.	GMT[f] of HAI antibodies showed a nine-fold increase following the second dose of the vaccine. This was observed in 74% of subjects receiving 100 μg GI.1 VLP vaccine.	El-Kamary et al. [39]
Challenge	None reported	Secretors had ≥4-fold increase in BT50[g]; the latter persisted until day 180 post-challenge with live NV. Fecal viral shedding was higher in individuals with no detectable pre-existing serum blocking antibodies compared to those with pre-existing blocking antibodies. HBGA-blocking antibodies were associated with protection following NV GI.1 challenge.	Reeck et al. [40]
Clinical	60% responded to at least one GI VLPs (GI.1 1968, GI.1 2001, GI.2 1999, GI.3 1999, GI.4 2000) stimulation by secreting IFN-γ from $CD4^+$ T cells.	GI VLPs were able to bind to HBGA including fucose, Lewis and A antigens. Strain-specific HBGA-blocking antibodies were detected against GII.4 NoVs. Cross-reactive blocking antibodies were also detected against GI NoVs belonging to different genotypes.	Lindesmith et al. [41]
Clinical	Following the second dose of the vaccine (GI.1 VLPs), 70% of vaccine recipients showed ≥4-fold increase in specific IgA. 50% and 32% of vaccine recipients showed ≥4-fold increase in NoV-specific IgG and IgM, respectively.	Serum HBGA-blocking antibodies were associated with significant reduction in the frequency of NoV infection and illness. Pre-challenge serum levels of antibodies blocking the binding of NV VLPs to HBGA H1 correlated with a lower risk of virus-associated illness post-challenge.	Atmar et al. [42]

(continued)

Table 6.1 (continued)

Study design	Immune responses (antibodies, T cell responses)	HBGA[a]/HAI[b]	References
Challenge	All infected individuals showed a ≥4-fold rise in anti-NoV ELISA titers between day 0 and day 28 post-infection with Norwalk virus (GI.1).	At day 28 post-challenge, all infected individuals showed a peak serum HAI GMT as well as HBGA-blocking antibody levels. There was no rise in HAI titer in uninfected subjects.	Czako et al. [43]
Challenge	Following challenge of humans with GII.4, IgG response was stronger among secretors compared to nonsecretors.	A four fold increase of HBGA-blocking antibodies was reported 30 days post-challenge. HBGA-blocking antibodies were not detected in nonsecretors pre- and post-challenge with GII.4.	Frenck et al. [44]
Preclinical	Single and trivalent VLP vaccines (GI-3, GII.4, and RV rVP6) administered to mice induced a strong homologous IgG response. The trivalent formulation induced cross-reactive IgG responses, mucosal and T cell responses.	Single NoV-VLP immunization did not elicit cross-reactive HBGA-blocking antibody response. Cross-reactive HBGA-blocking antibody responses were elicited against heterologous NoV-VLPs in mice immunized with the trivalent vaccine. The intensity of blocking activity against homologous GII.4 and GI.3 VLPs was similar in mice immunized with the single and trivalent vaccine formulas.	Tamminen et al. [45]
Clinical	Following the first dose of the 50/50 µg bivalent GI.1/GII.4c VLPs, a significant increase in serum GI.1 and GII.4 Pan-Ig, IgG, and IgA GMTs was observed. Serum antibody responses against GI.1 were higher than those mounted against GII.4.	Strong HBGA-blocking antibody response was elicited after the first dose of vaccine with blocking titers of ≥200 against GI.1 and GII.4 viruses detected in the majority of subjects 28 days post-vaccination.	Treanor et al. [46]
Challenge	By day 28 post-challenge, ≥4-fold increase in NV-specific salivary IgA was detected in 70% of infected subjects whereas all infected individuals showed >4-fold increase in fecal IgA. Similarly, all infected subjects showed a >4-fold increase in IgA and IgG ASC levels.	Higher HBGA-blocking antibody titers were detected in infected individuals who did not develop gastroenteritis. Pre- existing serum HBGA-blocking antibodies significantly correlated with pre-existing salivary but not fecal IgA.	Ramani et al. [47]
Clinical	Robust increase in serum IgG and IgA was reported following IM[h] vaccination with the bivalent GI.1/GII.4 VLP vaccine; greater responses were detected against GI.1.	Pre-challenge HBGA-blocking antibody levels of >1:500 in the vaccinees significantly decreased the likelihood of developing moderate-to-severe vomit or diarrhea following GII.4 challenge.	Atmar et al. [48]

Challenge	None reported	Human volunteers orally challenged with Norwalk virus elicited homotypic and heterotypic serum HBGA-blocking antibodies with higher titers detected against the homotypic challenge strain (Norwalk virus GI.1). Modest HBGA-blocking antibody response was detected against heterotypic NoV strains; GI.4, GI.7, and GII.4 (HOV and Sydney) VLPs.	Czako et al. [49]
Challenge	Sera from volunteers previously challenged with GII.4 demonstrated IgG cross-reactivities to GII.4 variants circulating between 1998 and 2012.	Cross-reactivity was observed to GII.4 variants circulating between 1998 and 2012. ≥4-fold increase in serum-blocking titers was detected in post-challenge sera from all secretors previously challenged with GII.4.	Dai et al. [50]
Clinical	Following immunization with multivalent GI.1/GII.4 VLPs, IgG GMFR[i] titers peaked at day 7 against GI and GII.4. Cross-reactive NoV-specific IgG antibodies were detected against GII.2, GII.3, and GII.14 VLPs. IgG antibodies were also detected against novel GII.4 variants emerging following the time of vaccination.	The blockade antibody titers against GI and GII.4 variants peaked at days 7 and 35 post-vaccination. Cross-reactive blocking titers were detected against non-vaccine genotypes including GII.3 and GII.14 as well as novel GII.4 strains. Interestingly, vaccination stimulated a boost in blocking titers against GII.4 variants circulating prior to vaccination.	Lindesmith et al. [51]
Clinical	An increase in Pan-Ig, IgG, and IgA to GI.1 and GII.4 was observed and peaked at 7–10 days post-vaccination with the 15/50 μg or 50/50 μg of GI.1/GII.4 VLP vaccine.	HBGA-blocking antibodies peaked at day 7–10 post-vaccination; the baseline titers were higher for GII.4 compared to GI.1. The magnitude of HBGA-blocking antibody response was less than that of Pan-Ig, IgG, and IgA.	Atmar et al. [52]
Preclinical	Mice immunized with six different VLPs (GII.4 1999, GII.4 2009, GII.4 2012, GII.12, GI.1, and GI.3) mounted type-specific and cross-reactive high serum IgG.	Robust genotype-specific blocking antibody titers were detected in all immune sera.	Malm et al. [53]
Preclinical	Both IM- and IN[j]-administered GII.4 VLPs induced serum and mucosal-specific IgG. IN immunization induced serum and mucosal specific IgA responses.	Similar blocking activity was observed in sera following IN or IM immunization. Intranasal immunization elicited mucosal IgA blocking antibodies; the latter strongly correlated with mucosal IgA but not IgG.	Tamminen et al. [54]

(continued)

Table 6.1 (continued)

Study design	Immune responses (antibodies, T cell responses)	HBGA[a]/HAI[b]	References
Preclinical	High GII.4-specific IgG titers with high avidity were detected in sera of mice immunized with GII.4-1999 VLPs alone, GII VLPs mix (GII.4 1999, GII.4 NO, GII.12), or GII VLPs mix and GI VLPs (GI.1, GI.3) regardless of the administration route (IM or ID[k]). The magnitude and functional avidity of GII.4-specific IFN-γ-producing T cells were similar in immunized mice. CD107 was detected on splenocytes of immunized mice stimulated with peptide 99–45 containing highly conserved epitope of GII NoVs.	Similar blocking antibody activity was detected in sera of immunized mice regardless of the administration route (IM or ID).	Malm et al. [55]

[a]Histo-blood group antigen
[b]Hemagglutination inhibition
[c]Virus replicon particles
[d]Virus-like particles
[e]Antibody-secreting cells
[f]Geometric Mean Titer
[g]Blocking Titer 50: percentage of sera necessary to block 50% of NoV-HBGA binding
[h]Intramuscular
[i]Geometric Mean Fold Rise
[j]Intranasal
[k]Intradermal

(binding and conformational changes) regulate the access of virus-specific antibodies to blockade epitopes and propose that this could be a frequent mechanism of GII.4 NoV to evade humoral immune responses [63]. These reasons could explain the persistence of GII.4 worldwide and the challenges facing the development of a cross-reactive and highly effective NoV vaccine.

6.4 Correlates of Protection

6.4.1 Histo-Blood Group Antigen (HBGA) and Norovirus

One of the most studied determinants of NoV susceptibility is HBGA in addition to HBGA-blocking antibodies identified as correlates of protection against NoV infection. Human experimental challenge studies using Norwalk virus revealed subjects that were resistant to infection with NV GI.1 yet lacking NV GI.1 specific serum antibodies [4, 64]. HBGA is recognized as a putative receptor for NoV. A number of studies suggested that susceptibility to NoV infection is associated with the individual's HBGA secretory status [40, 44, 48, 65–70]. Individuals lacking HBGA expression are called nonsecretors and those expressing HBGA are called secretors. Extensive studies suggested the potential role of HBGAs as mediators of NoV infection especially with nonsecretors showing resistance to infection with certain NoV genotypes, specifically GI.1 and GII.4 [66, 71–74]. Nonsecretor individuals have shown lower frequencies in developing NoV-associated gastroenteritis following challenge with GII.4 [44] and after a GII.17 outbreak [75].

Briefly, HBGAs are small carbohydrate molecules expressed on the surface of human red blood cells [76], epithelial cells of many tissues of vertebrate species (humans, rats, rabbits, baboons, old world monkeys) [77], cells of the intestinal tract as well as being secreted in fluids such as saliva [78, 79]. The interaction between NoVs and HBGAs occurs in a genotype- and strain-specific manner [4, 72, 78]. The synthesis pathway of HBGAs starts with a disaccharide precursor and gene encoding glycosyltransferases [72, 78]. Fucosyltransferase encoded by the Fucosyl-transferase 3 (*FUT3*) gene adds a fucose residue to the disaccharide precursor in either alpha-1, 3 or alpha-1, 4 linkages, generating the Lewis A (Le^a) nonsecretor phenotype. The *FUT2* gene encodes functional alpha-1,2-fucosyltranferase that adds additional monosaccharides to the precursor. The latter is responsible for the secretor phenotype leading to the expression of HBGA type H on the surface of the gastrointestinal tract and its secretion in saliva. Further modifications by FUT3, A, or B enzymes (encoded by the *ABO* gene) lead to the generation of HBGA types A and B in addition to Lewis x (Le^x) and Le^y [78, 79]. The genes involved in the HBGA synthesis pathway are highly polymorphic and their frequencies vary widely depending on ethnicity [80].

NoVs attach to HBGA through the P2 subdomain of the major capsid protein (VP1) [78]. The P2 subdomain is also the target of neutralizing antibodies due to its exposure on the surface of viral capsid. Due to the genetic variability of NoVs and the polymorphism in the *FUT2* gene responsible for the expression of HBGA, different HBGA binding profiles have been reported [78]. Some NoV strains are known to bind to A/B and/or H antigens (GII VA387, GII.4 Grimsbey strain (GrV), GI.1 Norwalk, GI.2 C59, GII.13 Parris island (PiV), GII.3 Mexico (MxV), GII.2 BUDS, GII.1 Hawaii (HV)). Other strains bind to Lewis and/or H antigens (GII.9 VA207, GI.8 Boxer virus, GII.21 Operation Iraqi Freedom (OIF)) [81]. The Farmington Hills 2002 binds to both A and Lewis antigens [78]. GI.1 virus-like particles (VLPs) bind to HBGA H1, H2, H3, Le^b, and Le^y and GII.4 (VA387) VLPs can bind to all these antigens except H2 [81, 82]. The newly circulating GII.17 NoV was described to have a wide spectrum of HBGA binding [75, 83].

Twenty percent of European and North American populations have a nonfunctional *FUT2* gene and hence categorized as nonsecretors [79]. Moreover, it has been estimated that 5–50% of different populations globally are nonsecretors [84]. A nonfunctional *FUT2* gene is mainly caused by the G428A nonsense mutation resulting in the lack of HBGA expression in both the gastrointestinal tract and in saliva [85]. This mutation is linked to high resistance to GI.1 Norwalk virus infection and respective illness [65, 66]. Importantly, resistance to GII.4 NoV was also reported among nonsecretors. The relationship between susceptibility to GII.4 infection and HBGA secretory status was evaluated by Frenck et al. [44]. In this study, GII.4, the predominant NoV genotype, was used to challenge 40 healthy adults (23 secretors and 17 nonsecretors). GII.4 infection occurred in 70% of secretor individuals regardless of their ABO blood group. Of the infected secretors, 57% experienced AGE as compared to 6% of nonsecretors. The impact of the secretor status was also investigated in different settings including outbreaks and sporadic infections [68]. The homozygous nonsense mutation (428G→A) in FUT2 was found to correlate with complete resistance to illness [70]. Of all symptomatic individuals included in this study, 49% were homozygous and 51% heterozygous secretors and none were secretor-negative. Saliva from secretor-positive and symptomatic patients but not secretor-negative and asymptomatic individuals was able to bind to the norovirus strain belonging to the GGII Lordsdale-like cluster, causing the respective outbreak. This report showed that the FUT2 nonsecretor genotype is associated with resistance to nosocomial and sporadic outbreaks of NoV infection. The role of secretor status and susceptibility to GII.4 [69] and non-GII.4 genotypes was further confirmed among pediatric patients [67]. More studies are needed to understand the association between secretor status and disease burden caused by NoV at the population level. It is yet to be determined whether being a secretor or not affects immune responses to vaccines.

6.4.2 HBGA-Blocking Antibodies

As most NoV strains bind to HBGAs, there has been an increasing interest in antibodies blocking the NoV–HBGA interaction and their effect on prevention of NoV infection and NoV-associated illness. HBGA-blocking antibodies have been identified as potential correlates of protection against NoV infection and illness. Table 6.1 summarizes the studies reporting on the latter. Reeck et al. [40] was the first group to report on the association between pre-existing HBGA-blocking antibodies and protection against NoV-gastroenteritis in a challenge study. In this study, following oral inoculation of 34 healthy adults (18–50 years old) with Norwalk GI.1 sera samples were collected at days 0, 2, 7, 14, 28, and 180 post-challenge. Measurable pre-existing HBGA-blocking titers were detected in infected individuals without gastroenteritis symptoms. The level of HBGA-blocking antibodies was higher in this group of individuals compared to infected subjects experiencing clinical symptoms. Moreover, fecal viral shedding was higher in individuals without detectable pre-existing serum-blocking antibodies compared to those with pre-existing blocking antibodies [40]. While the results of this study highlighted these blocking antibodies as correlates of protection, a number of issues were raised. These include the minimum threshold blocking titer protecting from NoV-gastroenteritis, the duration of immunity associated with these titers, cross-reactivity potential, and whether potential NoV vaccines would be able to stimulate protective serum-blocking antibodies. Similar results were observed in a randomized double-blind placebo-controlled trial using GI.1 NV VLPs, where a four fold increase in IgA-specific seroresponse was detected in the majority of immunized volunteers as well as the presence of higher levels of pre-challenge HBGA-blocking antibodies; the latter correlated with protection against Norwalk virus infection following challenge [42]. The pre-challenge HBGA-blocking antibody level of >200 was associated with >50% reduction in the frequency of NV infection and illness. This further suggests the important role of pre-existing HBGA-blocking antibodies in protection from NoV infection and reduction of NoV-associated illness.

The HBGA-blocking antibody assay was described by Harrington et al. [86]. Briefly, samples are typed for the presence of ABH antigens in relation to blood type and secretor status by using antibodies to Le^a, Le^b, A antigen, and B antigen. Samples are boiled to remove any potential anti-NoV antibody. Boiled samples are used to coat plates and VLP binding is detected by using sera to be tested from human volunteers and goat anti-human immunoglobulin G (IgG) alkaline phosphatase conjugate; plates are developed with *p*-nitrophenyl phosphate substrate (pNPP) and optical densities determined at 405 nm (OD_{405}).

Earlier studies evaluated the ability of antibodies isolated from GI- or GII-infected individuals and mice inoculated with virus replicon-particle VRP-expressing NoV VLPs to block viral binding to ABH-HBGAs [37] (Table 6.1). Pre-challenge and convalescent sera were collected from volunteers infected with GI (NV) and GII (SM, or HV) strains [86]. Acute and convalescent sera were also collected from human volunteers during outbreaks caused by NoV GI strains [87] including Desert Shield-like virus (DS), LV (Lordsdale-like virus), and Chiba-like virus. These sera were analyzed for their ability to block NV-VLP binding to synthetic H1 antigen. In general, pre-challenge sera did not block H1 binding to NV-VLP in contrast to convalescent sera showing >90% of blockade of H1 binding. Similarly, homotypic murine IgG blocked NV VLP-H type 1 binding. Convalescent sera from LV-infected individuals were able to block entirely LV-H type 3 binding whereas sera from other infection groups were able to partially block H3 binding. Heterotypic antibody responses following infection with GII strains were completely unable to block H1 binding to NV-VLP similar to results observed in mice. This study demonstrated that heterotypic antibody responses can be induced by NoV infection in humans; the former were stronger within the same genogroup rather than between genogroups. Murine homotypic responses were stronger than heterotypic antibody responses; the latter were in agreement with results observed in humans. Moreover, heterotypic antibodies blockade to ABH antigens were induced by GI variants but not following infection with GII strains.

High serum HBGA-blocking antibody titers were detected in secretors, indicating previous NoV infections and a possible correlate of protection against NoV illness and infection [40, 44]. GII.4, the predominant NoV genotype, was used to challenge 40 healthy adults (23 secretors and 17 nonsecretors) [44]. Thirty days post-challenge, 13 secretors had a four fold increase from pre-challenge titers of receptor-blocking antibodies with more frequent gastroenteritis symptoms observed in secretors than in nonsecretors. Similarly, secretors showed a higher IgG seroresponse (Table 6.1). Moreover, secretors including A, B, O, and Lewis positive individuals (i.e., various HBGA types) were more susceptible to NoV infection and showed AGE symptoms [75]. This human challenge study confirmed further the significance of the secretor status and susceptibility to GII.4 infection but also suggested the involvement of other factors associated with high risk of infection especially among secretors with low levels of pre-existing antibodies.

Induced HBGA-blocking antibodies against noroviruses have been shown to be reactive not only against homologous NoV strains but also to heterologous strains [41, 45, 49–51]. Archived pre-challenge and convalescent sera collected from ten individuals previously infected with GI.1-1968 have been used to investigate the immune responses against a panel of five GI VLPs: GI.1-1968, GI.1-2001, GI.2-1999, GI.3-1999, and GI.4-2000 [41]. Whereas most of the pre-challenge sera were not able to block HBGA–VLP binding despite the presence of GI-VLP-specific IgG, convalescent-phase sera were able to block the interaction between all the tested GI VLPs and HBGA. This further suggests that infection with NoV GI elicits cross-reactive HBGA-blocking antibodies. Sera from volunteers previously infected with GII.2-1976 or GII.4 were used to compare the results to those observed using sera from individuals previously infected with GI.1-1968. Interestingly, GII.4 VLP-HBGA blocking activity was not reported using sera from volunteers infected with GI.1-1968 and GII.2-1976. Consequently cross-blocking activity was detected between different GI genotypes suggesting the lack of variability in the P2 domain in contrast to variability detected in GII.4 strains [72]. Importantly, a substantial increase in virus-specific serum IgG was also detected in 90% of convalescent sera against all tested VLPs. In addition to humoral responses, GI-specific T-cell responses were also reported. Peripheral blood mononuclear cells (PBMCs) from 60% of samples responded to in vitro stimulation with at least one VLP by the production of IFN-γ from $CD4^+$ T cells. This supports earlier results which demonstrated an increase in IFN-γ secretion in human volunteers challenged with GII SMV in addition to secretion of other Th1 cytokines by PBMCs stimulated with VLPs [14]. These results suggest the role of cellular immune responses in NoV infection.

Sera from patients at pre- and post-challenge with GII.4/2002 cluster [44] were used to assess the antigenic relatedness of GII.4 variants [50]. Low antibody titers were detected in pre-challenge sera compared to more than four-fold increase in convalescent sera (i.e., 30 days post-challenge). These titers were detected against 14 GII.4 P particles with significant cross-reactivities between GII.4 variants circulating between 1998 and 2012. Similar results were observed when testing for HBGA-blocking activity with significant differences between pre- and post-challenge titers. Importantly, no significant differences were detected between the post-challenge titers against different P particles. This cross-reactivity among NoV variants circulating between 1998 and 2012 suggests that the HBGA-binding sites are conserved among GII.4 variants. While encouraging, these results raise further questions related to the evolution of NoV and the impact of host immunity as well as HBGAs [57]. Conserved and less conserved central binding pockets (CBPs) are known on the HBGA binding surfaces; amino acids spanning the conserved CBP are at 100% similarity among GII.4 strains [88, 89] in contrast to those surrounding this site. The former is suggested to function as major antigenic determinants for a broadly protective vaccine formulation against GII.4 variants. Consequently, HBGAs exert an important role in NoV viral evolution and host selective pressure.

Following vaccination of human volunteers (18–49 years old) with a bivalent VLP vaccine containing GI.1 and GII.4, Pan-Ig, NoV-specific IgG and IgA as well as HBGA-blocking antibody titers were determined [46]. One week following the first dose of the vaccine, the latter induced a large Pan-Ig, IgG, and IgA with similar patterns detected against GI and GII components of the vaccine. Interestingly, serum IgG geometric mean titers (GMTs) were lower in subjects aged 65–83 years compared to other age groups. A strong HBGA-blocking antibody response was elicited following the first dose of vaccine with blocking titers of $\geq$200 against GI.1 and GII.4 viruses; the former were detected in the majority of subjects 28 days post-vaccination. This study guided the use of 50 μg/VLP of the bivalent vaccine in a follow-up study using sera from vaccinated participants to analyze the cross-reactivity of IgG and blockade antibody titers against the vaccine as well as two GI VLPs, two GII VLPs, and five GII.4 VLPs that were not included in the candidate vaccine [51]. GI.3-, GI.4-, and GI.1-specific IgG seroresponses peaked 1 week following vaccination yet were variable among individuals and waned down with time despite two booster doses administered at 28 and 35 days following vaccination. The IgG seroresponses to GII.4 VLPs were fairly consistent peaking 1 week following vaccination and following the same trends observed with GI.1 across time. Similarly, the titers of the blockade antibodies against GI.1, GI.3, and GI.4 peaked at day 7 in 50–90% of participants. These results are similar to previously reported ones in mice immunized with trivalent (GI.1, GII.1, GII.2) or tetravalent NoV vaccine (GI.1, GII.1, GII.2, GII.4) [37] and humans infected with GI.1 [41]. The cross-blocking activity in sera peaked 7 days post-vaccination, remained detectable on day 35 and returned to baseline on day 180. Similar results were also observed for GII.4 VLPs blockade antibodies. Interestingly, blockade antibody geometric mean fold rise (GMFR), defined as the ratio of post-vaccination antibody titer to pre-vaccination titer, was also detected against GII.3 and GII.14 which were not included in the vaccine in addition to blockade antibodies against two novel strains (GII.4.2012 and GII.4.2006b.P.D302). These novel strains were not circulating at or before the time of serum samples collection. These results suggest the ability of multivalent NoV vaccine to induce robust HBGA-blocking antibodies against vaccine components, previously encountered NoV strains as well as novel strains. As NoV mainly infects children <5 years old and elderly, the ability of this multivalent vaccine to show similar results is to be assessed in these two populations especially since the results above were based on a sample size of ten adult volunteers (18–49 years old).

Archived sera from previously challenged individuals with NV-GI.1 [40] (previously used to evaluate the association between pre-existing HBGA-blocking antibodies and protection against illness) were used to evaluate the persistence and breadth of HBGA-blocking antibodies in the context of experimental HuNoV infection using NoV VLPs (GI.1, GI.4, GI.7, GII.4 HOV, and GII.4 Sydney). Infected individuals had ≥4-folds increase in homotypic HBGA-blocking antibody titers (compared to pre-challenge titers). These titers remained detectable for up to 6 months post-infection [40]. Heterotypic HBGA-blocking antibodies were also detected against NoVs GI.4, GI.7, and GII.4, albeit not as robust as homotypic blockade responses similar to previously reported results [14, 37, 90]. Moreover, immunization with single NoV-VLP did not elicit cross-reactive HBGA-blocking antibody response [45, 91]. Single and trivalent VLPs induced IgG production in mice; moreover, cross-reactive seroresponses were mounted in mice vaccinated with the trivalent formulation. The GII.4 single VLP and the trivalent VLPs induced a strong IFN-γ response following stimulation with capsid P-domain T-cell epitopes in vitro [92].

Cross-reactive HBGA-blocking antibody responses were elicited against heterologous NoV-VLPs in mice immunized with the trivalent vaccine containing VLPs of GI-3 and GII.4 and tubular rotavirus (RV) recombinant VP6 (rVP6) [45]. Importantly, sera from immunized mice were able to block binding to VLPs that were not included in the vaccine formulation yet belonging to the same genogroup, namely GII.4 NO and GI.1 VLPs. These results mirror previously reported ones [38, 40, 41]. RV rVP6-GII.4 was previously tested in mice and induced antibodies that blocked the binding of GII.4 VLPs to HBGA H type 3 [93]. Similarly, the coadministration of RV rVP6 with GII.4 VLPs induced strong HBGA-blocking antibodies (blocking titer 50 (BT50) of 100–200 as compared to BT50 of 25 in its absence) in recently immunized mice [94]. Similarly, the co-delivery of NV VLPs with Toll-like receptor (TLR) agonists 7 or 9 to mice induced a strong blocking activity to H type 1 [95], the latter previously defined as a correlate of protection for NV in human challenge studies [40, 86].

LoBue et al. immunized BALB/c mice with the following formulas: monovalent VLPs including Norwalk virus (NV) GI.1, Lordsdale-like virus (LV) GII.4, or MNV-1 (GV) VLPs [38]. Multivalent groups received GI-specific VLPs representing Southampton (SoV) GI.2, Desert Shield (DSV) GI.3, and Chiba (GI.4) strains with or without NV VLPs; GII-specific VLPs including Hawaii (GII.1), Toronto (TV) (GII.3), and M7 (GII.13) strains with or without LV VLPs; or complete VLP cocktails containing all GI and GII VLPs with or without NV and LV VLPs or all GI and GII VLPs with or without MNV VLPs (GV). Sera from mice immunized with monovalent and multivalent VLP vaccines with or without adjuvant induced IgG2a titers; the latter correlates with Th1 responses [38]. Moreover, cross-reactive and receptor-blocking IgG responses were also detected against heterologous strains. Multivalent genogroup-specific formulation (GI VLPs or GII VLPs) stimulated cross-reactive IgG seroresponses aligned with blockade response to NV or LV VLPs (not included in the formula) suggesting that vaccines should contain both GI and GII strains to stimulate a broader and cross-reactive response. Following MNV challenge, decreased viral load correlated with reduction in clinical disease without protection against infection, specifically in mice vaccinated with the multivalent formulation. Interestingly, the intensity of blocking activity against homologous GII.4 and GI.3 VLPs was similar in mice immunized with single and trivalent vaccines. Similar results were also reported in human volunteers orally challenged with Norwalk virus. Homotypic and heterotypic serum HBGA-blocking antibodies were detected; with higher titers observed against the homotypic challenge strain (Norwalk virus GI.1). Modest HBGA-blocking antibody response was detected against heterotypic NoV strains; GI.4, GI.7, and GII.4 (HOV and Sydney) VLPs [49]. This study also suggested that pre-existing HBGA-blocking antibodies are independent of the clinical outcomes of NoV challenge and thus offering little protection upon re-exposure. However, these results should be cautiously interpreted, especially due to the lack of statistical power as well as other factors that might have contributed to the differences in affinity of blocking antibodies and the lack of knowledge about the persistence of heterotypic blocking antibody responses. These results highlight the need to evaluate the role of pre-existing immune responses as well as blocking responses on the breadth and quality of NoV immunity.

Furthermore, HBGA-blocking antibodies have also been correlated with protection against norovirus gastroenteritis in a human volunteer challenge study [47]. This study was conducted in order to characterize mucosal and cellular immune responses following Norwalk virus challenge [96]. In addition to the increase in NoV-specific IgA in serum, virus-specific salivary and fecal IgA increased in subjects infected with Norwalk virus; salivary IgA increased ≥4-folds 28 days following infection with NV in 70% of subjects whereas 100% of volunteers showed ≥4-folds increase in fecal IgA level. Pre-existing salivary IgA correlated with protection from gastroenteritis whereas pre-existing fecal IgA correlated with a reduced viral load. It was suggested that the NV-specific pre-existing salivary IgA that correlated with protection against NV illness might be secretory rather than serum-derived given that serum IgA did not contribute to protection against NV-associated illness in this study. This study also analyzed the cellular immune responses among vaccinated participants. An increase in IgA and IgG antibody secreting cells (ASCs) was observed in all infected individuals 7 days post-challenge and declined dramatically at 2 weeks. Importantly, ACS responses were skewed toward IgA whereas memory B-cell responses were skewed toward IgG. The latter persisted for 180 days. It is yet to be explored whether similar results could be obtained following vaccination. Pre-challenge serum HBGA-blocking antibodies strongly correlated with pre-existing salivary IgA and serum IgG that correlated with protection against developing NV-associated gastroenteritis following Norwalk virus challenge. In an attempt to quantify salivary anti-NoV antibodies using a saliva-based ELISA, Tamminen et al. [97] recently reported the ability of purified salivary IgA and IgG to block the binding of NoV VLPs to HBGAs. While salivary IgG and IgA correlated with their serum counterpart as previously reported [47], these results were detected in healthy adult volunteers and consequently protection from NoV infection and/or disease could not be concluded. Nevertheless, these studies highlight the significance of mucosal, cellular, and humoral responses and suggests the potential role that pre-existing HBGA-blocking antibodies could play in protection following NoV infection.

Although HBGA-blocking antibodies are considered correlates of protection among NoV-infected individuals and recipients of experimental vaccines, they may not be broadly used as surrogate measure of protection against NoV infection post-vaccination. In a phase 1/2 double-blind, placebo-controlled trial [48], GI.1/GII.4 bivalent vaccine was administered intramuscularly in two doses. Both vaccine and placebo groups were then challenged with GII.4 strain, Farmington Hills-like variant. A robust increase in serum IgG and IgA was reported following intramuscular vaccination with the bivalent GI.1/GII.4 VLP vaccine in all vaccine recipients with higher GMFR and seroresponse frequency against GI.1. The high levels of pre-challenge HBGA-blocking antibodies, as well as serum IgA antibodies but not IgG, detected in the placebo group, rendered subjects less susceptible to NoV infection and illness. Interestingly, some individuals in the placebo group had no detectable levels of pre-challenge HBGA-blocking antibodies, yet showed protection following challenge with GII.4. This observation, however, did not apply to vaccine recipients. Among the vaccinees, pre-challenge HBGA-blocking antibody titer was higher than that observed in placebo subjects, yet this was not necessarily associated with decreased illness and infection frequency. This confirms again as previously reported [44] the need to further explore other factors affecting susceptibility to infection.

High levels of serum HBGA-blocking antibodies were also documented in healthy adults vaccinated with bivalent NoV-VLP vaccine containing either 15 μg/50 μg or 50 μg/50 μg of GI.1/GII.4 [52]. Higher seroresponses, Pan-Ig and IgA antibodies, were detected against the GII.4 component of the vaccine in the 15/50 formulation. HBGA-blocking antibody levels in vaccinated subjects receiving either one of the vaccine formulations showed peak levels at day 7–10 post-vaccination, following the same trend as Pan-Ig and IgA seroresponses. While still higher than the placebo, anti-GII.4 HBGA as well as antibody seroresponses declined by day 28. This result is in agreement with a volunteer vaccine study whereby the first dose of an intramuscular bivalent vaccine (GI.1/GII.4 VLPs) induced a robust HBGA-blocking antibody response [46].

HBGA-blocking antibodies were also detected in mice immunized intramuscularly with one of six NoV VLPs (GII.4 1999, GII.4 2009, GII.4 2012, GII.12, GI.1, and GI.3) [53]. HBGA-blocking antibodies were detected against homologous VLPs with BT90 $\geq$ 200 using synthetic HBGAs and BT90 $\geq$ 400 using salivary HBGAs. BT90 is defined as the percentage of sera necessary to block 90% of NoV–HBGA binding. Similarly, systemic blockade activities were elicited following intranasal or intramuscular immunization of Balb/c mice with NoV GII.4 VLPs whereas intranasal administration of the vaccine mounted IgA blocking titers that correlated with mucosal NoV-specific IgA; thus suggesting the former to be the main blocking antibody on mucosa [54]. Moreover, homologous antibodies were detected to all VLPs; cross-reactive ones were detected against heterologous VLPs within the same genogroup. This study also reported the mapping of $CD4^+$ and $CD8^+$ T-cell epitopes using matrix pools of peptides as well as individual peptides located in the S and P domains of VP1 [53]. Robust T-cell responses to peptides 99-45 and 99-46 (11mers) were reported; the latters contain the GII conserved PAPLGTPDF 9 aa stretch. When mice were immunized with monovalent NoV VLPs (GI.1-2001, GI.3-2002, GII.4-1999, or GII.4-2010 New Orleans), cross-reactive HBGA-blocking antibodies within genogroup were detected [98]. The same group previously reported that sera from mice immunized with GII.4 VLPs with or without GII.12 VLPs boost were able to mount a cross-reactive IgG response and to block the binding of GII.4 VLPs to HBGAs [99]. Previous studies reported the lack of correlation between fecal IgA and protection [47, 100] whereas pre-existing [48] and early post-infection salivary IgA [66] were associated with protection. Consequently, NoV vaccines inducing mucosal immune responses should be actively considered.

Malm et al. recently assessed the relationship between serum NoV-specific IgG and avidity as well as blocking ability [55]. Mice were immunized with single GII.4 1999 VLPs, a GII-mix VLPs (GII.4-1999, GII.4 NO, GII.12), or GI-mix VLPs (GI.1 and GI.3). GII.4, delivered alone or in combination with other VLPs, induced serum-specific IgG regardless of the mode of vaccination. Similar blocking activities developed regardless of the mode of immunization (intradermal or intramuscular). T-cell responses were also analyzed and results showed that the magnitude and functional avidity of GII.4-specific IFN-γ producing T-cells were similar in immunized mice. CD107, a marker of T cell activation, was detected on splenocytes of immunized mice stimulated with peptide 99-45 containing highly conserved epitope of GII NoVs and thus suggesting the potential role of cytotoxic T lymphocytes (CTL) directed against this epitope.

The induction of HBGA-blocking antibodies in mice was also reported in a series of preclinical vaccine studies using the P particle or chimeric vaccines containing P particle. The P domain (P Polypeptide) was detected in the stool of NoV-infected patients; however, the role of this truncated protein without a C-terminal arginine cluster is unknown. Consequently, a number of studies evaluated the immunogenicity domains (reviewed in Chap. 7) and the ability of these vaccine constructs to stimulate HBGA blockade activity [101–105].

6.4.3 Other Correlates of Protection

While serum antibodies blocking the binding of NoV to HBGAs have been used as correlates of protection, a number of challenges were associated with the assay including the effect of temperature, pH, availability and quality of purified HBGAs [40, 41]. Consequently, the hemagglutination inhibition (HAI) assay was explored as an alternative to the HBGA-blocking assay, especially since hemagglutination inhibition activity was reported to increase following challenge or vaccination of human subjects [39, 82]. Briefly, the assay consisted of preparing serially diluted VLPs and incubation with an equal volume of 0.5% human red blood cell suspension in 96-well V bottom plates. The amount of Norwalk VLP antigen corresponding to four hemagglutination units was determined and confirmed by back titration. Heat-inactivated test sera were incubated with an equal volume of Norwalk VLPs (LigoCyte, lot 45–06002) containing four hemagglutination units followed by addition of 0.5% red blood cells suspension. Plates were incubated at 4 °C and checked for agglutination every 30 min. HAI titers were calculated as the inverse of the highest dilution that inhibited hemagglutination, with a compact negative red blood cell pattern (button of red blood cells). El-Kamary et al. [39] used this assay to examine the ability of vaccine-induced antibodies to inhibit hemagglutination of O-type red blood cells NoV VLPs. The vaccine consisted of GI.1 VLPs and a monophosphoryl lipid A (MPL) adjuvant and a toll-like receptor 4 (TLR4) agonist and chistosan. The study was a two phase 1, double-blind clinical trial. Study 1 evaluated dosage and safety of three doses (5, 15, or 50 μg) of intranasally administered adjuvanted NoV VLPs administered to healthy adults (18–49 years); study 2 compared two dosages of the vaccine (50 and 100 μg). HAI antibodies increased in subjects receiving the highest dose (100 μg) of the vaccine. These results were in 72% and 75% agreement with IgG and IgA seroresponses, respectively. This study also reported on the NoV-specific IgG and IgA ACSs following intranasal immunization. IgG ACSs expressed $CD19^+$ $CD27^+$ integrin α_4/β_7^+ $CD62L^-$ and thus allowing homing to the gut mucosa in contrast to IgG ACSs with peripheral lymphoid tissues homing receptors (i.e., $CD19^+$ $CD27^+$ integrin α_4/β_7^+ $CD62L^+$).

HAI was also tested as a correlate of protection following human challenge with Norwalk virus GI.1 [43]. Participants showing ≥4-fold increase in anti-Norwalk virus antibodies by ELISA also showed an increase in HAI titer between day 0 and day 28 post-infection. Higher titers were detected among infected subjects who did not develop gastroenteritis. Importantly, the HAI titers correlated significantly with HBGA-blocking titers (Table 6.1). While these results suggest the suitability of serum HAI assay as alternative correlate of protection to HBGA-blocking antibodies, larger trials are needed to evaluate this measure and confirm its reliability in defining immune responses developed following vaccination.

In addition to using HBGA-blocking assays [37, 86], the HAI assay [43] and stem cell-derived human intestinal enteroids (HIE)-based assay [5] to evaluate vaccines in vitro, a new cell line-based model was developed as a surrogate neutralization assay [106]. Human embryonic kidney 293T cells over-expressing FUT2 (293T-FUT2) were generated. The 293T-FUT2 cell line was able to express H2 and Le^y antigens and consequently allowed the attachment of NoV VLPs. NoV GI.1, GII.4, and GII.17 VLPs were shown to bind to 293T-FUT2 cells in a dose-dependent manner. In order to assess the suitability of using NoV-specific mouse antisera to block VLP attachment, a cellular ELISA-based inhibition assay was generated. Antisera against the three tested VLPs showed similar dose-dependent inhibitory patterns as compared to control sera. The inhibition of attachment was strain-specific and not heterologous. When compared to the HBGA-blocking assay using sera of immunized mice with GI.1, GII.4, or GII.17 VLPs, this ELISA model significantly correlated with the latter suggesting its potential use as a surrogate neutralization assay to assess NoV vaccines. The authors suggested several advantages of the described model over HAI and HBGA as correlates of protection. These include: the expression of HBGA on 293T-FUT2 cells authentically simulates host cells; the stable expression of HBGA on these cells is not compromised as the case of the HBGA-blocking assay, the latter being affected by the HBGA lot whereas HAI assay is affected by the source of RBCs [107] and the B cell culture model [108] affected by the virus stock and the source of the fetal bovine serum [6]. Finally, these cells are more easily obtained as compared to HIEs [5].

6.5 Duration of Immunity

An important challenge for the development of vaccines is their ability to achieve a broad and long-lasting immunity. These aims are hampered mainly by the genetic variability of noroviruses and their impact on immune evasion. Moreover, there is lack of well-established data on protective immunity generated following primary and secondary NoV infections. The duration of immunity to noroviruses, estimated in early human challenge studies, was reported to range between 2 months and 2 years in early studies [64] and later reported to last at least 6 months [109]. These studies reported homologous protection against NoV infection in addition to the lack of correlation between pre-existing antibodies and protection against NoV infections. Mathematical modeling of community NoV transmission estimated a mean duration of immunity ranging between 4 and 8 years. This model was however implemented in the absence of strain variations while taking into consideration the natural history of NoV, its transmission and the natural waning of immunity [110]. The estimates provided by this study targeted children less than 5 years old and adults.

Avidity assays were developed to measure the strength of antigen binding with many antigenic determinants to multivalent antibodies [111]. Following natural infection with different NoV strains in the Netherlands, homologous as well as heterologous IgG and IgA responses were detected as well as significant rise in homologous HBGA-blocking antibody titers. The role of the latter was studied in human adult challenge studies [42, 51] with limited studies on immune responses in children. In an attempt to assess the role of antibodies developed against NoV GII.4, their duration, blocking ability and avidity, two healthy children were followed from birth and up to 8 years [1]. This study and others [112] estimated that a child would acquire one to three NoV infections during their first year of life. Interestingly, maternal antibodies were shown to influence susceptibility to NoV infection in early infancy; the higher the level of maternal antibodies with higher avidity, the longer the child was protected during early years of life. A homologous but not heterologous protective blocking antibody titer was detected among children and was associated with protection from infection [113, 114]. Protective GII.4-specific antibodies with high blocking capacity and avidity were generated at 2–3 years of age and were retained up to 8 years of follow-up. Although it is challenging to follow up a birth cohort to investigate the development and the evolution of NoV-specific antibodies, similar studies are critical to validate these findings and to address the potential age for the administration of candidate NoV vaccines in relation to duration of protection.

Recently, blocking antibodies to pandemic GII.4 NoV (MD145, Grimsby, Houston, and Sydney) were detected in pre-pandemic sera collected decades prior to GII.4 circulation; moreover, the IgG blocking titers against GII.4 were sustained for 6 years [115]. A long-lived serological memory against NoV infection was suggested without ruling out the possibility of reinfection during the study period leading to immunological memory. This result is in accordance with a mathematical model of NoV transmission suggesting that NoV immunity can last between 4 and 9 years [110].

Recently, the relative avidity of NoV-blocking antibodies was measured in sera [116] following vaccination of participants with a multivalent adjuvanted vaccine containing GI.1 and GII.4c VLPs [45, 51, 116]. A broadly reactive serum blockade response was mounted 1 week following the first dose of vaccine, whereas, booster vaccination 28 days following immunization did not induce any significant change in blockade titers. In this study, the avidity of GI.1-blocking antibodies increased at day 7 post-vaccination and remained elevated for 180 days in secretors individuals only. While similar results were observed against GII.4c whereby blocking antibody titers peaked at day 7 post-vaccination, and remained elevated until day 180, the former were independent of the secretor status. Importantly and while the avidity of blockade antibody to GI.1 increased over time, a decrease in blockade antibody titer was detected to GI.3 and GI.4. This was suggested to be largely associated with somatic hypermutation. In addition to the increased antibodies' avidity against GII.4c, an increase in the antibodies' avidity against GII.4.1997, 2002, and 2006b was observed. The avidity against GII.4c and GII.4.1997 was similar but higher than that observed against GII.4.2006b suggesting that exposure to GII.4 would shape further GII.4 blockade antibody responses [51]. Interestingly, some secretor volunteers showed elevated levels of antibodies' avidity against GII.4.2012, a novel strain circulating 1 year following collection of samples.

The duration of NoV-specific immunity is not well characterized yet. NoV infection is affected by the host genetic background (secretor vs. nonsecretor status), the viral variant (NoV strain), the host pre-existing immunity, as well as the level of exposure (dose). The pre-existing immunity to NoV strains and host genetics could be further studied and characterized by inclusion of blockade antibody avidity in vaccination and challenge studies.

6.6 In Vitro Culture Systems and Impact on NoV Correlates of Protection

The development of an effective vaccine is hampered by a number of factors. The antigenic diversity of circulating NoV strains, the differences in and duration of protective immunity, the lack of a specific correlate of protection as a surrogate of clinical end-point and until recently the lack of an in vitro culture system for HuNoV. A number of cell lines have been tested to culture HuNoV [117]. Unfortunately, these attempts were not successful in showing de novo infection.

Wobus et al. [24] were the first to report the successful replication of murine norovirus 1 (MNV-1) in vitro using murine bone marrow-derived DCs and macrophages. The replication of MNV-1 in vitro was supported in IFN-$\alpha\beta R^{-/-}$ and in the absence of the cellular transcription factor STAT1. While these results were encouraging, a major challenge exists in using MNV-1 in mice model. This is especially due to the inability of murine in contrast to human noroviruses to cause disease in immunocompetent hosts. Consequently, human macrophages and DCs isolated from PBMCs of NoV-susceptible individuals were next tested in vitro for their ability to grow HuNoV [23]. NoV RNA was not detected in macrophages, monocyte-derived DCs or CD11c^{+} DCs. The different results obtained in vitro between HuNoV and MNV could be explained by distinct tropism; MNV is known to bind to a ganglioside-linked sialic acid [118], and HuNV binds to HBGAs as primary receptors and susceptibility factors. Moreover, the resistance to norovirus might be explained by the strong IFN responses produced by macrophages and DCs [23].

Attempts continued to search for a system to grow noroviruses. Results confirmed the visualization of viral NoV antigen in B cells of Peyer's patches biopsies of mice infected with MNV [28] as well as intestinal B cells of chimpanzees infected with HuNoV [119]; this is in addition to lower MNV titers reported in B cell-deficient mice [120]. These data were naturally followed by infection of B cells with MNV and HuNoV which led to the generation of infectious viral particles [108, 121]. MNV-1 and MNV-3 replicated efficiently in B cells yet at a slower rate compared with macrophages [108]. While the majority of studies used GI.1, the HuNoV GII.4-Sydney was later used to infect BJAB cells (human EBV-negative B cell line) in the presence and absence of synthetic HBGA and resulted in moderate increase in viral genome copies. These results highlighted yet again the role of

HBGA in promoting infection but also the possible need/presence for another receptor or a co-receptor on B cells promoting viral entry. The latter was previously demonstrated in HBGA-expressing intestinal epithelial cells (IECs) shown to be nonpermissive for infection [122].

HIEs derived from stem cells isolated from the intestinal crypts in human intestines were successfully infected with multiple GII.4 variants [5]. In contrast to the B cell culture system described earlier, GII.4 was successfully propagated in HIEs in the absence of commensal bacteria previously suggested to act as a cofactor for viral replication [6, 108]. Interestingly, NoV strains other than GII.4 were also successfully propagated in HIEs culture. GI.1, GII.3, and GII.17 strains have been shown to grow in HIEs in the presence of human bile; the latter was not required for the propagation of GII.4 [5]. The growth of these variants was dose-dependent on human bile level. Moreover, replication of HuNoVs differed by viral variant and intestinal segment whereby GII.4 and GII.3 replicated in enteroids of the duodenum, jejunum, and ileum indicating that enterocytes are the focal site of infection and replication. In general, secretor-negative HIEs were not infected indicating the relevance of host-susceptibility factors.

Recently, a human intestinal organoids (HIO) model was developed for HuNoV culture [7]. HIOs were derived from human pluripotent stem cells (PSCs), the latter successfully differentiated into HIOs containing all intestinal cell types. Promising results were obtained whereby viral RNA was detected in HIOs expressing different types of HBGAs. When compared to the HIE culture system [5], viral replication in HIOs was lower. The major reason was suggested to be related to differences between the sensitivity of HIEs to HuNoV infection compared to HIOs, the former being enterocytes whereas HIOs require development from stem cells. In a recent study, attempts to grow HuNoVs, human Sapovirus (HuSaV), and bovine NoV (BoNoV) in various cells and culture conditions were unsuccessful with minimal increase in the level of viral RNA in all tested cell lines under different culture conditions [123].

Despite the successful development of B cell and HIEs culture systems, attempts to optimize and develop more efficient, cost-effective, and reproducible culture systems are still ongoing. The development of a cell culture system for human norovirus is critical for the understanding of the virus biology including pathogenicity and infectivity, antigenic determinants, viral evolution, and virus–host interaction. Importantly, it will facilitate assessing the ability of vaccines to prevent and control NoV infection.

6.7 Conclusion

Until recently, our understanding of NoV correlates of protection resulted from volunteer challenge studies, preclinical and clinical studies. Several surrogate measures of immune correlates of protection have been described: NoV-specific antibodies, antibodies blocking binding of NoV to HBGAs, and hemagglutination inhibition antibodies. While the humoral immune responses have been well studied, other arms of immunity are likely to play a significant role in protection against NoV infection. Consequently, it is critical to identify other immune correlates of protection. The newly developed NoV culture systems are considered a breakthrough in the field. These culture systems will help understanding the NoV-specific immune mechanisms and responses, as well as viral pathogenesis. These are critical for the development of effective vaccines with long-lasting systemic and mucosal responses directed against GI, GII, and newly emerging strains.

References

1. Blazevic V, Malm M, Honkanen H, Knip M, Hyoty H, Vesikari T. Development and maturation of norovirus antibodies in childhood. Microbes Infect. 2016;18(4):263–9.
2. Debbink K, Lindesmith LC, Donaldson EF, Baric RS. Norovirus immunity and the great escape. PLoS Pathog. 2012;8(10):e1002921.
3. Banyai K, Estes MK, Martella V, Parashar UD. Viral gastroenteritis. Lancet. 2018;392 (10142):175–86.
4. Ramani S, Estes MK, Atmar RL. Correlates of protection against norovirus infection and disease-where are we now, where do we go? PLoS Pathog. 2016;12(4):e1005334.
5. Ettayebi K, Crawford SE, Murakami K, Broughman JR, Karandikar U, Tenge VR, Neill FH, Blutt SE, Zeng X-L, Qu L, Kou B, Opekun AR, Burrin D, Graham DY, Ramani S, Atmar RL, Estes MK. Infectious disease: replication of human noroviruses in stem cell-derived human enteroids. Science. 2016;353(6306):1387–93.
6. Jones MK, Grau KR, Costantini V, Kolawole AO, de Graaf, Freiden P, Graves CL, Koopmans M, Wallet SM, Tibbetts SA, Schultz-Cherry S, Wobus CE, Vinje J, Karst SM. Human norovirus culture in B cells. Nat Protoc. 2015;10(12):1939–47.
7. Zhang D, Tan M, Zhong W, Xia M, Huang P, Jiang X. Human intestinal organoids express histo-blood group antigens, bind norovirus VLPs, and support limited norovirus replication. Sci Rep. 2017;7(1):12621.
8. Newman KL, Moe CL, Kirby AE, Flanders WD, Parkos CA, Leon JS. Norovirus in symptomatic and asymptomatic individuals: cytokines and viral shedding. Clin Exp Immunol. 2016;184(3):347–57.
9. Troeger H, Loddenkemper C, Schneider T, Schreier E, Epple HJ, Zeitz M, Fromm M, Schulzke JD. Structural and functional changes of the duodenum in human norovirus infection. Gut. 2009;58(8):1070.
10. Atmar RL, Opekun AR, Gilger MA, Estes MK, Crawford SE, Neill FH, Graham DY. Norwalk virus shedding after experimental human infection. Emerg Infect Dis. 2008;14(10):1553–7.
11. Murata T, Katsushima N, Mizuta K, Muraki Y, Hongo S, Matsuzaki Y. Prolonged norovirus shedding in infants $\leq$6 months of age with gastroenteritis. Pediatr Infect Dis J. 2007;26 (1):46–9.
12. Rockx B, de Wit, Vennema H, Vinjé J, de Bruin, van Duynhoven, Koopmans M. Natural history of human calicivirus infection: a prospective cohort study. Clin Infect Dis. 2002;35 (3):246–53.
13. Teunis PFM, Sukhrie FHA, Vennema H, Bogerman J, Beersma MFC, Koopmans MPG. Shedding of norovirus in symptomatic and asymptomatic infections. Epidemiol Infect. 2014;143(8):1710–7.
14. Lindesmith L, Moe C, LePendu J, Frelinger JA, Treanor J, Baric RS. Cellular and humoral immunity following snow mountain virus challenge. J Virol. 2005;79(5):2900–9.
15. Souza M, Azevedo MS, Jung K, Cheetham S, Saif LJ. Pathogenesis and immune responses in gnotobiotic calves after infection with the genogroup II.4-HS66 strain of human norovirus. J Virol. 2008;82(4):1777–86.

16. Souza M, Cheetham SM, Azevedo MS, Costantini V, Saif LJ. Cytokine and antibody responses in gnotobiotic pigs after infection with human norovirus genogroup II.4 (HS66 strain). J Virol. 2007;81(17):9183–92.
17. Gustavsson L, Skovbjerg S, Lindh M, Westin J, Andersson L-M. Low serum levels of CCL5 are associated with longer duration of viral shedding in norovirus infection. J Clin Virol. 2015;69:133–7.
18. Baldridge MT, Lee S, Brown JJ, McAllister N, Urbanek K, Dermody TS, Nice TJ, Virgin HW. Expression of Ifnlr1 on intestinal epithelial cells is critical to the antiviral effects of interferon lambda against norovirus and reovirus. J Virol. 2017;91(7):e02079–16.
19. Baldridge MT, Nice TJ, McCune BT, Yokoyama CC, Kambal A, Wheadon M, Diamond MS, Ivanova Y, Artyomov M, Virgin HW. Commensal microbes and interferon-lambda determine persistence of enteric murine norovirus infection. Science. 2015;347(6219):266–9.
20. McCartney SA, Thackray LB, Gitlin L, Gilfillan S, Virgin HW, Colonna M. MDA-5 recognition of a murine norovirus. PLoS Pathog. 2008;4(7):e1000108.
21. Nice TJ, Baldridge MT, McCune BT, Norman JM, Lazear HM, Artyomov M, Diamond MS, Virgin HW. Interferon-lambda cures persistent murine norovirus infection in the absence of adaptive immunity. Science. 2015;347(6219):269–73.
22. Thackray LB, Duan E, Lazear HM, Kambal A, Schreiber RD, Diamond MS, Virgin HW. Critical role for interferon regulatory factor 3 (IRF-3) and IRF-7 in type I interferon-mediated control of murine norovirus replication. J Virol. 2012;86(24):13515–23.
23. Lay MK, Atmar RL, Guix S, Bharadwaj U, He H, Neill FH, Sastry KJ, Yao Q, Estes MK. Norwalk virus does not replicate in human macrophages or dendritic cells derived from the peripheral blood of susceptible humans. Virology. 2010;406(1):1–11.
24. Wobus CE, Karst SM, Thackray LB, Chang K-O, Sosnovtsev SV, Belliot G, Krug A, Mackenzie JM, Green KY, Virgin HW. Replication of norovirus in cell culture reveals a tropism for dendritic cells and macrophages. PLoS Biol. 2004;2(12):e432.
25. Qu L, Murakami K, Broughman JR, Lay MK, Guix S, Tenge VR, Atmar RL, Estes MK. Replication of human norovirus RNA in mammalian cells reveals lack of interferon response. J Virol. 2016;90(19):8906–23.
26. Karst SM, Wobus CE, Lay M, Davidson J, Virgin HW. STAT1-dependent innate immunity to a Norwalk-like virus. Science. 2003;299(5612):1575–8.
27. Nishana M, Raghavan SC. Role of recombination activating genes in the generation of antigen receptor diversity and beyond. Immunology. 2012;137(4):271–81.
28. Mumphrey SM, Changotra H, Moore TN, Heimann-Nichols ER, Wobus CE, Reilly MJ, Moghadamfalahi M, Shukla D, Karst SM. Murine norovirus 1 infection is associated with histopathological changes in immunocompetent hosts, but clinical disease is prevented by STAT1-dependent interferon responses. J Virol. 2007;81(7):3251–63.
29. Henderson KS. Murine norovirus, a recently discovered and highly prevalent viral agent of mice. Lab Anim (NY). 2008;37(7):314–20.
30. Sadler AJ, Williams BRG. Interferon-inducible antiviral effectors. Nat Rev Immunol. 2008;8:559.
31. Taniguchi T, Takaoka A. The interferon-α/β system in antiviral responses: a multimodal machinery of gene regulation by the IRF family of transcription factors. Curr Opin Immunol. 2002;14(1):111–6.
32. Nice TJ, Osborne LC, Tomov VT, Artis D, Wherry EJ, Virgin HW. Type I interferon receptor deficiency in dendritic cells facilitates systemic murine norovirus persistence despite enhanced adaptive immunity. PLoS Pathog. 2016;12(6):e1005684.
33. Li XD, Sun L, Seth RB, Pineda G, Chen ZJ. Hepatitis C virus protease NS3/4A cleaves mitochondrial antiviral signaling protein off the mitochondria to evade innate immunity. Proc Natl Acad Sci U S A. 2005;102(49):17717–22.
34. Elinav E, Strowig T, Kau AL, Henao-Mejia J, Thaiss CA, Booth CJ, Peaper DR, Bertin J, Eisenbarth SC, Gordon JI, Flavell RA. NLRP6 inflammasome regulates colonic microbial ecology and risk for colitis. Cell. 2011;145(5):745–57.

35. Wang P, Zhu S, Yang L, Cui S, Pan W, Jackson R, Zheng Y, Rongvaux A, Sun Q, Yang G, Gao S, Lin R, You F, Flavell R, Fikrig E. Nlrp6 regulates intestinal antiviral innate immunity. Science. 2015;350(6262):826–30.
36. McFadden N, Bailey D, Carrara G, Benson A, Chaudhry Y, Shortland A, Heeney J, Yarovinsky F, Simmonds P, Macdonald A, Goodfellow I. Norovirus regulation of the innate immune response and apoptosis occurs via the product of the alternative open reading frame 4. PLoS Pathog. 2011;7(12):e1002413.
37. LoBue AD, Lindesmith L, Yount B, Harrington PR, Thompson JM, Johnston RE, Moe CL, Baric RS. Multivalent norovirus vaccines induce strong mucosal and systemic blocking antibodies against multiple strains. Vaccine. 2006;24(24):5220–34.
38. LoBue AD, Thompson JM, Lindesmith L, Johnston RE, Baric RS. Alphavirus-adjuvanted norovirus-like particle vaccines: heterologous, humoral, and mucosal immune responses protect against murine norovirus challenge. J Virol. 2009;83(7):3212–27.
39. El-Kamary SS, Pasetti MF, Mendelman PM, Frey SE, Bernstein DI, Treanor JJ, Ferreira J, Chen WH, Sublett R, Richardson C, Bargatze RF, Sztein MB, Tacket CO. Adjuvanted intranasal Norwalk virus-like particle vaccine elicits antibodies and antibody-secreting cells that express homing receptors for mucosal and peripheral lymphoid tissues. J Infect Dis. 2010;202(11):1649–58.
40. Reeck A, Kavanagh O, Estes MK, Opekun AR, Gilger MA, Graham DY, Atmar RL. Serological correlate of protection against norovirus-induced gastroenteritis. J Infect Dis. 2010;202(8):1212–8.
41. Lindesmith LC, Donaldson E, Leon J, Moe CL, Frelinger JA, Johnston RE, Weber DJ, Baric RS. Heterotypic humoral and cellular immune responses following Norwalk virus infection. J Virol. 2010;84(4):1800–15.
42. Atmar RLMD, Bernstein DIMD, Harro CDMD, Al-Ibrahim MSMBCB, Chen WHMD, Ferreira JSM, Estes MKPD, Graham DYMD, Opekun ARPAC, Richardson CPD, Mendelman PMMD. Norovirus vaccine against experimental human Norwalk virus illness. N Engl J Med. 2011;365(23):2178–87.
43. Czako R, Atmar RL, Opekun AR, Gilger MA, Graham DY, Estes MK. Serum hemagglutination inhibition activity correlates with protection from gastroenteritis in persons infected with Norwalk virus. Clin Vaccine Immunol. 2012;19(2):284–7.
44. Frenck R, Bernstein DI, Xia M, Huang P, Zhong W, Parker S, Dickey M, McNeal M, Jiang X. Predicting susceptibility to norovirus GII.4 by use of a challenge model involving humans. J Infect Dis. 2012;206(9):1386–93.
45. Tamminen K, Lappalainen S, Huhti L, Vesikari T, Blazevic V. Trivalent combination vaccine induces broad heterologous immune responses to norovirus and rotavirus in mice. PLoS One. 2013;8(7):e70409.
46. Treanor JJ, Atmar RL, Frey SE, Gormley R, Chen WH, Ferreira J, Goodwin R, Borkowski A, Clemens R, Mendelman PM. A novel intramuscular bivalent norovirus virus-like particle vaccine candidate-reactogenicity, safety, and immunogenicity in a phase 1 trial in healthy adults. J Infect Dis. 2014;210(11):1763–71.
47. Ramani S, Neill FH, Opekun AR, Gilger MA, Graham DY, Estes MK, Atmar RL. Mucosal and cellular immune responses to Norwalk virus. J Infect Dis. 2015;212(3):397–405.
48. Atmar RL, Bernstein DI, Lyon GM, Treanor JJ, Al-Ibrahim MS, Graham DY, Vinje J, Jiang X, Gregoricus N, Frenck RW, Moe CL, Chen WH, Ferreira J, Barrett J, Opekun AR, Estes MK, Borkowski A, Baehner F, Goodwin R, Edmonds A, Mendelman PM. Serological correlates of protection against a GII.4 norovirus. Clin Vaccine Immunol. 2015;22(8):923–9.
49. Czako R, Atmar RL, Opekun AR, Gilger MA, Graham DY, Estes MK. Experimental human infection with Norwalk virus elicits a surrogate neutralizing antibody response with cross-genogroup activity. Clin Vaccine Immunol. 2015;22(2):221–8.
50. Dai YC, Zhang XF, Xia M, Tan M, Quigley C, Lei W, Fang H, Zhong W, Lee B, Pang X, Nie J, Jiang X. Antigenic relatedness of norovirus GII.4 variants determined by human challenge sera. PLoS One. 2015;10(4):e0124945.

51. Lindesmith LC, Ferris MT, Mullan CW, Ferreira J, Debbink K, Swanstrom J, Richardson C, Goodwin RR, Baehner F, Mendelman PM, Bargatze RF, Baric RS. Broad blockade antibody responses in human volunteers after immunization with a multivalent norovirus VLP candidate vaccine: immunological analyses from a phase I clinical trial. PLoS Med. 2015;12(3): e1001807.
52. Atmar RL, Baehner F, Cramer JP, Song E, Borkowski A, Mendelman PM, NOR Study Group on Behalf of the, Al-Ibrahim MS, Bernstein DL, Brandon DM, Chu L, Davis MG, Epstein RJ, Frey SE, Rosen JB, Treanor JJ. Rapid responses to 2 virus-like particle norovirus vaccine candidate formulations in healthy adults: a randomized controlled trial. J Infect Dis. 2016;214 (6):845–53.
53. Malm M, Tamminen K, Vesikari T, Blazevic V. Type-specific and cross-reactive antibodies and T cell responses in norovirus VLP immunized mice are targeted both to conserved and variable domains of capsid VP1 protein. Mol Immunol. 2016;78:27–37.
54. Tamminen K, Malm M, Vesikari T, Blazevic V. Mucosal antibodies in duced by intranasal but not intramuscular immunization block norovirus GII. 4 virus like particle receptor binding. Viral Immunol. 2016;29(5):315–9.
55. Malm M, Tamminen K, Heinimaki S, Vesikari T, Blazevic V. Functionality and avidity of norovirus-specific antibodies and T cells induced by GII.4 virus-like particles alone or co-administered with different genotypes. Vaccine. 2018;36(4):484–90.
56. Liu W, Chen Y, Jiang X, Xia M, Yang Y, Tan M, Li X, Rao Z. A unique human norovirus lineage with a distinct HBGA binding interface. PLoS Pathog. 2015;11(7):e1005025.
57. Lindesmith LC, Donaldson EF, Baric RS. Norovirus GII.4 strain antigenic variation. J Virol. 2011;85(1):231–42.
58. Lindesmith LC, Beltramello M, Donaldson EF, Corti D, Swanstrom J, Debbink K, Lanzavecchia A, Baric RS. Immunogenetic mechanisms driving norovirus GII.4 antigenic variation. PLoS Pathog. 2012;8(5):e1002705.
59. Shanker S, Choi JM, Sankaran B, Atmar RL, Estes MK, Prasad BV. Structural analysis of histo-blood group antigen binding specificity in a norovirus GII.4 epidemic variant: implications for epochal evolution. J Virol. 2011;85(17):8635–45.
60. Shanker S, Czako R, Sankaran B, Atmar RL, Estes MK, Prasad BV. Structural analysis of determinants of histo-blood group antigen binding specificity in genogroup I noroviruses. J Virol. 2014;88(11):6168–80.
61. Singh BK, Leuthold MM, Hansman GS. Human noroviruses' fondness for histo-blood group antigens. J Virol. 2015;89(4):2024–40.
62. Eden J-S, Tanaka MM, Boni MF, Rawlinson WD, White PA. Recombination within the pandemic norovirus GII.4 lineage. J Virol. 2013;87(11):6270–82.
63. Lindesmith LC, Mallory ML, Debbink K, Donaldson EF, Brewer-Jensen PD, Swann EW, Sheahan TP, Graham RL, Beltramello M, Corti D, Lanzavecchia A, Baric RS. Conformational occlusion of blockade antibody epitopes, a novel mechanism of GII.4 human norovirus immune evasion. mSphere. 2018;3(1):e00518–7.
64. Parrino TA, Schreiber DS, Trier JS, Kapikian AZ, Blacklow NR. Clinical immunity in acute gastroenteritis caused by Norwalk agent. N Engl J Med. 1977;297(2):86–9.
65. Hutson AM, Atmar RL, Graham DY, Estes MK. Norwalk virus infection and disease is associated with ABO histo-blood group type. J Infect Dis. 2002;185(9):1335–7.
66. Lindesmith L, Moe C, Marionneau S, Ruvoen N, Jiang X, Lindblad L, Stewart P, LePendu J, Baric R. Human susceptibility and resistance to Norwalk virus infection. Nat Med. 2003;9:548.
67. Currier RL, Payne DC, Staat MA, Selvarangan R, Shirley SH, Halasa N, Boom JA, Englund JA, Szilagyi PG, Harrison CJ, Klein EJ, Weinberg GA, Wikswo ME, Parashar U, Vinje J, Morrow AL. Innate susceptibility to norovirus infections influenced by FUT2 genotype in a United States pediatric population. Clin Infect Dis. 2015;60(11):1631–8.
68. Le Pendu J, Rydell GE, Nasir W, Larson G. Human norovirus receptors. In: Viral gastroenteritis. Philadelphia, PA: Elsevier; 2016. p. 379–96.

69. Lopman BA, Trivedi T, Vicuna Y, Costantini V, Collins N, Gregoricus N, Parashar U, Sandoval C, Broncano N, Vaca M, Chico ME, Vinje J, Cooper PJ. Norovirus infection and disease in an Ecuadorian birth cohort: association of certain norovirus genotypes with host FUT2 secretor status. J Infect Dis. 2015;211(11):1813–21.
70. Thorven M, Grahn A, Hedlund KO, Johansson H, Wahlfrid C, Larson G, Svensson L. A homozygous nonsense mutation (428G→A) in the human secretor (FUT2) gene provides resistance to symptomatic norovirus (GGII) infections. J Virol. 2005;79(24):15351–5.
71. Bucardo F, Kindberg E, Paniagua M, Vildevall M, Svensson L. Genetic susceptibility to symptomatic norovirus infection in Nicaragua. J Med Virol. 2009;81(4):728–35.
72. Lindesmith LC, Donaldson EF, Lobue AD, Cannon JL, Zheng DP, Vinje J, Baric RS. Mechanisms of GII.4 norovirus persistence in human populations. PLoS Med. 2008;5 (2):e31.
73. Nordgren J, Nitiema LW, Ouermi D, Simpore J, Svensson L. Host genetic factors affect susceptibility to norovirus infections in Burkina Faso. PLoS One. 2013;8(7):e69557.
74. Ruvoen-Clouet N, Belliot G, Le Pendu. Noroviruses and histo-blood groups: the impact of common host genetic polymorphisms on virus transmission and evolution. Rev Med Virol. 2013;23(6):355–66.
75. Zhang XF, Huang Q, Long Y, Jiang X, Zhang T, Tan M, Zhang QL, Huang ZY, Li YH, Ding YQ, Hu GF, Tang S, Dai YC. An outbreak caused by GII.17 norovirus with a wide spectrum of HBGA-associated susceptibility. Sci Rep. 2015;5:17687.
76. Storry JR, Olsson ML. Genetic basis of blood group diversity. Br J Haematol. 2004;126 (6):759–71.
77. Marionneau S, Cailleau-Thomas A, Rocher J, Le Moullac-Vaidye, Ruvoen N, Clement M, Le Pendu. ABH and Lewis histo-blood group antigens, a model for the meaning of oligosaccharide diversity in the face of a changing world. Biochimie. 2001;83(7):565–73.
78. Donaldson EF, Lindesmith LC, Lobue AD, Baric RS. Norovirus pathogenesis: mechanisms of persistence and immune evasion in human populations. Immunol Rev. 2008;225(1):190–211.
79. Heggelund JE, Varrot A, Imberty A, Krengel U. Histo-blood group antigens as mediators of infections. Curr Opin Struct Biol. 2017;44:190–200.
80. de Mattos LC. Structural diversity and biological importance of ABO, H, Lewis and secretor histo-blood group carbohydrates. Rev Bras Hematol Hemoter. 2016;38(4):331–40.
81. Huang P, Farkas T, Zhong W, Tan M, Thornton S, Morrow AL, Jiang X. Norovirus and histo-blood group antigens: demonstration of a wide spectrum of strain specificities and classification of two major binding groups among multiple binding patterns. J Virol. 2005;79 (11):6714–22.
82. Hutson AM, Atmar RL, Marcus DM, Estes MK. Norwalk virus-like particle hemagglutination by binding to H histo-blood group antigens. J Virol. 2003;77(1):405–15.
83. Chan MCW, Lee N, Hung T-N, Kwok K, Cheung K, Tin EKY, Lai RWM, Nelson EAS, Leung TF, Chan PKS. Rapid emergence and predominance of a broadly recognizing and fast-evolving norovirus GII.17 variant in late 2014. Nat Commun. 2015;6:10061.
84. Nordgren J, Sharma S, Kambhampati A, Lopman B, Svensson L. Innate resistance and susceptibility to norovirus infection. PLoS Pathog. 2016;12(4):e1005385.
85. Rydell GE, Kindberg E, Larson G, Svensson L. Susceptibility to winter vomiting disease: a sweet matter. Rev Med Virol. 2011;21(6):370–82.
86. Harrington PR, Lindesmith L, Yount B, Moe CL, Baric RS. Binding of Norwalk virus-like particles to ABH histo-blood group antigens is blocked by antisera from infected human volunteers or experimentally vaccinated mice. J Virol. 2002;76(23):12335–43.
87. Becker KM, Moe CL, Southwick KL, MacCormack JN. Transmission of Norwalk virus during a football game. N Engl J Med. 2000;343(17):1223–7.
88. Tan M, Jiang X. Histo-blood group antigens: a common niche for norovirus and rotavirus. Expert Rev Mol Med. 2014;16:e5.

89. Tan M, Xia M, Chen Y, Bu W, Hegde RS, Meller J, Li X, Jiang X. Conservation of carbohydrate binding interfaces: evidence of human HBGA selection in norovirus evolution. PLoS One. 2009;4(4):e5058.
90. Farkas T, Thornton SA, Wilton N, Zhong W, Altaye M, Jiang X. Homologous versus heterologous immune responses to Norwalk-like viruses among crew members after acute gastroenteritis outbreaks on 2 US Navy vessels. J Infect Dis. 2003;187(2):187–93.
91. Xia M, Farkas T, Jiang X. Norovirus capsid protein expressed in yeast forms virus-like particles and stimulates systemic and mucosal immunity in mice following an oral administration of raw yeast extracts. J Med Virol. 2007;79(1):74–83.
92. LoBue AD, Lindesmith LC, Baric RS. Identification of cross-reactive norovirus CD4+ T cell epitopes. J Virol. 2010;84(17):8530–8.
93. Blazevic V, Lappalainen S, Nurminen K, Huhti L, Vesikari T. Norovirus VLPs and rotavirus VP6 protein as combined vaccine for childhood gastroenteritis. Vaccine. 2011;29 (45):8126–33.
94. Malm M, Heinimaki S, Vesikari T, Blazevic V. Rotavirus capsid VP6 tubular and spherical nanostructures act as local adjuvants when co-delivered with norovirus VLPs. Clin Exp Immunol. 2017;189(3):331–41.
95. Hjelm BE, Kilbourne J, Herbst-Kralovetz MM. TLR7 and 9 agonists are highly effective mucosal adjuvants for norovirus virus-like particle vaccines. Hum Vaccin Immunother. 2014;10(2):410–6.
96. Atmar RL, Opekun AR, Gilger MA, Estes MK, Crawford SE, Neill FH, Ramani S, Hill H, Ferreira J, Graham DY. Determination of the 50% human infectious dose for Norwalk virus. J Infect Dis. 2014;209(7):1016–22.
97. Tamminen K, Malm M, Vesikari T, Blazevic V. Norovirus-specific mucosal antibodies correlate to systemic antibodies and block norovirus virus-like particles binding to histo-blood group antigens. Clin Immunol. 2018;197:110–7.
98. Malm M, Tamminen K, Lappalainen S, Uusi-Kerttula H, Vesikari T, Blazevic V. Genotype considerations for virus-like particle-based bivalent norovirus vaccine composition. Clin Vaccine Immunol. 2015;22(6):656.
99. Tamminen K, Huhti L, Vesikari T, Blazevic V. Pre-existing immunity to norovirus GII-4 virus-like particles does not impair de Novo immune responses to norovirus GII-12 genotype. Viral Immunol. 2013;26(2):167–70.
100. Parra GI, Green KY. Sequential gastroenteritis episodes caused by 2 norovirus genotypes. Emerg Infect Dis. 2014;20(6):1016–8.
101. Tamminen K, Huhti L, Koho T, Lappalainen S, Hytonen VP, Vesikari T, Blazevic V. A comparison of immunogenicity of norovirus GII-4 virus-like particles and P-particles. Immunology. 2012;135(1):89–99.
102. Tan M, Huang P, Xia M, Fang P-A, Zhong W, McNeal M, Wei C, Jiang W, Jiang X. Norovirus P particle, a novel platform for vaccine development and antibody production. J Virol. 2011;85(2):753–64.
103. Wang L, Cao D, Wei C, Meng X-J, Jiang X, Tan M. A dual vaccine candidate against norovirus and hepatitis E virus. Vaccine. 2014;32(4):445–52.
104. Wang X, Ku Z, Dai W, Chen T, Ye X, Zhang C, Zhang Y, Liu Q, Jin X, Huang Z. A bivalent virus-like particle based vaccine induces a balanced antibody response against both enterovirus 71 and norovirus in mice. Vaccine. 2015;33(43):5779–85.
105. Xia M, Tan M, Wei C, Zhong W, Wang L, McNeal M, Jiang X. A candidate dual vaccine against influenza and noroviruses. Vaccine. 2011;29(44):7670–7.
106. Wang X, Wang S, Zhang C, Zhou Y, Xiong P, Liu Q, Huang Z. Development of a surrogate neutralization assay for norovirus vaccine evaluation at the cellular level. Viruses. 2018;10 (1):27.
107. Uusi-Kerttula H, Tamminen K, Malm M, Vesikari T, Blazevic V. Comparison of human saliva and synthetic histo-blood group antigens usage as ligands in norovirus-like particle binding and blocking assays. Microbes Infect. 2014;16(6):472–80.

108. Jones MK, Watanabe M, Zhu S, Graves CL, Keyes LR, Grau KR, Gonzalez-Hernandez MB, Iovine NM, Wobus CE, Vinje J, Tibbetts SA, Wallet SM, Karst SM. Norovirus: enteric bacteria promote human and mouse norovirus infection of B cells. Science. 2014;346 (6210):755–9.
109. Johnson PC, Mathewson JJ, DuPont HL, Greenberg HB. Multiple-challenge study of host susceptibility to Norwalk gastroenteritis in US adults. J Infect Dis. 1990;161(1):18–21.
110. Simmons K, Gambhir M, Leon J, Lopman B. Duration of immunity to norovirus gastroenteritis. Emerg Infect Dis. 2013;19(8):1260–7.
111. Kanno A, Kazuyama Y. Immunoglobulin G antibody avidity assay for serodiagnosis of hepatitis C virus infection. J Med Virol. 2002;68(2):229–33.
112. Blazevic V, Malm M, Salminen M, Oikarinen S, Hyoty H, Veijola R, Vesikari T. Multiple consecutive norovirus infections in the first 2 years of life. Eur J Pediatr. 2015;174 (12):1679–83.
113. Blazevic V, Malm M, Vesikari T. Induction of homologous and cross-reactive GII.4-specific blocking antibodies in children after GII.4 New Orleans norovirus infection. J Med Virol. 2015;87(10):1656–61.
114. Malm M, Uusi-Kerttula H, Vesikari T, Blazevic V. High serum levels of norovirus genotype-specific blocking antibodies correlate with protection from infection in children. J Infect Dis. 2014;210(11):1755–62.
115. Sharma S, Carlsson B, Czako R, Vene S, Haglund M, Ludvigsson J, Larson G, Hammarstrom L, Sosnovtsev SV, Atmar RL, Green KY, Estes MK, Svensson L. Human sera collected between 1979 and 2010 possess blocking-antibody titers to pandemic GII.4 noroviruses isolated over three decades. J Virol. 2017;91(14):e00567–17.
116. Lindesmith LC, Mallory ML, Jones TA, Richardson C, Goodwin RR, Baehner F, Mendelman PM, Bargatze RF, Baric RS. Impact of pre-exposure history and host genetics on antibody avidity following norovirus vaccination. J Infect Dis. 2017;215(6):984–91.
117. Ha S, Choi I-S, Choi C, Myoung J. Infection models of human norovirus: challenges and recent progress. Arch Virol. 2016;161(4):779–88.
118. Taube S, Perry JW, Yetming K, Patel SP, Auble H, Shu L, Nawar HF, Lee CH, Connell TD, Shayman JA, Wobus CE. Ganglioside-linked terminal sialic acid moieties on murine macrophages function as attachment receptors for murine noroviruses. J Virol. 2009;83 (9):4092–101.
119. Bok K, Parra GI, Mitra T, Abente E, Shaver CK, Boon D, Engle R, Yu C, Kapikian AZ, Sosnovtsev SV, Purcell RH, Green KY. Chimpanzees as an animal model for human norovirus infection and vaccine development. Proc Natl Acad Sci U S A. 2010;108(1):325–30.
120. Zhu S, Regev D, Watanabe M, Hickman D, Moussatche N, Jesus DM, Kahan SM, Napthine S, Brierley I, Hunter RN 3rd, Devabhaktuni D, Jones MK, Karst SM. Identification of immune and viral correlates of norovirus protective immunity through comparative study of intra-cluster norovirus strains. PLoS Pathog. 2013;9(9):e1003592.
121. Karst SM. Identification of a novel cellular target and a co-factor for norovirus infection – B cells and commensal bacteria. Gut Microbes. 2015;6(4):266–71.
122. Guix S, Asanaka M, Katayama K, Crawford SE, Neill FH, Atmar RL, Estes MK. Norwalk virus RNA is infectious in mammalian cells. J Virol. 2007;81(22):12238–48.
123. Oka T, Stoltzfus GT, Zhu C, Jung K, Wang Q, Saif LJ. Attempts to grow human noroviruses, a sapovirus, and a bovine norovirus in vitro. PLoS One. 2018;13(2):e0178157.

The Current Status of Norovirus Vaccine Development 7

Nada M. Melhem, Farouk F. Abou Hassan, and Mohammad Ramadan

7.1 Introduction

Norovirus (NoV) is the leading cause of epidemic gastroenteritis affecting all age groups [1–3]. While NoV infection is usually self-limited, NoV-associated gastroenteritis lasts for weeks or months among immunocompromised individuals and causes severe outcomes among children less than 5 years old. Despite the decline in annual deaths caused by diarrhea, the burden of disease caused by NoVs was estimated at 677–700 million cases [4, 5]. The incidence estimates of NoV-associated outcomes in different age groups are summarized in Chap. 1. The data clearly depict NoV as a public health problem resulting in >200,000 deaths and an economic burden of >$64 billion yearly [5]. 82% of NoV-associated disease and 97% of deaths worldwide occur in low- and middle-income countries. Until the write-up of this book, noroviruses were classified into seven genogroups (GI–GVII) based on the amino acid sequence of the viral capsid protien (VP1). This classification was recently updated by Chhabra et al. to include three new genogroups (GVIII–GX) [6]. An aged-structured dynamic transmission model predicted that norovirus pediatric vaccination averts 33 and 60% of cases among children under 5 years for low vaccine efficacy and high vaccine efficacy, respectively [7]. Currently, there are no licensed vaccines for the prevention of norovirus infection. Recently, norovirus was identified by the World Health Organization's (WHO) Product Development for Vaccines Advisory Committee as a priority disease for vaccine development [8]. This chapter summarizes the currently available data from challenge, preclinical and clinical studies as well as their impact on the prevention and control of NoV infection.

N. M. Melhem (✉) · F. F. Abou Hassan
Medical Laboratory Sciences Program, Division of Health Professions, Faculty of Health Sciences, American University of Beirut, Beirut, Lebanon
e-mail: melhemn@aub.edu.lb

M. Ramadan
Faculty of Medicine, American University of Beirut, Beirut, Lebanon

N. M. Melhem (ed.), *Norovirus*,
https://doi.org/10.1007/978-3-030-27209-8_7

7.2 Challenge Studies

Challenge studies have been useful in generating data defining the antigenic relatedness among NoV genotypes, genetic susceptibility as well as deciphering the role of immune responses against NoV infection. Table 7.1 summarizes the main results of these studies. Early reports suggested that challenge with NoV leads to short-term protection against homologous but not heterologous or heterotypic strains [18]. Other reports assessed the role of existing NoV-specific antibodies in protecting adults against infection whereby two groups of volunteers were challenged via ingestion of a water solution containing Norwalk inoculum 8FIIa [9]. These groups were previously included in challenge studies and selected based on their low and high levels of serum-specific anti-NoV titers. These groups received the same dosage of a single, two, three or four doses of virus inoculum. Volunteers with low and high antibody titers suffered from gastroenteritis following the first challenge. Authors concluded that existing high antibody titers did not protect volunteers against Norwalk-gastroenteritis. Following the second and third challenges, volunteers with low initial antibody titers developed resistance to experimental infection; the latter lasted for up to 6 months. The results of this study were among the early ones concluding that repeated exposure resulted in resistance to norovirus infection.

Challenge studies have also been used to understand the predictors of susceptibility and protection against noroviruses [10]. Following challenge, 62% of secretor-positive (Se^+) volunteers became infected whereas all secretor-negative (Se^-) volunteers were resistant to Norwalk virus (NV) infection; thus confirming the correlation between susceptibility to NoV and fucosyl transferase 2 (*FUT2*) expression (95% of Se^+ volunteers expressed *FUT2* gene). When salivary IgA titers were measured in 50 volunteers infected with NV, variable levels were detected among Se^+ infected and uninfected adults. A faster increase in salivary IgA (1–5 days post-challenge) was detected among uninfected participants suggesting previous exposure to NV. However, Se^- volunteers with pre-challenge NV-specific IgG were also resistant to infection indicating that susceptibility to NoV infection is complex and does not solely depend on the expression of a functional *FUT2* gene. This study highlighted the role of acquired immunity associated with the production of salivary IgA in Se^+ individuals.

Antigenically and morphologically similar to wild type NoV, virus-like particles (VLPs) self-assemble following in vitro expression of NoV major capsid protein VP1. Interestingly, serum IgG and salivary IgA were detected post-challenge with NV [19, 20]. Challenge studies examined the ability of VLPs to stimulate B- and T-cell responses against NoV in a group of volunteers infected with Snow Mountain virus (SMV), a GII strain [11]. These responses were cross-reactive within but not across genogroups. Specifically, anti-SMV IgG were cross reactive with Hawaii virus (HV, a GII NoV) but not against NV (GI NoV). When characterizing further

Table 7.1 Norovirus challenge studies

Challenge virus	Dosage (Route)	Schedule	Sample size[a] (Age)	Immune responses	References
Norwalk inoculum 8FIIa	20 μL (Oral)	Single dose; N = 42 two doses; N = 22 three doses; N = 19 four doses; N = 2	42 (≥18 years)	Already existing high antibody titers did not protect volunteers. Resistance to NoV infection lasted up to 6 months following the second challenge.	Johnson et al. [9]
Norwalk virus 8FIIa or 8FIIb	$10 - 3 \times 10^8$ PDU[b] (NA)	Single dose	77 (20–49 years)	Salivary IgA in infected secretor-positive individuals peaked at day 14 following challenge. Salivary IgA titers in uninfected secretor-positives were 2-folds higher than baseline 2–3 days following challenge.	Lindesmith et al. [10]
SMV[c]	$10–10^5$ PCR units (Oral)	Single dose	15 (21–54 years)	By day 14 post-challenge, ≥4-fold rise in serum IgG titers and anti-SMVs salivary IgA titers (67%) was detected. Increase in Th1 cytokines (IFN-γ, TNF-α, and IL-2) was also detected.	Lindesmith et al. [11]
GI.1-1968	–	–	10 (18–49 years)	Cross-reactive IgG antibodies detected against a panel of GI VLPs. IFN-γ specific T-cell responses were also detected.	Lindesmith et al. [12]
GI.1 NoV	0.48, 4.8, 48 or 4800 RT-PCR units (Oral)	Single dose	48 (18–50 years)	≥4-fold rise in NV-VP1 IgG and ≥4-fold increase in HOV[d] protease antibodies	Ajami et al. [13]

(continued)

Table 7.1 (continued)

Challenge virus	Dosage (Route)	Schedule	Sample size[a] (Age)	Immune responses	References
GII.4 (Cin-1 strain) NoV	5×10^4 RT-PCR (Oral)	Single dose	40 (18–49 years)	Stronger IgG response and 4-fold increase in HBGA-blocking titers among secretors	Frenck et al. [14]
GI.1 NoV	0.48, 4.8, 48 or 4800 RT-PCR units (Oral)	Single dose	34 (18–50 years)	≥4-fold increase in serum blocking antibodies to homologous GI.1 NoV and one or more heterologous NoVs (GI.4, GI.7, GII.4 HOV and GII.4 Sydney)	Czako et al. [15]
Norwalk virus 8FIIb inoculum	6.5×10^7 GEC[e] or 10^4 GEC (Oral)	Single dose	52 (mean: 26.7 years)	Increased cytokine levels in infected individuals (IFN-α, IL-2, IL-4, IL-5, IL-6, IL-8, IL-10, MCP-1, TNF-α and TNF-β)	Newman et al. [16]
Norwalk virus GI.1	0.48, 4.8, 48, or 4800 RT-PCR units (Oral)	Single dose	57 (18–50 years)	Increased levels of NoV-specific fecal IgA, salivary IgA and IgG and IgA ASCs[f]; higher HBGA-blocking titers in infected individuals who did not develop gastroenteritis	Ramani et al. [17]

[a]The sample size reported in this table represents the total number of volunteers (receiving challenge and placebo)
[b]PCR detectable units
[c]Snow Mountain virus
[d]Houston virus
[e]Genome equivalent copies
[f]Antibody-secreting cells

the properties of salivary IgA, results showed the lack of cross-reactive antibody titers with NV and less cross-reactive to HV as compared to IgG. Little was known at the time about the characteristics of the T helper type 1 (Th1) and Th2 cellular immune responses following NoV challenge. In an attempt to determine the ability of SMV challenge to stimulate T-cell responses, VLPs of three NoV strains (GI.1, GII.1, and GII.2) were used to stimulate in vitro peripheral blood mononuclear cells (PBMCs) isolated from volunteers pre- and post-challenge with SMV [11]. Post-challenge PBMCs of many infected and uninfected participants mounted an interferon (IFN)-γ response and produced other cytokines including tumor necrosis factor (TNF)-α and interleukin (IL)-2 (Th1 responses). These results are similar to previously reported ones following challenge of volunteers with NV VLPs [21]. When stimulated with HV VLPs, post-challenge PBMCs of infected and uninfected volunteers mounted a cross-reactive Th1 response (IFN-γ and IL-2). Following in vitro stimulation with NV VLPs, no significant difference was recorded in the levels of IFN-γ, TNF-α, IL-2, IL-10 and IL-4 between pre- and post-challenge PBMCs indicating lack of cross-reactivity. This study was the first to characterize T-cell responses in vitro following challenge.

The same group used archived sera and PBMCs from healthy adults infected with GI.1-1968 strain to assess the heterotypic humoral and cellular immune responses. Briefly, VLPs of five GI strains were used: G1.1-1968, GI.1-2001, GI.2-1999, GI.3-2000 and GI.4-2000 [12]. This study demonstrated the generation of cross-reactive IgG antibodies to the diverse panel of GI VLPs. The high degree of reactivity to GI VLPs was explained by: (1) the ability of GI.1 to induce high affinity antibodies; (2) the pre-existing memory B (B_M) cell responses with high affinity to other GI strains; or (3) the conserved capsid protein acting as a cross-reactive antibody-binding site. IFN-γ specific T-cell responses targeting homologous and heterologous GI NoV strains were also detected ex vivo in archived PBMCs of most volunteers. This bias of humoral and T-cell responses to GI VLPs was suggested as a potential mechanism of deceptive imprinting or original antigenic sin (OAS). OAS explains the ability of the host immune responses to stimulate memory cells to produce specific cellular responses to a similar antigen rather than inducing a primary response to a virus variant slightly different from the founder or infecting strain [22].

In an attempt to develop an assay to detect antibodies to NoV proteases, the seroresponses of individuals challenged with NV were tested against proteases from GI.1 NV and GII.4 Houston virus (HOV) [13]. Approximately 53% of volunteers challenged with GI.1 NV ($n = 48$) (shedding the virus, and showing a seroresponse to NV VLPs) exhibited ≥4-fold rise in NV-protease antibody titers; pre-challenge IgG titers directed against NoV protease were detected in 40% of volunteers. Interestingly, 74% of infected volunteers showed increased protease-specific antibody titers to homologous and heterologous protease (GI and GII). All infected volunteers with increased anti-protease antibodies also exhibited a ≥4-fold increase in IgG NV-VP1-specific antibodies (day 28 post-challenge). These results suggested the immunogenicity of NoV protease.

While the majority of challenge studies were performed using GI.1 NoV, Frenck et al. [14] performed the first challenge study using NoV GII.4, the predominant human Norovirus (HuNoV) genotype worldwide. The latter was administered orally to healthy adults 18–49 years old. This study recruited secretors and non-secretors in order to assess the association of histo-blood group antigens (HBGAs) and susceptibility to HuNoV infection. Low levels of pre-existing anti-GII.4 antibodies were required prior to participation to reduce the impact of previous exposure/immunity. A strong correlation between the secretor status and symptomatic disease was clearly shown. While this study did not report on the immune responses following challenge, it established a challenge model using GII.4 NoV.

The acute immune responses to HuNoV infection are not well understood. Consequently, archived sera from NoV challenge studies were used to characterize the acute serum cytokine responses following challenge with Norwalk virus 8FIIb [16]. Four days following challenge, a general increase in a number of cytokines was detected among infected secretor-positive individuals compared to uninfected volunteers. The levels of IFN-α, IL-2, IL-4, IL-5, IL-6, IL-8, IL-10, monocyte chemoattractant protein (MCP)-1, TNF-α and TNF-β peaked by day 2 with only IL-10 maintaining significantly high titers up to 4 days post-challenge. These findings demonstrated the ability of NoV infection to mount Th1 and Th2 responses. In another challenge study [11], IFN-γ, IL-2 and TNF-α also increased; similar results were also reported in animal models [23, 24]. The use of the NoV experimental challenge model was important to elucidate the comprehensive pattern of acute cytokine responses for the first time and confirmed results reported in animal models. Importantly, this study also demonstrated the lack of correlation between pre-challenge cytokine concentrations and infection following challenge.

Another experimental challenge study was performed whereby 29 individuals were infected with NoV GI.1 and 5 received placebo (sterile water) [15]. Although the highest responses targeted homologous strains as previously reported [11, 25, 26], ≥4-fold increase in serum-blocking antibodies was elicited post-challenge to one or more heterologous GI (GI.4, GI.7) as well as the GII divergent NoVs (GII.4 HOV and GII.4 Sydney). This broad cross-reactivity supports the presence of cross-reactive epitopes as previously reported [27]. Consequently, factors other than antigenic relatedness could play a role in the breadth of blockade responses. The quality of the generated immune responses following challenge and vaccination and its sustainability remain poorly studied.

Following inoculation with NV GI.1, the mucosal and cellular immune responses were assessed in 57 healthy adults [17]. Four days post-challenge, 58% of susceptible participants (i.e. expressing *FUT2*) showed clinical signs and symptoms. Moreover, the HBGA-blocking antibodies titer was significantly higher among infected volunteers who did not develop gastroenteritis compared to those who did. More than 4-fold increase in NV-specific fecal IgA was detected in 95% of participants by day 14 post-challenge; these high levels of fecal IgA titers were maintained in 85% of infected subjects until day 28 post-challenge. Similar increases were reported for NV-specific salivary IgA levels albeit in 65 and 35% of infected volunteers by days 14 and 28 post-challenge, respectively. Pre-challenge levels of salivary NoV-specific IgA and NoV-specific memory IgG were associated with protection from gastroenteritis whereas pre-existing levels of NoV-specific fecal IgA were not;

the latter correlated with reduced viral load. This study confirmed the role of mucosal immune responses in protection against the disease as well as control of viral load. Moreover, 95% of infected persons had robust increases in NoV-specific IgA and IgG antibody-secreting cells (ASCs) by day 7 post-challenge. By days 14 and 28, IgA ASCs were maintained among 90 and 67% of volunteers whereas IgG ASCs were detected among 74 and 67% of participants, respectively. Importantly, similar results were observed for NoV-specific IgG and IgA B_M cells whereby 75 and 50% of infected persons maintained more than 4-fold rise for 180 days post-challenge, respectively.

Challenge studies have been instrumental in elucidating and characterizing immune responses and correlates of protection following NoV infection. Most of these studies targeted adults and focused on outcomes following infection with limited number of NoV strains. It is yet to determine the trends of immune responses in children, i.e. unprimed population. With the advancement of vaccine studies, the breadth, magnitude and repertoire of immune responses against noroviruses will be further understood.

7.3 Preclinical Studies

Prior to banning research on non-human primates [28], chimpanzees were used as an animal model to study NoV infection [29]. Chimpanzees were vaccinated intramuscularly with two doses of GI VLPs or GII VLPs followed by intravenous challenge with GI and GII one month post-vaccination and GII eighteen months post-vaccination. Chimpanzees vaccinated with GI VLPs were protected against homologous challenge and mounted serum antibody responses for up to 24 months. While high titers of serum-specific GII antibodies were detected in animals immunized with GII.4 VLPs, the former did not protect against NV challenge.

Gnotobiotic pigs [30] and calves [23] were also found to be successfully permissive for infection with HuNoV GII.4 whereby low-to-moderate levels of serum-specific IgG and IgA were detected [23]. Moreover, gnotobiotic pigs and calves inoculated orally or intranasally with VLPs expressing HuNoV GII.4 mounted Th1 (IFN-γ, IL-12) and/or Th2 responses (IL-4) [23, 24]. IgM ASCs were detected in the intestines, spleen and blood followed by late IgG ASCs in tissues and blood. IgA ASCs were detected early in the intestine and blood at post-inoculation day (PID)-6 and later in spleen (PID-28). Following homologous challenge, a balanced Th1/Th2 response was reported in animals vaccinated with VLPs in the presence of mutant *Escherichia coli (E. coli)* heat-labile toxin (mLT) as adjuvant. This response was able to protect 100% of vaccinated animals. When VLPs were administered with immunostimulating complexes (ISCOM), animals mounted a Th2 response with 75% protection rate from diarrhea [31]. The same model was used to compare the efficacy of P particles, 24-mer particles containing the protruding (P) domain of NoV major capsid protein VP1 [32], and VLPs derived from GII.4 against NoV infection [33]. The results of this study were previously summarized [34]. Briefly, pigs were divided into three groups: groups 1 and 2 received GII.4/1997 NoV (VA387)-

derived P particle and VLPs, respectively, via the intranasal route; group 3 received GII.4/2006b NoV orally to simulate primary infection. Upon re-challenge with the homologous virus, oral NoV infection provided 83% protection rate from diarrhea compared to 47 and 60% protection rates provided by the intranasally administered P particle and VLP-based constructs, respectively. These results suggested that previous infection with NoV led to substantial protection. Importantly, NoV primary infection induced an increase in IFN-γ produced by $CD4^+$ T cells in the duodenum. The latter is in agreement with observations detected in humans [11]. This study showed an inverse correlation between the expansion of T cells, regardless of the subset, and protection post-challenge. P particles and VLPs induced similar protection rates with higher numbers of activated $CD4^+$ T cells in different tissues, $CD8^+$ T cells in the duodenum, regulatory T cells (Tregs) in peripheral blood and $CD4^+CD25^-FoxP3^+$ Tregs producing transforming growth factor (TGF)-β in the spleen following vaccination with the former. These results suggested that P particles are more immunogenic than VLPs. This study was the first to evaluate the protective efficacy of VLP- or P particle-based vaccine candidates against NoV infection. This study also assessed the role of different T-cell subsets following primary infection and vaccination with VLP or P particles pre- and post-challenge. While this study advances P particles as potential vaccine candidates, conflicting results were reported when comparing P particles to VLPs in mice model. Th2-skewed response was detected following use of P particles whereas a balanced Th1/Th2 response was detected following use of VLPs [35, 36].

The breadth of the immune response against diverse NoV variants was evaluated following immunization of rabbits with monovalent and bivalent vaccine formula containing consensus GII.4 (GII.4c) and/or GI.1 NV VLPs [37]. GII.4c VLPs were derived from GII.4 Houston 2002, Den Haag 2006, and Yerseke 2006 strains. High seroresponses were detected against VLPs derived from different wild-type GII.4 strains with low levels of cross-reactivity against GI strains. The bivalent form of the vaccine administered intramuscularly or intranasally in two doses (50 or 150 μg) with different adjuvants (Aluminum hydroxide or Alhydrogel ($Al(OH)_3$)), Monophosphoryl lipid A (MPLA), Toll-like receptor (TLR)-4 agonist, or Chitosan) was then tested in an attempt to investigate the ability of the rabbit model to mount better responses against heterologous strains. Intramuscular immunization elicited the highest serum antibody titers against homologous and heterologous strains in the vaccine. This response was slightly increased in the presence of $Al(OH)_3$. The development of consensus vaccine formulation may induce a broadly cross-reactive antibody response against different NoV genotypes, a model that successfully induced neutralizing antibodies against human immunodeficiency virus (HIV) [38–40].

While preclinical vaccine studies were performed using pigs and occasionally rabbits and chimpanzees, the majority of studies elucidating the role of different vaccine formulations were performed in the mouse model (Table 7.2). The use of the latter as a platform for vaccination with VLPs was instrumental in recording the generation of NoV-specific IgG and mucosal IgA responses. These responses were more efficiently mounted following intranasal administration of the vaccine rather

Table 7.2 Norovirus preclinical studies

Vaccine formulation	Animal, route of immunization	Dosage (Adjuvant)	Schedule	Immune responses post-vaccination	Challenge strain (Route)	IR[a] and protection post-challenge	References
GI.1 or GII.4 VLPs (Baculovirus expression system)	Chimpanzees, IM[b]	50 μg ($Al(OH)_3$)	2 doses, 30 days apart	Serum-specific IgM and IgG responses were detected following vaccination.	Homologous and heterologous challenge (IV[c])	Immunization with homologous GI VLPs protected animals following intravenous challenge.	Bock et al. [29]
GII.4 HS66 VLPs (Baculovirus expression system)	Gnotobiotic pigs, Oral and IN[d]	250 μg (mLT[e] or ISCOM[f])	3 doses (one oral dose, two IN doses), 10 days apart	Adjuvanted vaccine induced systemic and intestinal antibodies as well as cellular immune responses.	Homologous challenge (Oral)	Higher protection rates from diarrhea as well as Th1/Th2 responses were detected with mLT-adjuvanted vaccine. ISCOM-adjuvanted vaccine induced Th2 response with 75% protection rate.	Souza et al. [31]
GII.4 VA387 VLP/P particles (*E. coli* expression system)	Gnotobiotic pigs, IN or Oral	2.74×10^4 viral RNA copies (Chitosan, MPLA[g])	3 doses, 5, 10 and 21 days postpartum	Effector and memory T-cell expansion in tissues; stronger responses induced by P particles	Homologous NoV GII.4 2006b variant 092895 (Oral)	P particles and VLPs induced substantial cross protection following reinfection.	Kocher et al. [33]
GI.1 and/or GII.4 consensus VLPs (Baculovirus expression system)	Rabbits, IM or IN	50 or 150 μg (Al $(OH)_3$, MPLA, TLR[h]-4 agonist or chitosan)	2 doses, 3 weeks apart	The bivalent formulation (IM) induced higher serum antibody titers to GII.4, compared to	NA	NA	Parra et al. [37]

(continued)

Table 7.2 (continued)

Vaccine formulation	Animal, route of immunization	Dosage (Adjuvant)	Schedule	Immune responses post-vaccination	Challenge strain (Route)	IR[a] and protection post-challenge	References
				the IN route. Slightly higher titers were detected in sera of rabbits immunized with the $Al(OH)_3$-adjuvanted bivalent formulations via the IM route compared to monovalent VLP administered intranasally and in combination with other adjuvants.			
GI.1 VLPs (Baculovirus expression system)	Mice, OG[i]	5, 25, 50, 75, 100, 200, 300, 400 and 500 μg, (CT[j])	4 doses: 2 on d9, 1 on d26 and 1 on d45.	High serum IgG and intestinal IgA detected with or without CT	NA	NA	Ball et al. [41]
VEE[k] replicon-expressed GI.1 (VRP-NV1) and Norwalk VLPs (VEE-Nor)	Mice, VRPs via FP[l] inoculation; VLPs via OG	10^7 IU[m] in FP, 75 μg in OG	Single dose, booster on d23	Recombinant vaccine induced higher IgG, IgM, and IgA responses compared to VLPs.	NA	NA	Harrington et al. [42]

GII.4/Dijon 171/96 VLPs (Baculovirus expression system)	Mice, IN or OG	10 μg, (LT or mLT)	2 doses, 2 weeks apart	Seroresponses, fecal IgA and cellular responses were detected; the presence of adjuvants enhanced these responses.	NA	NA	Nicollier-Jamot et al. [43]
VEE replicons expressing GI.1/Norwalk, GII.1/Hawaii, GII.2/Snow Mountain, GII.4/Lordsdale VLPs (monovalent, trivalent or quadrivalent)	Mice, FP inoculation	2.5×10^6 IU	2 doses, 3 weeks apart	Trivalent vaccine (GI.1, GII.1, GII.2) induced a heterotypic IgG response against GII.4 though to a lesser degree than quadrivalent or monovalent preparations with GII.4.	NA	NA	LoBue et al. [26]
GII.4/VA387 VLPs (yeast expression system)	Mice, OG or IM	IM: 40 μg (Ribi adjuvant), Oral: 0.1 mg or 1 mg	IM: 3 doses, 2 weeks apart; OG: 5 doses, 2 weeks apart	Dose-dependent serum IgG and fecal IgA responses were detected following oral vaccination. HBGA-blocking activity was also detected.	NA	NA	Xia et al. [44]
VRPs expressing ORF1, ORF2 or ORF3 of MNV1.CW3; ORF2 of HuNoV Lordsdale (GII.4) or Chiba (GI.4)	Mice, SC[n]	2.5×10^6 IU	2 doses, 3 weeks apart	Vaccinated mice mounted cellular immune responses.	Homologous MNV1.CW3 (Oral)	The vaccine stimulated long-term protection against oral infection with MNV (up to 6 months).	Chachu et al. [45]

(continued)

Table 7.2 (continued)

Vaccine formulation	Animal, route of immunization	Dosage (Adjuvant)	Schedule	Immune responses post-vaccination	Challenge strain (Route)	IR[a] and protection post-challenge	References
rvAdGII.4[o]	Mice, IN	10^6 IU	3 doses, 2 weeks apart	The vaccine induced strong NoV-specific seroresponses (IgA, IgG and IgM) and mucosal responses (fecal IgG and IgA). Th1 and Th2 cytokines were also produced.	NA	NA	Guo et al. [46]
rvAdGII.4 and GII.4 VLPs	Mice, IN	10^6 IU of rAd and 10 μg of VLPs (CpG[p] and poly I:C[q])	3 doses, 2 weeks apart	Stronger humoral (IgG, IgA), mucosal (fecal IgA, IgG) and IFN-γ responses were detected in mice primed with rAd expressing GII.4 and boosted with VLPs compared to mice primed with VLPs and boosted with rAd or immunized with VLPs alone.	NA	NA	Guo et al. [47]

Monovalent VLP GI/GII or multivalent (VEE-replicon expression system)	Mice, FP inoculation	2 μg of each VLP (null VRP[r] or CpG)	2 doses, 4 weeks apart	Multivalent vaccines adjuvanted with null VRPs induced higher cross-reactive antibody responses as well as receptor-blocking IgG responses.	Heterologous, 3×10^7 PFU[s] MNV-1.CW.3 (Oral)	Decreased viral load following MNV challenge correlated with reduction in clinical disease.	LoBue et al. [48]
VEE replicons expressing GI.1-1968 or GII.4-2002 VLPs	Mice, Oral	2.5×10^6 IU	2 doses, 3 weeks apart	Cross-reactive IFN-γ ($CD4^+$ T cell-mediated) was detected following heterologous immunization.	NA	NA	LoBue et al. [49]
NV VLP (Tobacco mosaic virus expression system)	Mice, IN	5-25 μg (CT, TLR7, TLR7/8 agonists)	2 doses, 3 weeks apart	Higher serum IgG, IgG1, and IgG2a responses as well as mucosal (IgA) were induced by TLR7- or CT-adjuvanted vaccine formulations.	NA	NA	Velasquez et al. [50]
rVSV[t]-VP1 (HuNoV)	Mice, IN and Oral	10^6 PFU	Single dose	Stronger and robust humoral (IgG) and T-cell responses induced by rVSV-VP1 compared to immunization with Baculovirus-expressed VLPs.	NA	NA	Ma et al. [51]

(continued)

Table 7.2 (continued)

Vaccine formulation	Animal, route of immunization	Dosage (Adjuvant)	Schedule	Immune responses post-vaccination	Challenge strain (Route)	IR[a] and protection post-challenge	References
rVP6 of RV and/or NoV GII.4 VLPs (monovalent or bivalent, Baculovirus expression system)	Mice, ID[u] or IM	10 μg/antigen	2 doses, 3 weeks apart	Higher levels of NoV-specific IgG responses as well as cross-reactive responses to heterologous NoV were detected following bivalent vaccination compared to monovalent immunization. Similar trends were reported for HBGA-blocking antibodies.	NA	NA	Blazevic et al. [52]
NV VLPs (Tobacco mosaic virus expression system)	Guinea pigs, IN	25 μg (GARD[v])	2 doses, 3 weeks apart	NV-VLPs, with or without GARD, and in the presence of GelVac[w] induced significantly higher serum and mucosal responses than NV-VLPs.	NA	NA	Velasquez et al. [53]
P particle-VP8 chimeric vaccine (bivalent, *E. coli* expression system)	Mice, IN or SC	15 μg (Freund's adjuvant in SC)	3 or 4 doses, 2 weeks apart	Stimulation of homotypic NoV HBGA-blocking antibodies	NA	NA	Tan et al. [54]

P particle-M2e influenza chimeric vaccine (bivalent; *E. coli* expression system)	Mice, IN or SC	50 μg (Montanide ISA720)	3 doses, 2 weeks apart	Stimulation of antibodies blocking the binding of NoV-VLPs to HBGA	NA	NA	Xia et al. [55]
GII.4 VLP/P particles (Baculovirus expression system)	Mice, IM or ID	10 μg	2 doses, 3 weeks apart	VLP vaccine induced higher serum-specific IgG response with higher IgG avidity and balanced Th1/Th2 responses compared to those induced by the P particle vaccine. It also stimulated cross-reactive B- & T-cell responses and stronger HBGA-blocking antibody responses.	NA	NA	Tamminen et al. [35]
NoV strain VA387 GII.4 VLPs and P particles (Baculovirus expression system)	Mice, IN	30 μg	3 doses, 2 weeks apart	VLPs and P particles induced higher IgG and $CD4^+$ T-cell responses compared to P dimers.	NA	NA	Fang et al. [36]

(continued)

Table 7.2 (continued)

Vaccine formulation	Animal, route of immunization	Dosage (Adjuvant)	Schedule	Immune responses post-vaccination	Challenge strain (Route)	IR[a] and protection post-challenge	References
NoV VLPs (Tobacco mosaic virus expression system)	Mice, IN or Oral	25 μg intranasal or 100-200 μg oral (TLR 3,7,8 or 9 agonists used in both routes)	2 doses IN or 3 doses orally, 3 weeks apart	The nasal co-delivery of VLPs with TLR7 or TLR9 agonists induced the highest systemic (IgG, IgG1, IgG2a) and mucosal (IgA) immune responses. HBGA-blocking antibodies were also induced.	NA	NA	Hjelm et al. [56]
RV rVP6, NoV GII.4 VLPs and GI.3 VLPs (single or trivalent, Baculovirus expression system)	Mice, IM	10 μg	2 doses, 3 weeks apart	Single and trivalent vaccines induced a strong homologous IgG response. The trivalent formulation induced cross-reactive IgG responses and HBGA-blocking antibodies against homologous and heterologous VLPs. IFN-γ response (T cell-mediated) was also detected.	NA	NA	Tamminen et al. [57]

GII.4 or GII.12 VLPs (Baculovirus expression system)	Mice, IM or ID	1 μg ID or 10 μg IM	2 doses, 3 weeks apart (GII.12 boost at week 18)	Immunization with GII.4 VLP stimulated high levels and prolonged (27 weeks) NoV-specific IgG responses and HBGA-blocking antibodies. Heterologous boost with GII.12 induced similar GII.12-specific IgG and HBGA-blocking antibodies to the primary antigen without decreasing the levels of GII.4 specific responses (IgG or blocking antibodies).	NA	NA	Tamminen et al. [58]
GII.4 VLPs, chimeric (GII.4 VRPs), or multivalent VRPs (NV, SM, HV, LV) (VEE-replicon expressed)	Mice, FP injection	5×10^4 VRPs or VLPs	2 doses, 3 weeks apart	The chimeric and multivalent immunogens induced broader blocking immune responses against homologous and heterologous strains compared to monovalent immunogens.	NA	NA	Debbink et al. [27]

(continued)

Table 7.2 (continued)

Vaccine formulation	Animal, route of immunization	Dosage (Adjuvant)	Schedule	Immune responses post-vaccination	Challenge strain (Route)	IR[a] and protection post-challenge	References
NoV VA387 VLP or rNDV[x]-VP1 GII.4	Mice, IN	30 μl of 10^6 EID50[y]	3 doses, 2 weeks apart	Serum IgG levels were comparable in both groups. rNDV induced higher levels of serum IgG1 and fecal IgA.	NA	NA	Kim et al. [59]
Narita 103 (NoV GII) virus capsid protein (NaVCP) (Tobacco mosaic virus expression system)	Mice, IN	25 μg	2 doses, 3 weeks apart	Robust increase in VLP-specific serum IgG and mucosal IgA	NA	NA	Mathew et al. [60]
rVSV-VP1	Mice, IN and Oral	1 or 5×10^6 PFU (HSP 70[z])	Single dose	rVSV-HSP70-VP1 induced higher NoV-specific seroresponses as well as mucosal and T-cell responses compared to non-adjuvanted formulation.	NA	NA	Ma et al. [61]
GST[aa]-NoV P⁻HEV P fusion protein	Mice, IN	14.4 μg	3 doses, 2 weeks apart	High NoV-specific IgG titers and HBGA-blocking activity were reported.	NA	NA	Wang et al. [62]

NoV GII.4 Sydney-2012 VLP (Baculovirus expression system)	Mice, IM or OG	1 and 10 μg (Al $(OH)_3$, MPLA); 5 and 50 μg (CTB[ab])	2 doses, 3 weeks apart	Oral delivery enhanced the production of mucosal IgA in the presence of CTB whereas the IM route induced higher titers of VLP-specific IgG in groups receiving $Al(OH)_3$- and MPLA-adjuvanted vaccines. HBGA-blocking activity was highest in groups receiving the adjuvanted VLPs.	NA	NA	Huo et al. [63]
GI.1-2001, GI.3-2002, GII.4-1999, GII.4-2010 NO VLPs (monovalent) or bivalent GI.3/GII.4-1999 (Baculovirus expression system)	Mice, IM	10 μg	2 doses, 3 weeks apart	Both monovalent and bivalent vaccines induced genotype-specific serum IgG, IgG1, and IgG2a titers as well as strong HBGA-blocking activity.	NA	NA	Malm et al. [64]

(continued)

Table 7.2 (continued)

Vaccine formulation	Animal, route of immunization	Dosage (Adjuvant)	Schedule	Immune responses post-vaccination	Challenge strain (Route)	IR[a] and protection post-challenge	References
GII.4 and/or enterovirus 71 (EV71) VLPs (monovalent and bivalent) (Baculovirus expression system)	Mice, IP[ac]	5 μg ($Al(OH)_3$)	2 doses, 2 weeks apart	EV71- and GII.4-specific antibody responses were generated at comparable levels following bivalent and monovalent immunization. Sera from mice immunized with the bivalent vaccine blocked the binding of GII.4 VLPs to mucin.	NA	NA	Wang et al. [65]
NoV GII.4 or GI.3 VLPs with or without RV rVP6 (Baculovirus expression system)	Mice, IM	0.3 or 3 μg of NoV VLPs (10 or 30 μg of RV-VP6)	2 doses, 3 weeks apart	GII.4 or GI.3 VLPs with RV rVP6 induced higher NoV-specific IgG with considerable levels of cross-reactivity. HBGA-blocking antibodies as well as high levels of IFN-γ and IL-4 were produced.	NA	NA	Blazevic et al. [66]

NoV GI.1, GI.3, GII.4-1999, GII.4-2009 (New Orleans), GII.4-2012 (Sydney) and GII.12, GI.1 VLPs (Baculovirus expression system)	Mice, IM	10 μg each	2 doses, 3 weeks apart	Type-specific and cross-reactive high serum IgG antibody titers were detected with higher cross-reactivity within rather that across genogroups. GII.4 VLPs induced the highest and broadest cross-reactive responses. Splenocytes produced the highest levels of IFN-γ against GII.4 peptides.	NA	NA	Malm et al. [67]
VA387 GII.4 P dimer antigen (*E. coli* expression system)	Mice, IN or SL[ad]	0.1 μM/dose (FlaB[ae])	3 doses, 1 week apart	IN vaccination with FlaB stimulated higher serum-specific IgA and IgG compared to sublingual route. $CD4^+$ and $CD8^+$ T-cell responses were detected in spleen and mesenteric lymph nodes.	NA	NA	Verma et al. [68]
	Mice, SC	0.1 μM (CFA[af], alum or FlaB)	3 doses, 1 week apart	High serum and mucosal responses were induced in the presence of FlaB.			

(continued)

Table 7.2 (continued)

Vaccine formulation	Animal, route of immunization	Dosage (Adjuvant)	Schedule	Immune responses post-vaccination	Challenge strain (Route)	IR[a] and protection post-challenge	References
Trivalent fused AstV, P-HEV, P-NoV P (GII.4) (*E. coli* expression system) or free P domains of AstV, HEV and NoV	Mice, IN	15 μg (MPLA)	3 doses, 2 weeks apart	Higher IgG and HBGA-neutralizing titers with the trivalent vaccine compared to the responses detected to mixed vaccine	NA	NA	Xia et al. [69]
NoV GII.4 VLPs (Baculovirus expression system)	Mice, IN or IM	10 μg	2 doses, 3 weeks apart	IN administration stimulated serum and mucosal IgG and IgA responses.	NA	NA	Tamminen et al. [70]
Bivalent recombinant NoV GI (8K/1979/USA) and GII.4 (Minerva/2006/US) VLPs (Baculovirus expression system)	Guinea pigs, IN	5, 15, 50 or 100 μg	2 doses, 3 weeks apart	A dose-dependent increase in serum, vaginal and intestinal IgG titers was generated.	NA	NA	Ball et al. [71]
NoV (cluster 1: GII.17-1978, GII.17-2002, cluster II: GII.17-2005, cluster III: GII.17-2015) VLPs (VEE-replicon expression system)	Mice, IM	0.2 μg (MPLA and $Al(OH)_3$)	Single dose, boosters on d28 and d62	Detection of cross-reactive blockade activity between clusters I and II	NA	NA	Lindesmith et al. [72]

NoV (GII.4-1999) VLPs and RV VP6 (Baculovirus expression system)	BALB/c mice, IM	0.3 μg GII.4 VLP (1, 10 or 30 μg of rVP6)	2 doses, 3 weeks apart	VP6 exerted a dose-dependent adjuvant effect on GII.4-specific antibody responses.	NA	NA	Malm et al. [73]
NoV GI.1, GI.3, GII.4-1999, GII.4-2009 (New Orleans) and GII.12 VLPs (Baculovirus expression system)	BALB/c mice, IM, ID or both	10 or 50 μg of GII.4-1999 VLP; 60 μg GII-mix (GII.4-1999, GII.4-NO, GII.12); GII-mix at 60 μg and 40 μg of GI-mix (GI.1, GI.3)	2 doses, 3 weeks apart	IM and ID vaccination stimulated similar serum-specific IgG and T cell- ($CD4^+$ and $CD8^+$) specific responses against GII.4-1999 VLPs (administered alone or in combination with other VLPs). Similar results were detected for HBGA-blocking antibodies.	NA	NA	Malm et al. [74]

(continued)

Table 7.2 (continued)

Vaccine formulation	Animal, route of immunization	Dosage (Adjuvant)	Schedule	Immune responses post-vaccination	Challenge strain (Route)	IR[a] and protection post-challenge	References
VA387 GII.4 P dimer antigen (*E. coli* expression system)	BALB/c mice; IN or SL	0.5 μg Pd + FlaB, IN, SL, or IN/SL/SL or SL/IN/IN	3 doses, 2 weeks apart	IN/SL/SL induced the highest serum-specific IgG and IgA antibodies as well as the highest fecal secretory IgA response. Gut homing B cells, follicular helper T cells as well as systemic Pd-specific IFN-γ, IL-2, IL-4 and IL-5 responses were also the highest in the IN priming group.	NA	NA	Hwang et al. [75]
NoV (GII.4-1999) VLPs (Baculovirus expression system)	BALB/c mice; IM or IN	IM: GII.4 VLPs ($Al(OH)_3$ or MPLA); IN: 10 μg GII.4 VLPs	2 doses, 3 weeks apart	IN administration of GII.4 VLPs elicited systemic and mucosal IgA production.	NA	NA	Heinimäki et al. [76]

[a]Immune responses
[b]Intramuscular
[c]Intravenous
[d]Intranasal
[e]Mutant *E. coli* heat-labile toxin
[f]Immunostimulating complexes
[g]Monophosphoryl Lipid A

[h]Toll-like receptor
[i]Oral gavage
[j]Cholera toxin
[k]Venezuelan equine encephalitis
[l]Footpad
[m]Infectious units
[n]Subcutaneous
[o]Recombinant adenovirus (rAd) vector expressing capsid protein of NoV GII.4
[p]Oligodeoxynucleotide 1826 CpG DNA
[q]Polyinosinic:polycytidylic Acid
[r]VRPs lacking a transgene
[s]Plaque forming unit
[t]Vesicular stomatitis virus
[u]Intradermal
[v]TLR7-agonist gardiquimod
[w]Dry powder formulation
[x]Recombinant Newcastle disease virus
[y]Egg infective dose 50
[z]Heat shock protein-70
[aa]Glutathione-S-transferase
[ab]Cholera toxin subunit B
[ac]Intraperitoneal
[ad]Sublingual
[ae]Flagellar protein from *Vibrio vulnificus*
[af]Complete Freund's adjuvant

than the oral route [41, 77]. Similar results were reported following the intranasal delivery of recombinant NV-VLPs in BALB/c mice resulting in the stimulation of humoral [78, 79] and cellular NV-specific responses [79]. These studies provided the proof of concept that mucosal administration of VLP-based vaccines can generate immune responses in animal models.

Earlier studies tested the immunogenicity of recombinant Norwalk virus (rNV) VLPs administered orally to mice in the presence or absence of cholera toxin (CT) used as an adjuvant [41]. Serum-specific IgG was detected at all tested concentrations regardless of the presence or absence of the adjuvant. Moreover, higher intestinal IgA was detected following vaccination using high VLP dosages (200 and 500 μg without CT). In contrast, lower concentrations of VLPs stimulated mucosal IgA in the presence of CT. In addition to the easy delivery, oral immunization of non-replicating NV VLPs stimulated NoV-specific IgG and mucosal IgA responses. The immunogenicity of GII.4 (Dijon) (GII Grimsby-like strain (Lordsdale genotype)) VLPs was also assessed in mice.

The vaccine was administered alone (25 μg) or co-administered with the mucosal adjuvant heat-labile *E. coli* toxin (LT) or its non-toxic mutant (LT R192G) (10 μg VLPs + 10 μg LT) [43]. The vaccine was delivered in two doses either orally or intranasally. This vaccine induced high serum IgG and IgA responses as well as systemic IgG responses. Serum IgG1 and IgG2a subclasses increased in the absence of adjuvant regardless of the mode of administration. A robust fecal IgA response was also detected; the former was significantly lower following administration of VLPs alone. Significantly higher titers of serum IgG, IgA, IgG1 and IgG2a were detected following vaccination with VLPs and LT or its mutant. The intranasal administration induced better responses in the presence or absence of the adjuvant. Importantly, cells isolated from the spleen, cervical lymph nodes (CLNs), mesenteric lymph nodes (MLNs) and peyer's patches (PP) of mice immunized twice in the presence or absence of LT, regardless of the route of delivery, resulted in a mixed Th1/Th2-like cytokines production (IL-2, IFN-γ and IL-5). Similarly, the immunogenicity of another monovalent VLP-based vaccine expressing VP1 of GII.4-2012 was assessed in BALB/c mice. Mice were immunized intramuscularly, with and without adjuvants, and orally with VLPs only [63]. Serum-specific IgG was detected in all immunized mice in the presence of $Al(OH)_3$, MPLA or cholera toxin subunit B (CTB). MPLA and $Al(OH)_3$ are known humoral adjuvants stimulating Th1 and Th2 responses, respectively, translated into the production of VLP-specific IgG. Similar results were observed following oral immunization yet in 80% of mice in a dose-dependent pattern. Fecal IgA was also detected following oral immunization of mice in the presence or absence of CTB, the latter was previously reported by Ball et al. [41].

Several expression systems were tested for the production of immunogenic VLPs. These systems were compared to the standard baculovirus expression system. These include the Venezuelan equine encephalitis (VEE) [42, 49], yeast systems (*Pichia Pastoris*) [44], viral vectors (adenovirus, vesicular stomatitis virus, Newcastle disease virus) [46, 47, 59, 61, 80] and the plant systems (*Nicotiana benthamiana*) following inoculation with viral vectors derived from Tobacco Mosaic Virus (TMV)

[50, 56, 60]. The VEE replicon system is a promising backbone for NoV vaccine design in animal models. This system was used to clone and express the major capsid protein of Norwalk virus. The VEE replicon particles (VRPs) self-assemble to generate VLPs that are antigenically and morphologically identical to the wild type virus [42]. VRP-NV1 were used to immunize mice subcutaneously (i.e. footpad inoculation). Robust serum-specific IgM and IgG responses were detected following the first dose. A ten-fold increase in IgG titers was reported following the second dose. Importantly, serum IgG responses were cross-reactive with the capsid protein of a heterologous GI.1 strain (85% identical to Thistle Hall 1/91) [81]. Intestinal IgA antibodies also increased following the first dose of the VRP-NV1 with clear boosting effect following the second dose. The same group generated a multivalent VRP candidate vaccine expressing GI.1/Norwalk, GII.1/ Hawaii, GII.2/Snow Mountain, and GII.4/Lordsdale VLPs [26]. Mice were inoculated via footpad injection with monovalent, trivalent or quadrivalent form of the vaccine. Mice inoculated with monovalent VRPs mounted a robust homotypic serum IgG response. Moreover, heterotypic IgG responses were detected albeit 10–40 folds less than homotypic responses; heterotypic responses showed higher cross-reactivity within rather than across genogroups. The trivalent VRP formulation induced mucosal IgG and IgA responses in the gut and mucosal IgG in the spleen. On the other hand, the trivalent and quadrivalent VRPs induced heterotypic serum IgG responses. Interestingly, the trivalent formulation induced cross-reactive serum IgG to Lordsdale virus (LV) VLPs; the latter was not included in the vaccine. These results suggest the ability of a multivalent vaccine to generate a broader immune response compared to monovalent vaccines. The same group tested the immunogenicity of a VRP expressing NoV VLPs derived from GI.1-1968 or GII.4-2002 in mice [49]. Murine Norovirus (MNV)-1 and HuNoVs share many biochemical and genetic characteristics which led to the use of the former to understand the relationship between basic mechanisms of NoV replication in tissue culture, pathogenesis in a natural host and the role of innate and adaptive immune responses in the control of NoVs [82, 83]. Consequently, splenocytes isolated from mice immunized with live MNV were stimulated in vitro with a panel of VLPs containing overlapping 15-mer peptides spanning the capsids of NoV GI.1 and GII.4. $CD4^+$ T cells produced IFN-γ following in vitro stimulation with homotypic MNV but not heterotypic HuNoV VLPs. These results suggested that T cells are cross-reactive when NoV peptide sequences are conserved as previously reported [11]. Moreover, $CD4^+$-depleted splenocytes stimulated in vitro with MNV VLPs produced significantly lower levels of IFN-γ compared to $CD8^+$-depleted splenocytes. Other reports highlighted the role of $CD4^+$ T cells and IFN-γ following NoV and MNV challenge [11, 21, 43, 79]. Mapping and defining T-cell epitopes resulting in T-cell activation should be investigated further for the development of a successful vaccine against NoV.

LoBue et al. previously demonstrated that multivalent vaccination adjuvanted with alphavirus particles could be used as a tool to generate cross-reactive antibody immune responses to heterologous strains of NoV [48]. Mice were immunized with multivalent vaccine formulation: GI-specific VLPs representing Southampton (SoV) GI.2, Desert Shield (DSV) GI.3, and Chiba (GI.4) strains with or without NV VLPs;

GII-specific VLPs including Hawaii (GII.1), Toronto (TV) (GII.3), and M7 (GII.13) strains with or without LV VLPs; or complete VLP cocktails containing all GI and GII VLPs with or without NV and LV VLPs or all GI and GII VLPs with or without MNV VLPs (GV). The vaccine was administered alone or in conjugation with VRPs lacking a transgene (null VRPs) or oligodeoxynucleotide 1826 CpG DNA adjuvant. Multivalent genogroup-specific formulation (GI VLPs or GII VLPs) adjuvanted with null VRPs stimulated cross-reactive IgG seroresponses aligned with blockade response to NV or LV VLP (not included in the formula). IgG2a titers correlating with Th1 responses were also induced. Following MNV challenge, decreased viral load correlated with reduction in clinical diseases without protection against infection in mice vaccinated with the null VRPs-adjuvanted multivalent vaccine. The alphavirus replicon VRP was also used to form VLPs expressing a mix of GII.4 blockade epitopes [27, 84–86]. Debbink et al. [27] immunized mice with VRPs expressing the major capsid of GII.4-1987, GII.4-2002, GII.4-2006 or GII.4-2009 (parental strains). Following immunization, sera from animals were compared to those from mice immunized with chimeric VRPs expressing GII.4-2006.87A or GII.4-2006.87A.02E capsid genes. Moreover, the blocking ability of a multivalent VRP version of the vaccine (GII.4-1987, GII.4-2002 and GII.4-2006) was compared to that of chimeric VRPs expressing GII.4-2006.87A.02E. The latter blocked all the GII.4 VLPs (1987–2012). The results of this study demonstrated that single-strain VLPs are better than chimeric and multivalent VRPs at stimulating blockade activity against the parental strain. Moreover, stronger responses were detected against homologous strains without blocking heterologous ones. The authors suggested a number of mechanisms behind these results including the importance of including immunodominant epitopes (epitope A in this case) in vaccine constructs to induce broader blockade responses, in agreement with previously published data [84, 85, 87].

Other viral expression systems including recombinant Adenovirus (rAd), recombinant Newcastle disease virus (rNDV) and recombinant vesicular stomatitis virus (rVSV) systems were also evaluated [46, 47, 58, 61]. Mice receiving three doses of intranasal rAd vector expressing the capsid protein of GII.4 (rvAdGII4) mounted humoral (IgA, IgM and IgG) and mucosal (fecal IgA and IgG) responses [46]. Moreover, strong IgG and IgA responses were also reported in the intestinal homogenates, lung homogenates, and lung lavage. Th1 cytokines (IFN-γ, TNF-α, and IL-2) and Th2 cytokines (IL-4 and IL-5) increased as well in splenocytes of vaccinated mice stimulated in vitro with NoV VLPs [46]. The same construct was used in a prime boost regimen [47]. The rAd priming-VLP boosting induced better humoral, mucosal and Th1 cellular immune responses compared to the VLP-priming rAd boost or VLPs (Lordsdale virus, GII4 NV) alone. The protective role of these regimens and underlying potential mechanisms were not assessed. Moreover, the use of rAd vector in the generation of a NoV vaccine is limited by the pre-existing immunity to adenoviruses [47].

Similarly, the use of two subcutaneous doses of live MNV or VEE VRPs expressing the MNV VP1 capsid protein induced long-term protection against oral infection with MNV. The depletion of $CD4^+$ and/or $CD8^+$ T cells was associated with significant increase of viral titer in the intestines. Moreover, perforin but not IFN-γ was found to be important for the clearance of MNV infection. The adaptive immune responses, even though in a homologous system of vaccination and challenge, are clearly important in prevention of NoV infection. This proves again the shared responsibility between the humoral and the cellular immune responses especially since mice lacking anti-MNV antibodies were protected [88].

rNDV-expressing NoV VP1 was also tested [59]. This model is not hampered by pre-existing immunity in humans, is cheaper than VLPs and induces mucosal immune responses. The results in mice receiving three doses of live-attenuated rNDV expressing the VP1 protein of GII.4 NoV (strain VA387) via the intranasal route were compared to mice receiving baculovirus-expressed VLPs (ORF2 of NoV strain VA387). Both the rNDV-VP1 and the baculovirus-expressed VLPs induced high titers of NoV-specific serum IgG. rNDV-VP1 also induced a Th1 response (serum IgG2a, IFN-γ, IL-2, and TNF-α) compared to a Th2-skewed responses induced by the baculovirus-expressed VLPs. Significantly higher titers of intestinal IgA were also detected in mice immunized with the rNDV-VP1. This study advanced the rNDV system as an efficient immunogenic vaccine platform.

rVSV-expressing HuNoV capsid protein (rVSV-VP1) was successfully generated and administered intranasally and orally to BALB/c mice [51]. This model induced NoV-specific mucosal and T-cell immune responses. NoV-specific serum IgG responses were also generated and were ten times higher than responses mounted against the VLP-based vaccine. Similarly, fecal and vaginal IgA as well as T-cell proliferative responses to NoV antigens were detected in mice following oral and intranasal administration. This study demonstrated the ability of the rVSV backbone as a potential replacement for the traditional baculovirus-expressed VLPs for NoV vaccine formulations. An important limitation of this construct is the neurotropic nature of VSV. However, attenuation of the virus was achieved via the double insertion of heat shock protein-70 (HSP70) into rVSV- VP1 [61]. VSV has been used to express the *env* and/or *gag* genes of HIV [89, 90] and the capsid, E1, E2 and P7 genes of hepatitis C virus (HCV) [91]. The same construct was evaluated with HSP-70 co-expressed in the vector (rVSV-HSP70-VP1) to investigate the ability of the former to enhance VP-1-specific immune responses [61]. rVSV-HSP70-VP1 and rVSV-VP1 were superior to VLP-based vaccines at inducing serum-specific IgG responses, fecal IgA as well as T-cell immune responses measured using T-cell proliferation assays (using splenocytes). rVSV-HSP70-VP1 also stimulated the production of significantly higher levels of vaginal IgA compared to other constructs.

In addition to the baculovirus and the VEE expression systems, the production of VP1 protein using the yeast expression system has been successful and was used to test the immunogenicity of VA387 strain (GII.4) capsid protein. BALB/c mice were immunized via the intramuscular route or orally [44]. Dose-dependent high antibody titers were detected following administration of each vaccine. Moreover, VLP-specific serum IgA and fecal IgA were detected in mice following oral administration of the VLPs.

Similarly, the TMV expression system was also used to express NoV VLPs [50, 56, 60]. The immunogenicity of tobacco-derived NV VLPs was evaluated in mice following intranasal administration [50]. A number of adjuvants were used: Toll-like receptor 7 (TLR-7) agonist, imidazoquinoline-based Toll-like receptor 7 and/or 8 (TLR 7/8) agonist, and CT. The vaccine induced higher serum IgG and IgG isotypes (IgG1 and IgG2a) responses as well as mucosal IgA (fecal, vaginal, nasal, and bronchoalveolar) when CT and TLR7 agonist were compared to TLR7/8 agonists. The same vaccine construct was later administered to mice via the oral or intranasal route with or without a panel of TLR agonists (TLR-3, TLR-5, TLR-7, TLR7/8, and TLR-9) [56]. The intranasal administration of this vaccine in the presence or absence of an adjuvant stimulated the generation of IgG, IgG1 and IgG2a responses. Importantly, adjuvanted-vaccine with TLR-7 or TLR-9 agonist induced significantly higher serum IgG as well as IgG1 and IgG2a titers compared to other TLR agonists. Moreover, this formulation induced the highest mucosal immune responses (vaginal IgG and IgA, fecal IgA, bronchoalveolar IgA, and intranasal IgA). Salivary and gastrointestinal IgA titers were also significantly higher in mice receiving the vaccine with these adjuvants as compared to the mock-immunized group. In addition, serum and mucosal antibodies were able to block the binding of NV-VLPs to H type 1. On the other hand, oral immunization did not elicit intranasal, salivary, gastrointestinal or bronchoalveolar IgA antibodies; thus suggesting more potent systemic and mucosal responses induced through intranasal immunization in the presence of TLR-7 or TLR-9 agonist. Similar results were reported in mice following intranasal administration with two doses of the plant-derived expression system (*Nicotiana benthamiana*) encoding the capsid protein of Norovirus Narita 104 (genogroup II) VLP (NaVLP) [60]. Robust VLP-specific serum IgG and nasal IgA were detected in immunized mice; serum IgG levels kept increasing until day 56 following the first dose of the vaccine. Serum IgG isotype levels also showed a similar trend as serum IgG with higher levels of IgG1 compared to IgG2a isotype; thus suggesting a predominant Th2 response as previously reported [53, 92]. Collectively, these results confirmed the immunogenicity of the plant-derived expression system in mice. While the baculovirus expression system is effective, the yeast and plant-derived expression systems are cheaper models for the generation of NoV vaccine formulations.

The mucociliary activity in the nasal cavity rapidly clears foreign particles such as NoV-VLPs. The latter might lead to a decrease in the amount of absorbed antigens resulting in reduction of the immune response [53]. In order to increase the adherence of the antigen in the nasal cavity and antigen uptake, a vaccine formulation of NoV VLPs containing a muco-adhesive dry powder (GelVac) was developed and its immunogenicity was tested in guinea pigs [53, 71]. GelVac powder was prepared by drying the liquid formulation of GelSite. GelVac powder, GelVac NV VLP powder, GelVac gardiquimod (GARD) powder, or GelVac NV VLP + GARD powder were prepared. GARD is an imidazoquinoline-based, TLR7 agonist. The in situ gelation of this powder formulation was tested in rats and led to an increase in the adherence time of the antigen (up to 3 h compared to 10 min with liquid formulation) in the nasal cavity. The intranasal administration of NV-VLP with GelSite induced robust serum IgG, IgG1 and IgG2a as well as mucosal (salivary, intestinal, nasal, bronchoalveolar, vaginal and uterine) IgA responses in vaccinated guinea pigs. Vaginal and uterine IgG antibodies were also detected. When compared to animals immunized with the liquid formulation of the NV-VLP vaccine containing GelSite, the vaccine induced 20-, 114-, and 40-folds higher serum IgG, IgG1, and IgG2a, respectively. While GelSite is not immunogenic, it stabilizes the maintenance of antigens at the mucociliary interface. GelSite was used as well with recombinant GI (8K/1979/USA) and GII.4 (Minerva/2006/USA) bivalent VLPs expressing VP1 [71]. Different doses of the vaccine were administered to guinea pigs via the intranasal route. Results showed the generation of dose dependent VLP-specific serum IgG as well as IgG isotypes (IgG1 and IgG2) responses against both VLPs with higher IgG2 titers compared to IgG1. Vaginal and intestinal IgG titers were detected. Importantly, VLP-specific serum neutralizing antibodies inhibiting the binding of GI and GII.4 VLPs to porcine gastric mucin were also detected.

The P domain of VP1 capsid protein is divided into P1 and P2 subdomains. P1 subdomain contains residues 226–278 and 406–520. The P2 subdomain on the surface of the capsid plays a role in the interaction between HBGA and specific blockade antibodies. The P polypeptide, a truncated P domain without the C-terminal arginine cluster, was detected in stool of NoV-infected patients [93]. Upon expression in *E. coli*, the P domain generates the P domain dimer, 12-mer P particle, 24-mer P particle and polyvalent P domain complexes with many P dimers [94]. A number of studies compared the ability of these domains to VLPs in stimulating cell-mediated immune responses in the BALB/c mice model. The ability of the P domain complexes to induce a polyfunctional memory $CD4^+$ T-cell response was studied following the intranasal immunization of mice with three doses of P particle, P dimer or NoV VLPs (GII.4 VA387) [36]. Seroresponses were determined and splenocytes of immunized mice were used in vitro in intracellular cytokine staining assays. The P domain and VLP complexes induced strong antibody responses in immunized mice. Moreover, P domain complexes and VLPs induced $CD4^+$ central memory responses. The lower immunogenicity of the P dimer compared to the P domain and VLP vaccine was suggested to be caused by its small size and the lower valence of antigenic structures. Moreover, P domain complexes and VLP complexes were able to induce the maturation of bone marrow-derived dendritic cells (DCs), a finding supporting the existence of T-cell epitopes efficiently presented by DCs. It remains unknown whether these particles are able to induce cytotoxic T-cell responses. Similar studies elucidated further the role of DCs in controlling MNV infection by facilitating the generation of anti-MNV antibodies. DC-depleted mice suffered from increased viral titers in the intestinal tract and a significant decrease in the generation of antibody responses 2 weeks following infection [95]. These results confirmed the role of DCs in the generation of an adaptive immune response against NoVs in this model. While histological data from patients infected with NoV showed the expansion of perforin-positive cytotoxic T cells [96], the cellular immune responses mounted against NoVs were further studied as previously described [11, 21, 48, 49].

NoV P particles were also used in a dual vaccine combination with VP8, the major neutralizing antigen of rotavirus (RV) and matrix protein 2 (M2e) peptide of influenza virus [54, 55]. The P particle VP8 chimeric vaccine induced high titers of specific anti-NoV antibodies as well as higher VP8-specific antibodies when compared to titers induced following vaccination with the RV VP8 particle alone. This bivalent vaccine blocked the binding of NoV VLPs to HBGAs. Similarly, the bivalent vaccine containing NoV P particle and the M2e influenza epitope induced a strong and protective Th2 response against lethal challenge of mice with mouse adapted H1N1 influenza virus [55]. This vaccine candidate also stimulated strong anti-NoV antibodies inhibiting the binding of NoV VLPs to corresponding receptors. While the NoV P particle was used as a candidate platform for vaccine development against RV and influenza virus, it also showed a substantial stimulation of NoV-specific responses. The role of these dual vaccines is yet to be evaluated in humans.

More work on potential vaccine candidates of NoV and RV combination was reported during the past decade [52, 57, 66, 73]. Blazevic et al. [52] used rVP6 of RV and NoV GII.4 VLPs to immunize BALB/c mice parenterally. Mice received two doses, via the intradermal or intramuscular route. This combination vaccine induced cross-reactive NoV-GII.4, GII.12, and GI.3 specific antibodies as well as RV-specific antibody responses. Importantly, sera from immunized mice (with the single or bivalent form) blocked the binding of GII.4 VLPs to HBGA H type 3 of mice and induced the production of fecal IgG (NoV- and RV-specific). Similarly, the VLP GII.4 monovalent vaccine stimulated a cross-reactive antibody response against heterologous VLPs within the same GII genogroup (GII.4 New Orleans (NO) and GII.12) as well as GI.3 VLPs. Tamminen et al. later included GI.3 VLPs in this vaccine construct, in an attempt to induce neutralizing antibodies against both GI and GII NoVs [57]. The trivalent vaccine (GI.3, GII.4 VP1-derived VLPs as well as recombinant RV VP6, the most conserved and abundant RV protein) was administered via the intramuscular route. The latter was compared to single administration of GI.3 and GII.4 VP1-derived VLPs as well as recombinant VP6 of rotavirus, separately. The results showed the generation of cross-reactive IgG antibodies against heterologous NoV VLPs, specifically GI.1, GII.4 and GII.12. These cross-reactive responses lasted for 6 months. The administration of monovalent vaccines induced stronger homologous responses. Importantly, the combination vaccine stimulated robust IFN-γ specific responses. These responses were detected against 15-mer peptides representing T-cell epitopes of the capsid P domain of homologous (GII.4) and heterologous noroviruses (GII.4 NO, GII.12). These responses lasted for 24 weeks. Recently, mice were immunized with GI.3 or GII.4 VLPs with or without RV VP6. NoV- and RV-specific seroresponses and T-cell responses were assessed [66]. RV VP6 exerted an adjuvant effect and induced higher NoV-specific IgG responses. Moreover, sera from mice immunized with GII.4 or GI.3 VLPs with RV VP6 produced cross-reactive IgG against heterologous strains (GII.4 NO, GII.4 Sydney and GI.1). VP6-specific IL-4 and IFN-γ were detected in vitro in splenocytes of immunized mice. Similarly, GII.4 VLP-specific serum IgG as well as IgG1 and IgG2a were significantly higher when mice were immunized with GII.4 VLPs in the presence of VP6 as compared to its absence [73]. Moreover, a dose-dependent effect was reported. VP6 was shown to induce the activation and maturation of antigen presenting cells (APCs) in vitro; thus supporting its ability to exert an adjuvant effect [97]. These results entertain the idea of developing combination vaccines targeting enteric viruses.

Several studies evaluated the immunogenicity of polyvalent vaccine complexes [62, 98]. The polyvalent complexes were generated using a glutathione S-transferase (GST) gene fusion system in pGEX-4 T-1 vector to generate GST-NoV P^-. NoV P^- is a P domain with four amino acids deletion at the end [93]. The former was derived from GI.3 VA115, GII.4 VA387 and GII.9 VA207. BALB/c mice were immunized with different dual combinations [98]. The control mice received free NoV P dimers of the NoV strains stated above. High IgG antibody titers and T-cell responses were detected against individual antigens of the polyvalent complexes. Specifically, the levels of IL-2, IFN-γ and TNF-α produced by $CD4^+$ T cells were higher following polyvalent complexes immunization. Importantly, sera from mice immunized with the polyvalent complexes exhibited significantly higher blocking activity than sera from mice immunized with the free dimers. These results suggest that the polyvalent complex vaccine expressing the P domain is immunogenic and induces higher humoral and cellular immune responses compared to the formulation containing free P dimer. Wang et al. [98] suggested that the increased immunogenicity of the polyvalent complexes could be explained by the polyvalent antigenic structures of large complexes and thus could be used as platforms for multivalent NoV vaccine development. Consequently, the same group generated GST-NoV P^- -HEV P protein (protruding domain of hepatitis E virus (HEV)) [62]. Both HEV and NoV have protruding domains playing a role in attachment and stimulation of neutralizing antibodies [36, 99–106] and thus are attractive targets for vaccine generation. Following intranasal immunization of mice, polyvalent GST-NoV P^- -HEV P and oligomeric NoV P^- -HEV P induced significantly higher titers of NoV P-specific IgG compared to those induced by NoV P^- antigen; the latter also resulted in lower binding blockade activity.

With the increased interest in the development of combination vaccines, the immunogenicity of monovalent and bivalent vaccines containing GII.4 VLP and Enterovirus 71 (EV71) VLPs was tested in mice [65]. BALB/c mice received the monovalent (GII.4-VLP or EV71-VLP) or the bivalent (GII.4-VLP and EV71-VLP) form of the vaccine. Both formulations induced comparable antibody responses indicating the lack of any potential interference on the immunogenicity of NoV and EV71 VLPs. Recently, a trivalent fused vaccine construct, Astrovirus (AstV) P-HEV P-NoV P, containing the dimeric P domains (viral neutralizing antigens) of AstV, NoV (GII.4) and HEV, was designed [69]. The immune responses generated following vaccination with this construct were compared to those induced by a mix of free P domains of AstV, HEV and NoV. Both constructs induced high IgG titers to the P domains of all three viruses while the highest IgG titers targeted NoV P domains. The trivalent fused vaccine induced higher IgG titers against NoV P domain compared to the free P domains mix formulation albeit the difference was not statistically significant. The use of the adjuvant MPLA enhanced seroresponses to both vaccine formulations. More studies explored the ability of recombinant VA387 GII.4 NoV specific P domain (Pd) vaccine in the presence or absence of mucosal adjuvant FlaB (flagellar protein of *Vibrio vulnificus*), a TLR5 agonist [68]. Mice were immunized via the intranasal (IN) or the sublingual (SL) route. The intranasal immunization of mice with Pd or Pd + FlaB induced significantly higher serum IgG and IgA responses than SL immunization. The presence of FlaB enhanced these responses. Fecal IgA titers were also higher following SL immunization. $CD4^{+}IFN\gamma^{+}$ T-cell levels were high in cell suspensions from spleen, MLNs and peyer's patches specifically following IN immunization with Pd + FlaB. Similar results were also reported for $CD8^{+}IFN\gamma^{+}$ levels. TLR5-deficient ($TLR5^{-/-}$) mice immunized with the same construct (i.e. Pd+FlaB) did not produce antigen-specific antibodies. The results of this study suggested the ability of a vaccine containing a TLR5 agonist to stimulate NoV-specific immune responses through mucosal administration even though other studies have refuted the role of FlaB in inducing cytotoxic T lymphocyte (CTL) responses [107]. The same study compared the effect of alum and complete Freund's adjuvant (CFA) following subcutaneous (SC) immunization of a group of mice to the results obtained using FlaB. The results showed that CFA and FlaB induced higher serum IgG while fecal IgA was higher in the presence of FlaB. This study suggested the superiority of FlaB as a mucosal adjuvant. The same group explored combining the IN and SL routes in an attempt to stimulate systemic and mucosal immune responses [75]. Pd + FlaB were administered intranasally or sublingually alone or combined in prime-boost regimens: IN priming followed by two SL boosts or SL priming followed by two IN boosts. Serum-specific IgA and IgG were highest following IN/IN/IN and lowest following SL/SL/SL. On the other hand, fecal secretory IgA (sIgA) was highest in mice following the IN/SL/SL prime-boost regimen suggesting the superiority of the IN route in priming systemic responses while the SL route stimulating enteric responses. Moreover, Th1 and Th2 Pd-specific immune responses were detected with higher antigen-specific production of IFN-γ, IL-2, IL-4 and IL-5 in splenocytes of IN priming groups (IN/IN/IN and IN/SL/SL). This study emphasizes the impact of the priming site on the generation of balanced immune responses.

The immune responses following immunization of mice with GII.4-1999 VLPs via the IN route were also compared to the intramuscular (IM) administration [70]. High NoV serum-specific IgG antibodies were detected along with HBGA-blocking activity following both methods of vaccine delivery. Serum-specific IgA antibodies were, however, detected only following IN administration. The latter also induced mucosal (nasal wash) IgA with significant blocking activity. The same group administered GII.4-1999 VLPs parenterally with or without adjuvant (Al $(OH)_3$ or MPLA) [76]. Strong GII.4-specific IgG responses were detected following IM vaccination with $Al(OH)_3$ or MPL. IgA responses were low to inexistent following IM vaccination in the presence or absence of an adjuvant, respectively. This study compared immune responses generated following IM vaccination to the IN route; both routes induced similar IgG responses while the intranasal administration mounted higher IgA responses. The same group also investigated mucosal immunity following IM immunization and reported that the latter does not stimulate the production of mucosal IgA (MLN) in contrast to IN immunization which correlated with mucosal blocking activity [70]. This study emphasizes again the role of the mucosal delivery of NoV antigens in the development of strong mucosal immune responses.

Intramuscular immunization of mice with monovalent GI.1-2001, GI.3-2002, GII.4-1999, or GII.4-2010 NO VLPs or bivalent GI.3 and GII.4-1999 VLPs was also tested in BALB/c mice. Monovalent and bivalent vaccines induced comparable levels of IgG, IgG1 and IgG2a. Cross-reactive antibodies were detected following monovalent vaccination within genogroups but not between genotypes [64]. The same group previously immunized mice intramuscularly with a large number of VP1-derived VLPs containing GII.4 or GII.12 [58], following the same procedure previously reported (immunization with single VLPs and boost at d21). This vaccine induced serum-specific IgG responses against GII.4 and GII.12. When mice were immunized and boosted with GII.4 and GII.12, respectively, results showed that boosting with GII.12 did not decrease the IgG titers to GII.4 suggesting the lack of original antigenic sin.

When mice were immunized intramuscularly with GI.1, GI.3, GII.4-1999, GII.4 NO, GII.4 SYD, and GII.12, strong type-specific and cross-reactive serum IgG titers were detected within genogroups [67]. Importantly, this study reported the production of strong T-cell responses (IFN-γ) by splenocytes of immunized mice stimulated with synthetic peptides; the latter map the entire GII.4-1999 VP1 capsid. The same group compared the magnitude and the specificity of T-cell responses in mice immunized with GII.4-1999 VLPs (intramuscularly or intradermally), a mixture of GII.4-1999, GII.4 NO and GII.12 (intramuscularly) or a mixture of GII VLPs and GI VLPs (GI.1 and GI.3) (intramuscularly and intradermally simultaneously) [74]. Robust genotype-specific IgG antibodies were generated against GII.4 VLPs regardless of the mode of administration. Importantly, the GII.4-1999 peptide pool as well as the 18-mer peptides, 99–54 and 99–50 induced strong IFN-γ responses in splenocytes of immunized mice. The 99–45 and 99–50 peptide sequences were previously identified to contain $CD8^+$ and $CD4^+$ T-cell epitopes, respectively [67]. Moreover, CD107, a marker of T-cell activation, was detected on splenocytes of immunized mice stimulated with peptide 99–45 containing highly conserved epitope of GII NoVs and thus suggesting the potential role of CTLs directed against this epitope. It is yet to identify conserved human CTL epitopes. These studies underscore the importance of using multivalent vaccines to induce humoral and cellular immune responses.

The evolution of the capsid gene of GII.17 has been associated with its potential to cause pandemic outbreaks as the case with GII.4. Consequently, mice were immunized with GII.17 clusters I (GII.17-1978, 2002), II (2005) and IIIb (2015) VLPs [72]. Following intramuscular delivery, the vaccine induced higher cross-reactive blockade activity between clusters I and II and less with cluster IIIb. This suggests the presence of similar antigenic epitopes in clusters I and II. Amino acid residues 393–396 or epitope D were identified as variant epitopes affecting antigenicity and cross-reactive blockade activity thus suggesting the inclusion of GII.17 in multivalent NoV VLP vaccines. This epitope has also been identified in GII.4 strains [84, 108].

The diversity and variability of the reported data in animal models are the result of many factors mainly the administered dose of the vaccine, its mode of delivery, the number and genotypes included in vaccine constructs as well as the lack of complete understanding of standardized correlates of protection. Nevertheless, these studies generated promising data enhancing the interest in human vaccine trials to prevent and control NoV infection.

7.4 Clinical Studies

Several clinical studies evaluated the safety and immunogenicity of GI.1 VLPs in healthy volunteers (Table 7.3). The first of the series was a phase 1 clinical trial [109]. Two doses of the vaccine (100 or 250 μg) were administered orally to 20 healthy adults (18–46 years). 60% of the volunteers receiving the 100 μg dose and all those receiving 250 μg developed greater than four-fold increase of serum IgG levels, mostly following the first dose. The vaccine induced serum IgA responses in 40 and 50% of volunteers receiving the 100 μg and 250 μg, respectively. The vaccine was well tolerated and vaccinees did not report any adverse event. This study was the first to show that a non-replicating NoV vaccine administered orally could elicit immune responses in healthy adults. Similarly, GI.1 VLPs [110] were tested in 20 healthy adults receiving two or three doses of Norwalk capsid protein (NVCP); the latter was expressed in transgenic potatoes and assembled successfully into VLPs. Almost all volunteers in both groups showed an increase in IgA ASCs and 10% of volunteers receiving two doses of the vaccine showed serum-specific IgG responses. Following a third dose, 30 and 40% of volunteers developed serum IgG and IgM responses, respectively. The vaccine was well tolerated with no difference in adverse events rate between control and vaccine groups. This study introduced the plant expression system as a potential model for the development of NoV vaccines. However, more studies are needed to enhance the immunogenicity of this vaccine candidate in humans.

Further studies assessed the safety and immunogenicity of GI.1 NoV VLPs in healthy adults 18–40 years old receiving increasing doses of NoV VLPs on day 1 and day 21 [21]. Following oral vaccination and regardless of the dosages, volunteers developed significant increases in serum IgA ASCs with transient rise in IFN-γ produced by PBMCs in response to HuNoV. The latter observation was recorded at lower VLP dosages and lasted for up to 56 days. Low-to-moderate mucosal (fecal, vaginal and salivary) IgA responses were detected in 40% of vaccinees. A four-fold rise in serum IgG geometric mean peak titer was detected in 90% of the volunteers receiving 250 μg of the vaccine construct; higher dosages of the vaccine did not result in higher immune responses. The vaccine was generally well tolerated. This study marked the beginning of a whole new series of vaccine studies in an attempt to elucidate the correlates of protection following infection with HuNoV [34].

Table 7.3 Norovirus clinical vaccine studies

Vaccine formulation	Mode of immunization	Dosage (Adjuvant)	Schedule	Sample size[a], (Age)	Immune responses	HAI[b]-/HBGA[c]-blocking	Challenge	Immune responses post-challenge	References
GI.1 VLPs (Baculovirus expression system)	Oral	100 μg	2 doses, 3 weeks apart	5 (18–46 years)	≥4-fold rise in serum IgA and IgG in 40 and 60% of vaccine recipients, respectively.	Not tested	None	NA	Ball et al. [109]
		250 μg	2 doses, 3 weeks apart	15 (18–46 years)	≥4-fold rise in serum IgA and IgG titers in 50 and 100% of recipients, respectively.				
GI.1 VLPs (Transgenic potato expression system)	Oral	215–751 μg	2 doses, 3 Weeks apart	10 (≥18 years)	Serum-specific IgG response was detected in 10% of vaccine recipients. All participants showed an increase in IgA ASCs[d].	Not tested	None	NA	Tacket et al. [110]
			3 doses (d0, d7 & d21)	10 (≥18 years)	Serum-specific IgG and IgM responses were detected in 30 and 40% of volunteers, respectively. An increase in IgA ASCs was observed in 90% of volunteers.				
GI.1 VLPs (Baculovirus expression system)	Oral	250, 500, and 2000 μg	2 doses, 3 weeks apart	36 (18–40 years)	≥4-fold rise in serum IgG was detected in 70–90% of volunteers. Low-to-moderate mucosal IgA and significant rise in IgA ASCs were detected in all participants. A transient rise of IFN-γ was reported in volunteers receiving the lower doses.	Not tested	None	NA	Tacket et al. [21]
GI.1 VLPs (Baculovirus expression system)	IN[e]	**Study 1** 5 μg (Chitosan, MPLA [f])	2 doses, 3 weeks apart	28 (18–49 years)	0.9- and 1.2-fold increase in specific IgG and IgA GMFR[g], respectively. 20% of vaccinees showed IgA ASCs on d28 post-vaccination.	Not tested	None	NA	El-Kamary et al. [111]

(continued)

Table 7.3 (continued)

Vaccine formulation	Mode of immunization	Dosage (Adjuvant)	Schedule	Sample size[a], (Age)	Immune responses	HAI[b]-/HBGA[c]-blocking	Challenge	Immune responses post-challenge	References
		15 µg (Chitosan, MPLA)			1.9- and 2.5-fold increase in IgG and IgA GMFR, respectively. IgA ASCs were reported on d28 post-vaccination among 80% of participants.	Not tested	None	NA	
		50 µg (Chitosan, MPLA)			4.7- and 4.5-fold increase in IgG and IgA GMFR, respectively. IgA ASCs were detected by d28 post-vaccination among 56% of participants.	Not tested	None	NA	
	IN	**Study 2** 50 µg (Chitosan, MPLA)	2 doses, 3 weeks apart	61 (18–49 years)	4.6- and 7.6-fold increase in IgG and IgA GMFR, respectively. IgA ASCs were detected in all subjects (days 7 and 28).	33% of participants showed ≥4-fold rise in HAI titers (d21).	None	NA	
		100 µg (Chitosan, MPLA)			4.8- and 9.1-fold increase in IgG and IgA, respectively. IgA ASCs detected in all subjects (days 7 and 28).	63% of participants showed ≥4-fold rise in HAI titers (d21).	None	NA	
GI.1 VLPs (Baculovirus expression system)	IN	100 µg (Chitosan, MPLA)	2 doses, 3 weeks apart	98 (18–50 years)	Following the second dose, a ≥4-fold rise of NoV-specific serum IgA, IgG and IgM was detected among 70%, 50% and 32% of participants, respectively.	Presence of HBGA-blocking antibodies (32% of participants)	Homologous GI.1 NoV	Protection and prevention of infection (70%) and gastroenteritis (37%)	Atmar et al. [112]
GI.1 VLPs (Baculovirus expression system)	IN	**Study 1** 5, 15, 50 µg (Chitosan and MPLA)	2 doses, 3 weeks apart	28 (18–49 years)	Dose-dependent NoV-specific IgG and IgA seroresponses were detected. Mucosal ASCs as well as IgG and IgA memory B cell (B_M) responses were detected following vaccination with 50 µg and 100 µg GI.1 VLP.	IgA B_M and IgG B_M titers significantly correlated with serum HAI.	GI.1 virus (Oral)	62% of recipients were protected against infection.	Ramirez et al. [113]
		Study 2 50, 100 µg (Chitosan and MPLA)	2 doses, 3 weeks apart	61 (18–49 years)					

GI.1 and consensus GII.4 VLPs (Baculovirus expression system)	IM[b]	**Stage 1** 5, 15, 50, or 150 μg (MPLA and $Al(OH)_3$)	2 doses, 4 weeks apart	48 (18–49 years)	≥4-fold increase in GI.1 Pan-Ig titers in all participants following the first dose (independent of dosage); ≥4-fold increase in Ig titers against GII.4 in 56% of participants in the 5 μg group; 88%, 80% and 75% of participants receiving 15, 50 and 150 μg showed a ≥4-fold increase of total seroresponses to GII.4. Titers did not increase following the second dose.	Most subjects in different groups achieved HBGA-blocking antibody titers of ≥200 (a level considered protective against infection) against both GI.1 and GII.4 by day 28 post-vaccination.	None	NA	Treanor et al. [114]
		Stage 2 50 μg (MPLA and $Al(OH)_3$)		8 (18–49 years) 9 (50–64 years) 10 (65–85 years)	High Pan-Ig response was detected following the first dose in all age groups. Titers peaked by day 7; seroresponses did not increase following the second dose.				
GI.1 and consensus GII.4 VLPs (Baculovirus expression system)	IM	50 μg each (MPLA, Al $(OH)_3$)	2 doses, 4 weeks apart	127 (18–50 years)	Robust increase in serum IgG and IgA following the first dose with greater responses detected against GI.1	Increase in HBGA-blocking titers	Heterologous GII.4 NoV, a 2002 Framington-Hills-like variant (Oral)	Lower frequency of moderate-to-severe vomiting or diarrhea among vaccinated volunteers	Atmar et al. [115]; Bernstein et al. [116]
GI.I and consensus GII.4 VLPs (VEE-replicon expression system for GI VLPs and Baculovirus expression system for consensus GII.4 VLPs)	IM	50 μg both (MPLA and $Al(OH)_3$)	2 doses, 4 weeks apart	10 (18–49 years)	A cross-reactive serum IgG response was reported against non-vaccine strains.	Detection of HBGA cross-blocking antibodies against non-vaccine strains	None	NA	Lindesmith et al. [117]
GI.1 and GII.4 VLPs (Baculovirus expression system)	IM	5, 15, 50 or 150 μg (MPLA or Al $(OH)_3$)	2 doses, 4 weeks apart	20 (18–49 years)	IgG and IgA specific responses (GI.1 and GII4) peaked 7 days post-vaccination and were generally maintained (d35).	Not tested	None	NA	Sundararajan et al. [118]

(continued)

Table 7.3 (continued)

Vaccine formulation	Mode of immunization	Dosage (Adjuvant)	Schedule	Sample size[a], (Age)	Immune responses	HAI[b]-/HBGA[c]-blocking	Challenge	Immune responses post-challenge	References
					High frequency of IgA-specific ASC responses were detected against vaccine strains.				
GI.1 and GII.4 VLPs (Baculovirus expression system)	IM	15 μg or 50 μg GI.1 VLPs and 50 μg of GII.4 VLPs (MPLA, Al $(OH)_3$)	Single dose	454 (18–49 years)	Increase in Pan-Ig, IgG and IgA against GI.1 and GII.4	Stimulation of strong HBGA-blocking antibody responses	None	NA	Atmar et al. [119]
GI.1 and consensus GII.4 VLPs (2006a (Yerseke), 2006b (Den Haag), and 2002 (Houston) (Baculovirus expression system)	IM	5, 15, 50, or 150 μg (MPLA, Al $(OH)_3$)	2 doses, 4 weeks apart	42 (18–49 years)	≥4-fold increase in IgG and IgA ASCs. A bias was observed towards IgA ASCs to GI.1 VLPs. Higher B_M cell responses were detected against GI.1 VLPs compared to GII.4 with a bias towards IgG responses.	HBGA-blocking antibodies against GI.1 and GII.4	None	NA	Ramani et al. [120]
rAd5-GI.1[i] Norwalk VP1	Oral	1×10^{10} or 1×10^{11} IU[j] (dsRNA hairpin)	Single dose	66 (18–49 years)	Volunteers receiving either doses mounted high serum IgG and IgA as well as fecal IgA responses. Circulating IgA and IgG ASCs were also induced 7 days post-vaccination.	61% and 78% of volunteers receiving the low and high dose-vaccine, respectively, showed >2-fold rise in BT50[k] titers in Le[b] and H1 assays.	None	NA	Kim et al. [121]
GI.1 and consensus GII.4 VLPs (Baculovirus expression system)	IM	15, 50 or 150 μg of each VLP (0, 15 or 50 μg of MPLA, Al $(OH)_3$)	1 or 2 doses, 4 weeks apart	420 (18–64 years)	By day 56 post-vaccination, all vaccine formulations elicited similar Pan-Ig and IgA responses. Higher seroresponses were detected to GI.1 (80–100%) compared to GII.4c (33.3–83.3%). These titers	Similar HBGA-blocking titers across groups	None	NA	Leroux-Roels et al. [122]

					persisted (day 393) mostly following the first dose. No evident benefit of MPLA was reported.				
GI.1 and consensus GII.4 VLPs (Baculovirus expression system)	IM	15 µg GI.1/ 50 µg GII.4c VLPs (Al $(OH)_3$)	Single dose	50 (18–49)	By day 8 post-vaccination, robust increases in serum Pan-Ig and IgA were detected. Salivary IgA responses were detected against both VLPs in >87% of volunteers.	Higher serum HBGA-blocking antibody titers detected against GII.4c rather than GI.1.	None	NA	Atmar et al. [123]

[a]Sample size includes total number of participants
[b]Hemagglutination inhibition assay
[c]Histoblood group antigens
[d]Antibody-secreting cells
[e]Intranasal
[f]Monophosphoryl lipid A
[g]Geometric Mean Fold Rise
[h]Intramuscular
[i]Recombinant Adenovirus vector
[j]Infectious unit
[k]Blocking Titer 50: percentage of sera necessary to block 50% of NoV-HBGA binding

As previously summarized [34], the ability of GI.1 VLPs, adjuvanted with MPLA and in the presence of chitosan acting as a mucoadherent, was tested in two phase 1 double-blind controlled human trials [111]. Healthy volunteers (18–49 years) tolerated increasing doses of the vaccine administered via the intranasal route replacing oral administration as an alternative mucosal delivery. Dose-dependent IgG and IgA responses were observed in response to increasing doses of the vaccine with more than four-fold increase in geometric mean fold-rise (GMFR) in the group receiving the 50 μg dose. Following the dose escalation study, IgA and IgG ASCs were detected among all study participants following the administration of 50 μg and 100 μg of the vaccine. IgA ASCs highly expressed the gut homing receptors $CD19^{+}CD27^{+}$ integrin α_4/β_7^{+} $CD62L^{-}$ and thus homing to peripheral and mucosal tissues; whereas IgG ASCs were $CD19^{+}CD27^{+}CD62L^{+}$ supporting homing to peripheral lymphoid tissues only. Based on the results of the hemagglutination inhibition (HAI) functional assay, immunized volunteers mounted a high HAI titer at higher doses of the vaccine with 72 and 75% agreement between the HAI and the IgG and IgA recorded data, respectively. The ability of this adjuvanted vaccine to induce mucosal and systemic immunity is noteworthy; however, its cross-reactive capabilities were not tested nor its ability to prevent infection with HuNoVs. The same vaccine construct was used to further evaluate the ability of GI.1 VLP to induce B_M responses [113]. Following primary immunization of volunteers with 50 μg or 100 μg NV VLPs, B_M cells expressing CD38 (B cell activation marker) and CD27 (memory marker) were detected in peripheral blood for up to 6 months. These cells are essential for reactivation upon antigenic re-exposure. In addition to B_M responses, dose-dependent functional IgG and IgA responses were detected following intranasal administration of the vaccine to healthy adults. The frequencies of B_M cells in blood correlated with the level of antibodies in sera; moreover, high HBGA-blocking antibody titers were associated with a decrease in the rate of infection. The results of this study showed the maintenance of NoV-specific antibody responses ensuring a rapid and strong anamnestic humoral response. These reports proved the ability of this VLP-based vaccine to generate a strong humoral immune response with potential long-term protection [111, 113]. It is yet to demonstrate the ability of any vaccine targeting HuNoVs to induce a similar impact on T-cell immune responses.

A randomized, double-blind, placebo-controlled, multicenter trial was conducted to assess the safety, immunogenicity, and efficacy of GI.1 VLP vaccine in preventing gastroenteritis [112]. Two doses of the vaccine, with chitosan and MPLA as adjuvants, or placebo were administered intranasally to healthy 18–50 year old adults. Homologous challenge with HuNoVs took place 3 weeks following the administration of the second dose of the vaccine. The participants enrolled in this study were likely to develop NoV-associated gastroenteritis due to the expression of O or A blood groups and a functional *FUT2* gene. Following the second dose of the vaccine, ≥four-fold increase in serum-specific IgA, IgG and IgM was reported among 70%, 50% and 32% of participants in the vaccine arm, respectively. The results of this study also showed that vaccine recipients were less likely to develop illness as compared to placebo-recipients; this was clear following challenge with homologous GI.1 whereby 82% of placebo recipients developed infection whereas the onset of illness was delayed and the severity of the disease was reduced among vaccinated participants who developed gastroenteritis. The correlation detected between increased pre-challenge serum level of HBGA-blocking antibodies and protection against infection and illness confirmed previous results reporting lower risk of NoV-associated gastroenteritis following challenge [124]. This study, like previous ones, did not test the immunogenicity of the suggested vaccine and its protective efficacy among young children and elderly. Nevertheless, it proved the ability of a two-dose intranasally-administered vaccine to stimulate homologous protection against NoV and to prevent infection.

Previous challenge studies demonstrated the ability of a diverse panel of GI VLPs to induce high levels of IgG with cross-reactive affinities [12]; thus suggesting the ability of GI.1 NoV to stimulate high-affinity antibodies to heterologous strains. The same construct induced an IFN-γ response to homologous and heterologous GI strains in vitro. A common capsid protein was suggested to provide a common epitope recognized by cross-reactive antibodies. The ability of a multivalent NoV VLP vaccine to induce a broad protective immunity was recently evaluated [117]. Study participants, 18–49 years old, were immunized intramuscularly with two doses of GI.I and GII.4c VLPs. Serum samples from vaccinated participants and placebo recipients were tested for VLP-specific antibody responses to a variety of HuNoV strains (GI, GII.4 and non-GII.4). Moreover, an antibody blockade assay was used to determine the cross-protection potential of this vaccine. The vaccine induced the production of IgG to NoV strains that were not included in the vaccine. Moreover, the results of the blockade assay positively correlated with the vaccine-induced immune responses in vaccinated adults. The production of cross-blocking antibodies and protection were detected against a newly emergent GII.4 strain not a constituent of the vaccine construct to which pre-existing immunity was reported to be unlikely. The authors suggested that the activation of pre-existing cross-reactive B_M cells directed against GII.4c, a vaccine component, contributed to antibody responses targeting the GII strains that were not included in the vaccine. This study supports the use of a multivalent vaccine to stimulate broad blockade responses to multiple epitopes; however, the evolution of the described antibody responses following virus challenge was not investigated.

In an earlier study, a bivalent adjuvanted-VLPs formulation of GI.1 and GII.4c was administered intramuscularly to healthy adults in the presence of 50 μg MPLA and 0.5 mg $Al(OH)_3$ [114]. GII.4 used in this construct is based on consensus sequence between GII.42006a (Yerseke), 2006b (Den Haag), and 2002 (Houston). Following the dose escalation phase of the study ($n = 48$ participants), 50 μg of each VLP component was selected to assess its immunogenicity ($n = 54$ participants). This concentration was used due to its ability to induce higher responses during the dose escalation phase. A high Pan-Ig response as well as strong HBGA-blocking antibody response were detected following the first dose. Seroresponses peaked at day 7 and were higher against GI.1 compared to those generated against GII.4. Lower IgG responses were detected in the older age groups (65–83 years). This study showed the ability of IM administration to induce faster and higher serum antibody responses compared to the oral or intranasal route albeit to different vaccine formulations [21, 109–113]. PBMCs from vaccinated volunteers [114] were collected to measure IgA and IgG ASCs as well as B_M cells [120]. IgA and IgG ASCs peaked at day 7 post-vaccination followed by a decline to pre-vaccination levels detected at day 28. There was minimal boosting effect observed following the second dose of the vaccine. IgA ASCs were stronger to GI.1 than GII.4 as previously reported [114, 118, 119]. This study also assessed the ability of the vaccine construct to induce B_M cells. Results showed variable responses with higher IgA B_M-cell responses to GI.1 compared to GII.4 at day 28 following the first dose of the vaccine. In contrast, IgG responses were higher than memory IgA responses to GII.4 at days 28 and 56 following the first dose. VLP-specific IgA B_M-cell responses returned to baseline values by day 180 while VLP-specific IgG memory responses were maintained. This study compared the magnitude of B-cell responses following experimental challenge with GI.1 [17] and vaccination using this bivalent vaccine. Similar ASC responses and IgG B_M cells were reported (at day 28 and 180) whereas a lower IgA memory B-cell response was reported at day 180 post-vaccination. It is yet to be determined whether these memory responses could be used as correlates of protection. Intramuscular administration of this bivalent formulation induced yet again strong B_M-cell responses.

Another bivalent vaccine was tested in a randomized, double-blind, placebo-controlled trial performed at five sites in the United States. This trial was performed to evaluate the safety and tolerability of two doses of intramuscularly administered vaccine containing GI.1 and GII.4c VLPs (consensus VLP derived from three GII.4 strains: Houston/TCH186/2002/US (ABY27560), DenHaag89/2006/NL (ABL74395), and Yerseke38/2006/NL (ABL74391) [37]) [116]. Healthy adults with functional *FUT2* gene were eligible to participate in the study. The data from this study revealed the ability of this construct to induce robust antibody responses following a single dose of the vaccine to both strains with a clear bias toward GI.1. Undetectable-to-little increase in these responses was recorded following the second dose. Vaccination appeared to decrease the incidence of acute illness, i.e. diarrhea and/or vomiting, following oral challenge with a heterologous GII.4 NoV strain. This is the first efficacy evaluation of bivalent vaccines confirming HuNoV infection at a lower rate than those previously reported in other human challenge studies [14, 112]. Similarly, the IM administration of another bivalent vaccine (GI.1, GII.4 VLPs) adjuvanted with MPLA and $Al(OH)_3$ induced strong immune responses [118]. The immunogenicity of GI.1 and GII.4 VLPs was evaluated in healthy adults in a phase 1 clinical trial. As previously summarized [34], this vaccine induced high seroresponses to GI.1 and GII.4 (VLP-specific IgG, IgA and Pan-Ig) albeit higher to the former. IgA-specific ASCs were predominantly detected to GI.1 VLPs. One week following the first immunization, the vaccine recipients mounted high ASC; higher VLP doses did not correlate with higher frequencies of ASCs. There was no boosting effect detected following the second immunization whereby lower ASC frequencies were reported to both genotypes compared to a rapid and robust B cell response detected following the first vaccination. The expression of mucosal homing phenotype ($CD27^+CD38^+\beta7^+$ or $CD27^+CD38^+CCR10^+$) was detected on the surface of activated B cells following IM vaccination with high proportion of IgA ASCs. The authors suggested the activation of pre-existing vaccine-specific B_M cells, i.e. a recall response rather than activation of naïve B cells, indicating previous exposure to HuNoV infection. More studies are needed to investigate the immunogenicity of this bivalent vaccine in NoV-naïve individuals. In conclusion, the findings of this study suggest the ability of a non-mucosal mode of vaccine delivery to induce ASCs with mucosal homing properties ($CD27^+CD38^+B7^+CD62L^-$ cells). The safety and immunogenicity of two formulations of the adjuvanted bivalent vaccine described above (15/50 μg and 50/50 μg of GI.1 and GII.4 VLPs, respectively) were tested in a randomized, double-blinded, placebo-controlled phase 2 study [119]. The vaccines were well-tolerated by healthy adults receiving a single dose intramuscularly. Pan-Ig, IgA and HBGA-blocking antibody responses were detected 7–10 days following vaccination. These responses declined by day 28 as previously reported [114, 116]. The responses detected against GI.1 were higher than those detected against GII.4. While HBGA-blocking antibody titers induced by the bivalent vaccine were lower than those previously reported [115], the results are not comparable due to the use of different assays. While higher seroresponses to GI.1 than GII.4 were reported when using equal concentrations of VLPs [114, 115], data from the current study showed that the 15/50 μg form stimulated higher GII.4 responses compared to the 50/50 μg form. Further studies are planned to assess the sustainability of the immune responses among participants, the impact of adjuvant (MPLA vs. $Al(OH)_3$), the impact of a booster dose, and response following challenge.

Recently, the safety and immunogenicity of another bivalent VLP (GI.1/GII.4c) vaccine was evaluated in a phase 2 clinical trial [122]. The vaccine was administered intramuscularly to healthy adults. Volunteers received variable concentrations of each VLP (GI.1/GII.4c) (15 μg, 50 μg, or 150 μg) with $Al(OH)_3$ with or without MPLA. The vaccines were well-tolerated and did not result in vaccine-related adverse events. By day 56 following the first dose and regardless of the vaccine formulation, an increase in serum Pan-Ig responses, IgA responses and HBGA-blocking antibodies was detected against both GI.1 and GII.4c VLPs. Similar to previous studies [114, 119], the best formulation resulting in robust antibody responses contained 15 μg and 50 μg of GI.1 and GII.4c VLPs, respectively, and adjuvanted with 500 mg $Al(OH)_3$. Moreover, higher seroresponses were detected to GI.1 (80–100%) compared to GII.4c (33–83%). These titers persisted up to 1 year following vaccination without any obvious impact on the magnitude or persistence of the immune responses following a second dose as previously reported [114, 118, 119]. The mucosal and humoral immune responses induced by the same vaccine construct [122] were recently evaluated by Atmar et al. [123]. 50 voluntary participants received a candidate vaccine containing: 15 μg GI.1 VLPs, 50 μg GII.4c VLPs and 500 μg $Al(OH)_3$. This vaccine was also well-tolerated with mild to moderate adverse events reported during the first 3 days post-vaccination. These included fatigue, headache, myalgia, and diarrhea. Following the intramuscular administration of the candidate vaccine, serum Pan-Ig, IgA and HBGA-blocking antibodies against GI.1 and GII.4c VLPs peaked by day 8 post-vaccination and slightly decreased by day 29. Salivary IgA was also produced and followed parallel trends as serum IgA, albeit at lower levels. The results of this study are in agreement with previously reported data [119] and suggested the ability of a single dose bivalent vaccine delivered via the intramuscular route to induce rapid and robust humoral and mucosal immune responses. Further studies are needed to determine the impact of these vaccines in preventing norovirus infection.

Recently, the safety and immunogenicity of an oral vaccine have been tested in a randomized, double-blind, placebo-controlled clinical trial [121]. The vaccine contained a non-replicating adenovirus-based vector expressing VP1 of NoV GI.1. The vaccine was well-tolerated and generated a significant increase in serum IgG and IgA following vaccination with 10^{10} or 10^{11} infectious units (IU). Similarly, VP1-specific IgA and IgG ASCs increased following vaccination. Fecal IgA increased in more than 40 and 30% of vaccinated participants with the low and high dose vaccine, respectively. VP1-specific IgG and IgA B_M cells peaked at day 7 and declined 28 days following vaccination. Importantly, IgA^+ B_M cells expressed $\alpha4\beta7$, the gut homing receptor. The oral administration of this vaccine induced gut immune activation and thus is potentially capable of inducing mucosal responses and protection following NoV infection.

In summary, clinical studies proved to date the safety and immunogenicity of NoV VLP vaccines. Oral and intranasal immunization induced seroresponses [21, 109–113]. The intramuscular administration of bivalent vaccines also stimulated strong immune responses [116, 118–120, 122]. Human challenge studies as well as clinical vaccine studies (Tables 7.1 and 7.3) tested the ability of different vaccine formulations to generate cross-reactive immune responses to account for the diversity of circulating HuNoV strains. The majority of these studies recruited healthy adults when HuNoV infection leads to acute gastroenteritis among children. Consequently, it is yet to determine how well these vaccines will perform in children, the duration of protection among different age groups, the effect of previous infection on the frequency and the magnitude of the immune responses and protection, the impact of genetic variation and escape mutants on the efficacy of these vaccines. The breadth of immune responses, the impact of pre-existing immunity, the relationship between these responses and prevention of NoV infection require further investigation.

7.5 The Future of Norovirus Vaccines

The disease and public health burden caused by NoV infections across age groups underscore the urgent need for preventive intervention strategies. The development of NoV vaccines has been on-going for more than a decade. Vaccine strategies have focused on a number of candidates tested in preclinical and clinical settings. VLP-based vaccine candidates expressing capsid proteins, P particle-based vaccines, VRPs as well as vaccine constructs containing NoV genotypes and other enteric viruses have been tested. Clinical studies proved the safety and the immunogenicity of many formulations administered intranasally, intramuscularly or orally. Until recently, the lack of methods for culture of noroviruses has been a significant and major challenge. The recently developed system of HuNoV culture and replication in stem cell-derived human enteroids [125] will help evaluating the existing correlates of protection and defining new ones following infection or vaccination.

A number of challenges hamper the development of a protective NoV vaccine. While the majority of clinical studies targeted healthy adults, NoV infection causes acute gastroenteritis requiring hospitalizations among children less than 5 years old [126]. Moreover, adults more than 65 years old and immunosuppressed individuals suffer from severe and extended episodes of gastroenteritis [127–129]. Consequently, efforts should be made to test vaccines among children as well as elderly and immunocompromised individuals. The incomplete understanding of the dynamics of virus shedding in symptomatic and asymptomatic individuals and its heterogeneity, and the debatable estimated duration of protection following infection or vaccination are major factors contributing to delaying the development of a successful vaccine.

Importantly, the viral antigenic diversity, whereby NoV escape mutants are periodically selected by herd immunity [130], exerts a toll on the ability of noroviruses to bind different HBGAs and consequently impact the immune responses. The epochal evolution of noroviruses leads to generation of new strains. The evolution of the capsid epitope A, an immunodominant and hypervariable epitope, correlates with the emergence of new NoV variants [108, 131–133]. Recently, variation within epitope D led to escape from neutralizing antibodies while conserving binding to specific carbohydrates ligands [134]. Lindesmith et al. [134] suggested that this mechanism could contribute to prolonged strain (GII.4) circulation. The ability of noroviruses to evade host immunity and establish prolonged infections have been recently related to nonstructural proteins p48 and p22 of noroviruses suggested to block the host secretory pathway [135]; the latter plays a role in cytokine secretion and expression of antigen presentation molecules on the cell surface (MHC class I, MHC class II, CD40, CD80 and CD86).

A successful NoV vaccine should induce long-term protection against homologous and heterologous viral strains. In addition to cross-protection between distantly related NoVs, an ideal vaccine would stimulate both humoral and T-cell mediated immune responses. Defining the correlates of protection through vaccine studies is crucial to predict and estimate the immunogenicity and effectiveness of these potential vaccines. Consequently, identifying specific T-cell epitopes of NoV capsids are important to understand the antigenic and immunogenic relationships of different circulating viral strains and their significant impact on vaccine design. More recently, NoV-specific T-cell responses to GII.4 and GI.3 were reported among children under 5 years, albeit in a small sample size [136]. Broadly neutralizing antibodies have been successfully generated to different viruses such as HIV [137] and influenza virus [138]. The recent development of HuNoV culture systems is a potential platform to generate such antibodies in an attempt to protect against extensive viral genetic diversity. An annual process of strain selection for vaccine formulation, as the case with influenza virus, could be a possible solution targeting the predominant circulating strains.

Currently, there are two vaccines in clinical trials [139]. The first one is a replication defective Adenovirus type 5 expressing the GI.1 capsid protein (Vaxart). This vaccine completed two phase I clinical trials, was safe and immunogenic [121]. The safety and immunogenicity of the second vaccine, a bivalent GI and GII.4 VLP vaccine, previously evaluated in young adults and elderly [114, 122], is currently being evaluated in children in a phase IIb study (Takeda). NoV causes the majority of deaths and illnesses in low-middle-income countries [5, 140]. While scarce data are available on the burden of disease and circulating strains of noroviruses in low- and low-middle income countries [141, 142], a vaccine will have a significant impact in these countries. Consequently, it is critical to understand the potential factors affecting the efficacy of a vaccine against noroviruses in resource-limited countries to prevent the hardships related to rotavirus vaccines [143]. These factors include environmental enteropathy, genetic and epigenetic determinants, nutritional deficiencies, gastrointestinal normal flora and microbial infections. Vaccine effectiveness studies in resource-limited settings are important to define immunological determinants of protection.

References

1. Banyai K, et al. Viral gastroenteritis. Lancet. 2018;392(10142):175–86.
2. Hall AJ, et al. Norovirus disease in the United States. Emerg Infect Dis. 2013;19(8):1198–205.
3. Patel MM, et al. Noroviruses: a comprehensive review. J Clin Virol. 2009;44(1):1–8.
4. Pires SM, et al. Aetiology-specific estimates of the global and regional incidence and mortality of diarrhoeal diseases commonly transmitted through food. PLoS One. 2015;10(12): e0142927.
5. Bartsch SM, et al. Global economic burden of norovirus gastroenteritis. PLoS One. 2016;11 (4):e0151219.
6. Chhabra P, de Graaf M, Parra GI, Chan MC, Green K, Martella V, Wang Q, White PA, Katayama K, Vennema H, Koopmans MPG, Vinjé J. Updated classification of norovirus genogroups and genotypes. J Gen Virol. 2019;100(10):1393–406.
7. Steele MK, et al. Targeting pediatric versus elderly populations for norovirus vaccines: a model-based analysis of mass vaccination options. Epidemics. 2016;17:42–9.
8. Giersing BK, et al. Report from the World Health Organization's third Product Development for Vaccines Advisory Committee (PDVAC) meeting, Geneva, 8–10th June 2016. Vaccine. 2017; https://doi.org/10.1016/j.vaccine.2016.10.090.
9. Johnson PC, et al. Multiple-challenge study of host susceptibility to Norwalk gastroenteritis in US adults. J Infect Dis. 1990;161(1):18–21.
10. Lindesmith L, et al. Human susceptibility and resistance to Norwalk virus infection. Nat Med. 2003;9:548.
11. Lindesmith L, et al. Cellular and humoral immunity following Snow Mountain virus challenge. J Virol. 2005;79(5):2900–9.
12. Lindesmith LC, et al. Heterotypic humoral and cellular immune responses following Norwalk virus infection. J Virol. 2010;84(4):1800–15.
13. Ajami NJ, et al. Antibody responses to norovirus genogroup GI.1 and GII.4 proteases in volunteers administered Norwalk virus. Clin Vaccine Immunol. 2012;19(12):1980–3.
14. Frenck R, et al. Predicting susceptibility to Norovirus GII.4 by use of a challenge model involving humans. J Infect Dis. 2012;206(9):1386–93.

15. Czako R, et al. Experimental human infection with Norwalk virus elicits a surrogate neutralizing antibody response with cross-genogroup activity. Clin Vaccine Immunol. 2015;22(2):221–8.
16. Newman KL, et al. Human norovirus infection and the acute serum cytokine response. Clin Exp Immunol. 2015;182(2):195–203.
17. Ramani S, et al. Mucosal and cellular immune responses to Norwalk virus. J Infect Dis. 2015;212(3):397–405.
18. Wyatt RG, et al. Experimental infection of chimpanzees with the Norwalk agent of epidemic viral gastroenteritis. J Med Virol. 1978;2(2):89–96.
19. Green KY, et al. Comparison of the reactivities of baculovirus-expressed recombinant Norwalk virus capsid antigen with those of the native Norwalk virus antigen in serologic assays and some epidemiologic observations. J Clin Microbiol. 1993;31(8):2185–91.
20. Jiang X, et al. Expression, self-assembly, and antigenicity of the Norwalk virus capsid protein. J Virol. 1992;66(11):6527–32.
21. Tacket CO, et al. Humoral, mucosal, and cellular immune responses to oral Norwalk virus-like particles in volunteers. Clin Immunol. 2003;108(3):241–7.
22. Muller S. Avoiding deceptive imprinting of the immune response to HIV-1 infection in vaccine development. Int Rev Immunol. 2004;23(5–6):423–36.
23. Souza M, et al. Pathogenesis and immune responses in gnotobiotic calves after infection with the genogroup II.4-HS66 strain of human norovirus. J Virol. 2008;82(4):1777–86.
24. Souza M, et al. Cytokine and antibody responses in gnotobiotic pigs after infection with human norovirus genogroup II.4 (HS66 strain). J Virol. 2007;81(17):9183–92.
25. Farkas T, et al. Homologous versus heterologous immune responses to Norwalk-like viruses among crew members after acute gastroenteritis outbreaks on 2 US Navy vessels. J Infect Dis. 2003;187(2):187–93.
26. LoBue AD, et al. Multivalent norovirus vaccines induce strong mucosal and systemic blocking antibodies against multiple strains. Vaccine. 2006;24(24):5220–34.
27. Debbink K, et al. Chimeric GII.4 norovirus virus-like-particle-based vaccines induce broadly blocking immune responses. J Virol. 2014;88(13):7256–66.
28. Cohen J. Committee recommends NIH continue chimpanzee research, with new limits. Science. 2011;334(6062):1484.
29. Bok K, et al. Chimpanzees as an animal model for human norovirus infection and vaccine development. Proc Natl Acad Sci U S A. 2011;108(1):325–30.
30. Cheetham S, et al. Pathogenesis of a genogroup II human norovirus in gnotobiotic pigs. J Virol. 2006;80(21):10372–81.
31. Souza M, et al. A human norovirus-like particle vaccine adjuvanted with ISCOM or mLT induces cytokine and antibody responses and protection to the homologous GII.4 human norovirus in a gnotobiotic pig disease model. Vaccine. 2007;25(50):8448–59.
32. Ramani S, Estes MK. And RL Atmar. Norovirus vaccine development, In Lennart Svensson, Ulrich Desselberger, Harry B. Greenberg, Mary K. Estes (eds) Viral gastroenteritis: molecular epidemiology and pathogenesis; 2016.Academic London p. 447–469.
33. Kocher J, et al. Intranasal P particle vaccine provided partial cross-variant protection against human GII. 4 norovirus diarrhea in gnotobiotic pigs. J Virol. 2014;88(17):9728–43.
34. Melhem NM. Norovirus vaccines: correlates of protection, challenges and limitations. Hum Vaccin Immunother. 2016;12(7):1653–69.
35. Tamminen K, et al. A comparison of immunogenicity of norovirus GII-4 virus-like particles and P-particles. Immunology. 2012;135(1):89–99.
36. Fang H, et al. Norovirus P particle efficiently elicits innate, humoral and cellular immunity. PLoS One. 2013;8(4):e63269.
37. Parra GI, et al. Immunogenicity and specificity of norovirus consensus GII.4 virus-like particles in monovalent and bivalent vaccine formulations. Vaccine. 2012;30(24):3580–6.
38. Meyerhoff RR, et al. HIV-1 consensus envelope-induced broadly binding antibodies. AIDS Res Hum Retrovir. 2017;33(8):859–68.

39. Qin Y, et al. Eliciting neutralizing antibodies with gp120 outer domain constructs based on M-group consensus sequence. Virology. 2014;462-463:363–76.
40. Qin Y, et al. Detailed characterization of antibody responses against HIV-1 group M consensus gp120 in rabbits. Retrovirology. 2014;11:125.
41. Ball JM, et al. Oral immunization with recombinant Norwalk virus-like particles induces a systemic and mucosal immune response in mice. J Virol. 1998;72(2):1345–53.
42. Harrington PR, et al. Systemic, mucosal, and heterotypic immune induction in mice inoculated with Venezuelan equine encephalitis replicons expressing Norwalk virus-like particles. J Virol. 2002;76(2):730–42.
43. Nicollier-Jamot B, et al. Recombinant virus-like particles of a norovirus (genogroup II strain) administered intranasally and orally with mucosal adjuvants LT and LT(R192G) in BALB/c mice induce specific humoral and cellular Th1/Th2-like immune responses. Vaccine. 2004;22 (9–10):1079–86.
44. Xia M, Farkas T, Jiang X. Norovirus capsid protein expressed in yeast forms virus-like particles and stimulates systemic and mucosal immunity in mice following an oral administration of raw yeast extracts. J Med Virol. 2007;79(1):74–83.
45. Chachu KA, et al. Immune mechanisms responsible for vaccination against and clearance of mucosal and lymphatic norovirus infection. PLoS Pathog. 2008;4(12):e1000236.
46. Guo L, et al. Intranasal administration of a recombinant adenovirus expressing the norovirus capsid protein stimulates specific humoral, mucosal, and cellular immune responses in mice. Vaccine. 2008;26(4):460–8.
47. Guo L, et al. A recombinant adenovirus prime-virus-like particle boost regimen elicits effective and specific immunities against norovirus in mice. Vaccine. 2009;27(38):5233–8.
48. LoBue AD, et al. Alphavirus-adjuvanted norovirus-like particle vaccines: heterologous, humoral, and mucosal immune responses protect against murine norovirus challenge. J Virol. 2009;83(7):3212–27.
49. LoBue AD, Lindesmith LC, Baric RS. Identification of cross-reactive norovirus CD4(+) T cell epitopes. J Virol. 2010;84(17):8530–8.
50. Velasquez LS, et al. An intranasally delivered toll-like receptor 7 agonist elicits robust systemic and mucosal responses to Norwalk virus-like particles. Clin Vaccine Immunol. 2010;17(12):1850–8.
51. Ma Y, Li J. Vesicular stomatitis virus as a vector to deliver virus-like particles of human norovirus: a new vaccine candidate against an important noncultivable virus. J Virol. 2011;85 (6):2942–52.
52. Blazevic V, et al. Norovirus VLPs and rotavirus VP6 protein as combined vaccine for childhood gastroenteritis. Vaccine. 2011;29(45):8126–33.
53. Velasquez LS, et al. Intranasal delivery of Norwalk virus-like particles formulated in an in-situ gelling, dry powder vaccine. Vaccine. 2011;29(32):5221–31.
54. Tan M, et al. Norovirus P particle, a novel platform for vaccine development and antibody production. J Virol. 2011;85(2):753–64.
55. Xia M, et al. A candidate dual vaccine against influenza and noroviruses. Vaccine. 2011;29 (44):7670–7.
56. Hjelm BE, Kilbourne J, Herbst-Kralovetz MM. TLR7 and 9 agonists are highly effective mucosal adjuvants for norovirus virus-like particle vaccines. Hum Vaccin Immunother. 2014;10(2):410–6.
57. Tamminen K, et al. Trivalent combination vaccine induces broad heterologous immune responses to norovirus and rotavirus in mice. PLoS One. 2013;8(7):e70409.
58. Tamminen K, et al. Pre-existing immunity to norovirus GII-4 virus-like particles does not impair de novo immune responses to norovirus GII-12 genotype. Viral Immunol. 2013;26 (2):167–70.
59. Kim S-H, et al. Newcastle disease virus vector producing human norovirus-like particles induces serum, cellular, and mucosal immune responses in mice. J Virol. 2014;88 (17):9718–27.

60. Mathew LG, Herbst-Kralovetz MM, Mason HS. Norovirus Narita 104 virus-like particles expressed in Nicotiana benthamiana induce serum and mucosal immune responses. Biomed Res Int. 2014;2014:807539.
61. Ma Y, et al. Heat shock protein 70 enhances mucosal immunity against human norovirus when coexpressed from a vesicular stomatitis virus vector. J Virol. 2014;88(9):5122–37.
62. Wang L, et al. A dual vaccine candidate against norovirus and Hepatitis E virus. Vaccine. 2014;32(4):445–52.
63. Huo Y, et al. Prevailing Sydney like norovirus GII. 4 VLPs induce systemic and mucosal immune responses in mice. Mol Immunol. 2015;68(2):367–72.
64. Malm M, et al. Genotype considerations for virus-like particle-based bivalent norovirus vaccine composition. Clin Vaccine Immunol. 2015;22(6):656–63.
65. Wang X, et al. A bivalent virus-like particle based vaccine induces a balanced antibody response against both enterovirus 71 and norovirus in mice. Vaccine. 2015;33(43):5779–85.
66. Blazevic V, et al. Rotavirus capsid VP6 protein acts as an adjuvant in vivo for norovirus virus-like particles in a combination vaccine. Hum Vaccin Immunother. 2016;12(3):740–8.
67. Malm M, et al. Type-specific and cross-reactive antibodies and T cell responses in norovirus VLP immunized mice are targeted both to conserved and variable domains of capsid VP1 protein. Mol Immunol. 2016;78:27–37.
68. Verma V, et al. Norovirus (NoV) specific protective immune responses induced by recombinant P dimer vaccine are enhanced by the mucosal adjuvant FlaB. J Transl Med. 2016;14:135.
69. Xia M, et al. A trivalent vaccine candidate against hepatitis E virus, norovirus, and astrovirus. Vaccine. 2016;34(7):905–13.
70. Tamminen K, et al. Mucosal antibodies induced by intranasal but not intramuscular immunization block norovirus GII.4 virus-like particle receptor binding. Viral Immunol. 2016;29 (5):315–9.
71. Ball JP, et al. Intranasal delivery of a bivalent norovirus vaccine formulated in an in situ gelling dry powder. PLoS One. 2017;12(5):e0177310.
72. Lindesmith LC, et al. Emergence of novel human norovirus GII.17 strains correlates with changes in blockade antibody epitopes. J Infect Dis. 2017;216(10):1227–34.
73. Malm M, et al. Rotavirus capsid VP6 tubular and spherical nanostructures act as local adjuvants when co-delivered with norovirus VLPs. Clin Exp Immunol. 2017;189(3):331–41.
74. Malm M, et al. Functionality and avidity of norovirus-specific antibodies and T cells induced by GII.4 virus-like particles alone or co-administered with different genotypes. Vaccine. 2018;36(4):484–90.
75. Hwang HS, et al. More robust gut immune responses induced by combining intranasal and sublingual routes for prime-boost immunization. Hum Vaccin Immunother. 2018;14 (9):2194–202.
76. Heinimaki S, et al. Parenterally administered norovirus GII.4 virus-like particle vaccine formulated with aluminum hydroxide or monophosphoryl lipid A adjuvants induces systemic but not mucosal immune responses in mice. J Immunol Res. 2018;2018:3487095.
77. Richardson C, et al. Norovirus virus-like particle vaccines for the prevention of acute gastroenteritis. Expert Rev Vaccines. 2013;12(2):155–67.
78. Guerrero RA, et al. Recombinant Norwalk virus-like particles administered intranasally to mice induce systemic and mucosal (fecal and vaginal) immune responses. J Virol. 2001;75 (20):9713–22.
79. Periwal SB, et al. A modified cholera holotoxin CT-E29H enhances systemic and mucosal immune responses to recombinant Norwalk virus–virus like particle vaccine. Vaccine. 2003;21(5–6):376–85.
80. Garaicoechea L, et al. Llama nanoantibodies with therapeutic potential against human norovirus diarrhea. PLoS One. 2015;10(8):e0133665.
81. Becker KM, et al. Transmission of Norwalk virus during a football game. N Engl J Med. 2000;343(17):1223–7.

82. Wobus CE, Thackray LB, Virgin HW. Murine norovirus: a model system to study norovirus biology and pathogenesis. J Virol. 2006;80(11):5104–12.
83. Wobus CE, et al. Animal models of norovirus Infection. In: Svensson L, Desselberger U, Greenberg HB, Estes MK, editors. Viral gastroenteritis: molecular epidemiology and pathogenesis. London: Academic; 2016. p. 397–422.
84. Lindesmith LC, et al. Immunogenetic mechanisms driving norovirus GII.4 antigenic variation. PLoS Pathog. 2012;8(5):e1002705.
85. Lindesmith LC, et al. Emergence of a norovirus GII.4 strain correlates with changes in evolving blockade epitopes. J Virol. 2013;87(5):2803–13.
86. Lindesmith LC, et al. Monoclonal antibody-based antigenic mapping of norovirus GII.4-2002. J Virol. 2012;86(2):873–83.
87. Debbink K, et al. Genetic mapping of a highly variable norovirus GII.4 blockade epitope: potential role in escape from human herd immunity. J Virol. 2012;86(2):1214–26.
88. Chachu KA, et al. Antibody is critical for the clearance of murine norovirus infection. J Virol. 2008;82(13):6610–7.
89. Rose NF, et al. An effective AIDS vaccine based on live attenuated vesicular stomatitis virus recombinants. Cell. 2001;106(5):539–49.
90. Tan GS, et al. Strong cellular and humoral anti-HIV Env immune responses induced by a heterologous rhabdoviral prime–boost approach. Virology. 2005;331(1):82–93.
91. Buonocore L, et al. Characterization of vesicular stomatitis virus recombinants that express and incorporate high levels of hepatitis C virus glycoproteins. J Virol. 2002;76(14):6865–72.
92. Jackson EM, Herbst-Kralovetz MM. Intranasal vaccination with murabutide enhances humoral and mucosal immune responses to a virus-like particle vaccine. PLoS One. 2012;7(7):e41529.
93. Tan M, Meller J, Jiang X. C-terminal arginine cluster is essential for receptor binding of norovirus capsid protein. J Virol. 2006;80(15):7322–31.
94. Huang Z, Wang X, Liu Q. Vaccine development. The Norovirus. 2017;13:187–206.
95. Elftman MD, et al. Multiple effects of dendritic cell depletion on murine norovirus infection. J Gen Virol. 2013;94(Pt 8):1761–8.
96. Troeger H, et al. Structural and functional changes of the duodenum in human norovirus infection. Gut. 2009;58(8):1070.
97. Malm M, et al. Rotavirus recombinant VP6 nanotubes act as an immunomodulator and delivery vehicle for norovirus virus-like particles. J Immunol Res. 2016;2016:9171632.
98. Wang L, et al. Polyvalent complexes for vaccine development. Biomaterials. 2013;34(18):4480–92.
99. He S, et al. Putative receptor-binding sites of hepatitis E virus. J Gen Virol. 2008;89(1):245–9.
100. Li SW, et al. A bacterially expressed particulate hepatitis E vaccine: antigenicity, immunogenicity and protectivity on primates. Vaccine. 2005;23(22):2893–901.
101. Tan M, Jiang X. Norovirus and its histo-blood group antigen receptors: an answer to a historical puzzle. Trends Microbiol. 2005;13(6):285–93.
102. Tan M, Jiang X. Norovirus-host interaction: implications for disease control and prevention. Expert Rev Mol Med. 2007;9(19):1–22.
103. Tan M, Jiang X. Norovirus gastroenteritis, carbohydrate receptors, and animal models. PLoS Pathog. 2010;6(8):e1000983.
104. Tan M, Jiang X. Norovirus-host interaction: multi-selections by human histo-blood group antigens. Trends Microbiol. 2011;19(8):382–8.
105. Tan M, Jiang X. Norovirus P particle: a subviral nanoparticle for vaccine development against norovirus, rotavirus and influenza virus. Nanomedicine. 2012;7(6):889–97.
106. Zhang J, et al. Analysis of hepatitis E virus neutralization sites using monoclonal antibodies directed against a virus capsid protein. Vaccine. 2005;23(22):2881–92.
107. Didierlaurent A, et al. Flagellin promotes myeloid differentiation factor 88-dependent development of Th2-type response. J Immunol. 2004;172(11):6922.

108. Lindesmith LC, et al. Mechanisms of GII.4 norovirus persistence in human populations. PLoS Med. 2008;5(2):e31.
109. Ball JM, et al. Recombinant Norwalk virus-like particles given orally to volunteers: phase I study. Gastroenterology. 1999;117(1):40–8.
110. Tacket CO, et al. Human immune responses to a novel Norwalk virus vaccine delivered in transgenic potatoes. J Infect Dis. 2000;182(1):302–5.
111. El-Kamary SS, et al. Adjuvanted intranasal Norwalk virus-like particle vaccine elicits antibodies and antibody-secreting cells that express homing receptors for mucosal and peripheral lymphoid tissues. J Infect Dis. 2010;202(11):1649–58.
112. Atmar RL, et al. Norovirus vaccine against experimental human Norwalk virus illness. N Engl J Med. 2011;365(23):2178–87.
113. Ramirez K, et al. Intranasal vaccination with an adjuvanted Norwalk virus-like particle vaccine elicits antigen-specific B memory responses in human adult volunteers. Clin Immunol. 2012;144(2):98–108.
114. Treanor JJ, et al. A novel intramuscular bivalent norovirus virus-like particle vaccine candidate—reactogenicity, safety, and immunogenicity in a phase 1 trial in healthy adults. J Infect Dis. 2015;210:1763–71. https://doi.org/10.1093/infdis/jiu337.
115. Atmar RL, et al. Serological correlates of protection against a GII.4 norovirus. Clin Vaccine Immunol. 2015;22(8):923–9.
116. Bernstein DI, et al. Norovirus vaccine against experimental human GII.4 virus illness: a challenge study in healthy adults. J Infect Dis. 2015;211(6):870–8.
117. Lindesmith LC, et al. Broad blockade antibody responses in human volunteers after immunization with a multivalent norovirus VLP candidate vaccine: immunological analyses from a phase I clinical trial. PLoS Med. 2015;12(3):e1001807.
118. Sundararajan A, et al. Robust mucosal-homing antibody-secreting B cell responses induced by intramuscular administration of adjuvanted bivalent human norovirus-like particle vaccine. Vaccine. 2015;33(4):568–76.
119. Atmar RL, et al. Rapid responses to 2 virus-like particle norovirus vaccine candidate formulations in healthy adults: a randomized controlled trial. J Infect Dis. 2016;214 (6):845–53.
120. Ramani S, et al. B-cell responses to intramuscular administration of a bivalent virus-like particle human norovirus vaccine. Clin Vaccine Immunol. 2017;24:e00571–16.
121. Kim L, et al. Safety and immunogenicity of an oral tablet norovirus vaccine, a phase I randomized, placebo-controlled trial. JCI Insight. 2018;3(13):121077.
122. Leroux-Roels G, et al. Safety and immunogenicity of different formulations of norovirus vaccine candidate in healthy adults: a randomized, controlled, double-blind clinical trial. J Infect Dis. 2018;217(4):597–607.
123. Atmar RL, et al. An exploratory study of the salivary immunoglobulin a responses to 1 dose of a norovirus virus-like particle candidate vaccine in healthy adults. J Infect Dis. 2018;219 (3):410–4.
124. Reeck A, et al. Serologic correlate of protection against norovirus-induced gastroenteritis. J Infect Dis. 2010;202(8):1212–8.
125. Ettayebi K, et al. Replication of human noroviruses in stem cell-derived human enteroids. Science. 2016;353(6306):1387–93.
126. Shah MP, Hall AJ. Norovirus illnesses in children and adolescents. Infect Dis Clin N Am. 2018;32(1):103–18.
127. Koo HL, DuPont HL. Noroviruses as a potential cause of protracted and lethal disease in immunocompromised patients. Clin Infect Dis. 2009;49(7):1069–71.
128. Florescu DF. The evaluation of critically ill transplant patients with infectious diarrhea. Curr Opin Crit Care. 2017;23(5):364–71.
129. van Beek J, et al. Chronic norovirus infection among solid organ recipients in a tertiary care hospital, the Netherlands, 2006–2014. Clin Microbiol Infect. 2017;23(4):265 e9–265 e13.

130. Donaldson EF, et al. Viral shape-shifting: norovirus evasion of the human immune system. Nat Rev Microbiol. 2010;8(3):231–41.
131. Siebenga JJ, et al. Epochal evolution of GGII.4 norovirus capsid proteins from 1995 to 2006. J Virol. 2007;81(18):9932–41.
132. Debbink K, et al. Within-host evolution results in antigenically distinct GII.4 noroviruses. J Virol. 2014;88(13):7244–55.
133. Lindesmith LC, et al. Conformational occlusion of blockade antibody epitopes, a novel mechanism of GII.4 human norovirus immune evasion. mSphere. 2018;3(1):e00518–7.
134. Lindesmith LC, et al. Human norovirus epitope D plasticity allows escape from antibody immunity without loss of capacity for binding cellular ligands. J Virol. 2019;93(2):e01813–8.
135. Roth AN, Karst SM. Norovirus mechanisms of immune antagonism. Curr Opin Virol. 2016;16:24–30.
136. Malm M, et al. Development of T cell immunity to norovirus and rotavirus in children under five years of age. Sci Rep. 2019;9(1):3199.
137. McCoy LE, Burton DR. Identification and specificity of broadly neutralizing antibodies against HIV. Immunol Rev. 2017;275(1):11–20.
138. Cho A, Wrammert J. Implications of broadly neutralizing antibodies in the development of a universal influenza vaccine. Curr Opin Virol. 2016;17:110–5.
139. Atmar RL, Ramani S, Estes MK. Human noroviruses: recent advances in a 50-year history. Curr Opin Infect Dis. 2018;31(5):422–32.
140. Kirk MD, et al. World Health Organization estimates of the global and regional disease burden of 22 foodborne bacterial, protozoal, and viral diseases, 2010: a data synthesis. PLoS Med. 2015;12(12):e1001921.
141. Kreidieh K, et al. The epidemiology of norovirus in the Middle East and North Africa (MENA) region: a systematic review. Virol J. 2017;14(1):220.
142. Nguyen GT, et al. A systematic review and meta-analysis of the prevalence of norovirus in cases of gastroenteritis in developing countries. Medicine (Baltimore). 2017;96(40):e8139.
143. Clarke E, Desselberger U. Correlates of protection against human rotavirus disease and the factors influencing protection in low-income settings. Mucosal Immunol. 2015;8(1):1–17.

Index

N. M. Melhem (ed.), *Norovirus*,
https://doi.org/10.1007/978-3-030-27209-8

D

J

L